NANOTECHNOLOGY

ETHICAL AND SOCIAL IMPLICATIONS

Nano and Energy

Series Editor: Sohail Anwar

Computational Nanotechnology: Modeling and Applications with MATLAB®
Sarhan M. Musa

Nanotechnology: Business Applications and Commercialization
Sherron Sparks

Nanotechnology: Ethical and Social Implications
Ahmed S. Khan

NANOTECHNOLOGY
ETHICAL AND SOCIAL IMPLICATIONS

AHMED S. KHAN

CRC Press
Taylor & Francis Group
Boca Raton London New York

CRC Press is an imprint of the
Taylor & Francis Group, an **informa** business

CRC Press
Taylor & Francis Group
6000 Broken Sound Parkway NW, Suite 300
Boca Raton, FL 33487-2742

Printed in the United States of America on acid-free paper
Version Date: 20120313

International Standard Book Number: 978-1-4398-5953-7 (Hardback)

Library of Congress Cataloging-in-Publication Data

Nanotechnology : ethical and social implications / editor, Ahmed S. Khan.
p. cm. -- (Nano and energy series)
Includes bibliographical references and index.
ISBN 978-1-4398-5953-7 (hardback)
1. Nanotechnology--Social aspects. 2. Nanotechnology--Moral and ethical aspects. I. Khan, Ahmed S.

T14.5.N343 2012
303.48'3--dc23 2012008934

Visit the Taylor & Francis Web site at
http://www.taylorandfrancis.com

and the CRC Press Web site at
http://www.crcpress.com

Dedication

To Tasneem, my parents, my students, and all seekers of truth and wisdom.

Ahmed S. Khan

Love and Knowledge

By love, the bitter becomes sweet;
by love, copper becomes gold;
by love, the dregs become clear;
by love, the dead become living;
by love, the king becomes a slave.
From knowledge, love grows.
Has stupidity ever placed someone on such a throne?*

Rumi

---

* *Rumi: Daylight—A Daybook of Spiritual Guidance (365 Selections from the Mathnawi),* II, 1542–3, translated by Camille and Kabir Helminski (Aptos, CA: Threshold Books, 1991), 142.

Epigraph

In an evolving universe, who stands still moves backward.

R. Anton Wilson

Empires of the future will be empires of the mind.

Winston Churchill

The principles of physics, as far as I can see, do not speak against the possibility of maneuvering things atom by atom.

Richard Feynman, Nobel laureate

The saddest aspect of society right now is that science gathers knowledge faster than society gathers wisdom.

Isaac Asimov

Contents

Preface

Historically, every new technological advance and innovation remake the world. The time to remake the world has become shorter with every new technological revolution. The Industrial Revolution took almost two centuries to reshape the world, the electronics revolution around seventy years, the information revolution two decades, and innovations in biotechnology and nanotechnology to reshape the world could be just a matter of less than a decade.

The projected impact of nanotechnology has been touted as a second industrial revolution—not the third, fourth, or fifth, because despite similar predictions for technologies such as computers and robotics, nothing has yet eclipsed the first. Society is at the threshold of a revolution that will transform the ways in which materials and products are created. How will this revolution develop? The opportunities that will develop in the future will depend significantly upon the ways in which a number of challenges are met. As we design systems on a nanoscale, we develop the capability to redesign the structure of all materials—natural and synthetic—along with rethinking the new possibilities of the reconstruction of any and all materials. Such a change in our design power presents tremendous social and ethical questions. To enable our future leadership to make decisions for sustainable ethical, economic nanotechnological development, it is imperative that we educate all nanotechnology stakeholders about the short-term and long-term benefits, limitations, and risks of nanotechnology. The social implications of nanotechnology encompass many fundamental areas such as ethics, privacy, environment, and security.

It is estimated that by 2020, nanotechnology will be a global industry with an annual investment of $3 trillion per year. In 1997, the investment in nanotechnology stood at $430 million, rising to more than $9 billion worldwide in 2004. At present, more than 800 products have been developed using nanotechnology. The development of nanotechnology requires multidisciplinary teams of highly trained researchers with backgrounds in biology, medicine, mathematics, physics, chemistry, material science, and engineering. This book focuses on stakeholders' awareness of the challenges posed by nanotechnology in terms of its risks and benefits to society and the demands of new skill sets on the workforce of the future.

Unique in its depth, breadth, and variety of viewpoints, *Nanotechnology: Ethical and Social Issues* stimulates, inspires, and builds awareness of nanotechnology's impact on society. Spanning the full range of nanotechnology areas, the book's selection of chapters focuses on how advances in science and the convergence of multiple technological disciplines in nanotechnology will impact society at personal, community, national, and international levels. The text explores both the positive and negative aspects of this rapidly developing technology amid the complexity of today's ethical and social issues. The book is written in an interactive format using case studies, scenarios, and Web exercises to promote critical thinking and problem-solving skills through the exploration of the following areas of nanotechnology:

- Introduction to nanotechnology and applications
- Nanotechnology ethical issues
- Nanotechnology social implications
- Nanotechnology and environmental and ecological issues
- Nanotechnology and genetics
- Nanotechnology and food and agriculture
- Nanotechnology and education
- Nanotechnology: international perspective
- Nanotechnology: predicting the future

The text provides an introduction to the technical aspects of nanotechnology while focusing on the social and ethical ramifications of nanotechnology. The book is targeted for all stakeholders of nanotechnology: undergraduate and graduate students in the sciences, engineering, business, and public policy; professors in the same disciplines; engineers; business and finance managers; and policymakers. The text is intended to:

- Provide an introduction to nantechnology and nanoscience
- Clarify the social and ethical ramifications of nanotechnology
- Analyze the impact of nanotechnology in medicine, health care, food, agriculture, military, and other areas
- Present new and emerging nanotechnologies in an interactive manner with the help of scenarios, case studies, flow charts, and Internet exercises
- Present materials in a format that promotes teaching and learning, using brain-based teaching theories and Bloom's taxonomy

The chapters contain numerous data tables, statistics, and figures to supplement and enhance the topics covered in each chapter. The book is a must-read for technical and nontechnical seekers of knowledge and wisdom. The text presents a wide spectrum of issues and insights. Chapter 1 provides an introduction to nanotechnology, its applications, ethical and social issues, and associated risks and uncertainties. Chapter 2 presents a comparison of ethics and nanoethics. It addresses the question: "Is there a need to start 'nanoethics' as a new subfield of ethics?" Chapter 3 presents a comparison of the ethical systems of Socrates and Confucius. Chapter 4 presents a discussion on ethics from a traditional Christian perspective. Chapter 5 presents the ethical and social dimension of nanotechnology. Chapter 6 discusses the anticipated and unanticipated consequences of technology. Chapter 7 presents portraits of carbon and describes its positive and negative sides to ecology and the environment. Chapter 8 discusses nanotechnology and ecological and environmental issues. Chapter 9 presents futuristic scenarios dealing with the impact of biogenetic advances on society. Chapter 10 discusses the impact of nanotechnology on food and agriculture. Chapter 11 deals with the ethical and social issues of genetically modified food. Chapter 12 present an introduction to titanium dioxide nanoparticles, their applications, and potential risks. Chapter 13 discusses the social implications of delaying the teaching of nanoscience education to K–12 students in the United States. Chapter 14 presents an international perspective by presenting the state and status of nanotechnology research, science, technology, and society education at Bilkent University, Turkey. Chapter 15 discusses the use of simulation in teaching nanotechnology and emerging technologies in academic programs. Chapter 16 describes the limitations of predicting the future by gauging public perceptions of nanotechnology. And finally, the book concludes by presenting a snapshot of the present size of human knowledge compared to the size of the known cosmos.

I hope that this text will enable all stakeholders of nanotechnology to develop a multifaceted understanding so that they can anticipate and address the complex ethical and social issues in a wise manner.

Ahmed S. Khan

Acknowledgments

I would like to thank Tasneen, Robert Lawrence, Fidel M. Salinas, Denise M. Smith, Shekar Viswanathan, Rose Chu, Gene Anderson, Anwar Sohail, and Maris Roze for their help and support. Special thanks to Nora Konopka, Prudy Taylor Board and Amy Blalock at Taylor & Francis Group/CRC Press for their support.

About the Editor

Dr. Ahmed S. Khan is professor of Electronics and Electrical Engineering in the College of Engineering and Information Sciences at DeVry University, Addison, Illinois. Dr. Khan has thirty years of experience in research; instruction; curricula design, development, evaluation, implementation and program accreditation; management, and supervision.

Dr. Khan received an MSEE from Michigan Technological University, an MBA from Keller Graduate School of Management, and his PhD from Colorado State University. His research interests are in the areas of nanotechnology, fiber optic communications, faculty development, outcomes assessment, application of telecommunications technologies in distance Education, and Social and Ethical Implications of Technology. He teaches Wireless Engineering, Network Engineering, Fiber Optic Communications, Science Technology and Society (STS), and Project Management. He also advises students on their senior design projects. He is the author of many educational papers and presentations. He has authored/coauthored *The Telecommunications Fact Book and Illustrated Dictionary 2E* (2006), *Technology and Society: Issues for the 21st Century and Beyond 3E,* (2008), *Fiber Optic Communication: An Applied Approach,* Prentice Hall, N.J. (2002), *Technology and Society: A Bridge to the 21st Century* (2002), *Technology and Society: Crossroads to the 21st Century* (1996), *Technology and Society: A Spectrum of Issues for the 21st Century* (1994), and *The Telecommunications Fact Book and Illustrated Dictionary* (1992).

Dr. Khan is a senior member of the Institute of Electrical and Electronics Engineering (IEEE), and a member of the American Society of Engineering Education (ASEE), and has been listed in Who's Who among America's Teachers. Dr. Khan has served as the faculty adviser to the student chapter of IEEE at DeVry-Addison since its inception in 1986. Dr. Khan also serves as a program evaluator for the Accreditation Board for Engineering and Technology (ABET).

Contributors

Gene Anderson
Kids Alive International
Taitung, Taiwan

Judith Light Feather
President and Founder
The NanoTechnology Group Inc.
Nacogdoches, Texas

Richard Hayes
Center for Genetics and Society
Berkeley, California

Tim Healy
Markkula Center for Applied Research
Santa Clara University
Santa Clara, California

Ahmed S. Khan
College of Engineering & Information Sciences
DeVry University
Addison, Illinois

Robert E. McGinn
Department of Management Science and Engineering
School of Engineering
Stanford University
Stanford, California

Margaret R. McLean
Markkula Center for Applied Research
Santa Clara University
Santa Clara, California

Todd Neff
Daily Camera
Boulder, Colorado

Deb Newberry
Director
Nanoscience Program
Dakota County Technical College
Rosemount, Minnesota

Haldun M. Ozaktas
Department of Electrical Engineering
Bilkent University
Ankara, Turkey

Fidel M. Salinas
CEO
LearningSolutions.3rdPlanet
Santa Clara, California
Adjunct Faculty
Eastern Illinois University
Charleston, Illinois
Associate Professor
National University
San Diego, California

M. A. Shah
Electron Microscopy Centre
Department of Physics
Faculty of Science
National Institute of Technology
Srinagar, India
Department of Physics
Faculty of Science
King Abdul Aziz University
Jeddah, Kingdom of Saudi Arabia

Michael Siegrist
ETH, Institute for Environmental Decisions (IED) Consumer Behavior
Zurich, Switzerland

Denise M. Smith
President
LearningSolutions.3rdPlanet Santa Clara, California
Adjunct Faculty
National University
San Diego, California

Shekar Viswanathan
Chair, Engineering and Technology
Department of Applied Engineering
National University
San Diego, California

Jiyan Yu
Department of Philosophy
State University of New York at Buffalo, New York

1

Nanotechnology: An Introduction to Applications and Ethical and Social Issues

Ahmed S. Khan

CONTENTS

KEY WORDS: nanotechnology, nanoparticle, nanotube, nanosize material, quantum dot, buckyball, nanoshell, MEMS, ethical and social issues, nanomedicine, risk assessment, uncertainties, lab safety

Introduction

The role of the infinitely small is infinitely large.

Louis Pasteur

As the second decade of the twenty-first century sets in, remarkable achievements are being reported at the frontiers of science and technology. Physicists at CERN (European Organization for Nuclear Research [Conseil Européen pour la Recherche Nucléaire]) have reported neutrinos traveling faster than the speed of light (Overby, 2011). Advances in neuroscience have led scientists to turn visual memories into video (Schroeder, 2011). Researchers have gained the ability to measure, manipulate, and organize matter on the nanoscale—1 to 100 billionths of a meter. At the nanoscale, physics, chemistry, material science, biology, and engineering are converging toward common principles, mechanisms, and tools. This convergence of multiple disciplines will have a significant impact on science, technology, and society.

Historically, every major technological advance and innovation has remade the world. The time to remake the world has become shorter with every new technological revolution. The Industrial Revolution took almost two centuries to reshape the world, the electronics revolution around seventy years, the information revolution two decades; and the time for innovations in biotechnology and nanotechnology to reshape the world could be a matter of less than a decade. In the post–WW II era the world was divided into the first world and the third world, but the information revolution revealed the "digital-divide," and advances in nanotechnology will divide the world into the nano-haves and the nano-have nots (Hjorth, Eichler, & Khan, 2007).

The nanoscale is not just another jump toward miniaturization, but a qualitatively new scale. Nanoscale behavior is dominated by quantum mechanics, material confinement in small structures, large interfacial volume fractions, and other unique properties, phenomena, and processes. Many current theories of matter at the microscale will be inadequate to describe the new phenomena at the nanoscale (Roco, 2011).

As the global economy continues to be transformed by new technology, an intense competition will grow for intellectual capital and intellectual property. Technology will continue to drive the global and domestic GDP (gross domestic product) (Khan & Qadeer, 2010).

The National Science Foundation predicts that the global marketplace for goods and services using nanotechnologies will grow to $1 trillion by 2015 and employ two million workers. It is estimated that by 2020 nanotechnology will be a $3-trillion-per-year global industry (Roco, 2011). In 1997 the investment in nanotechnology stood at $430 million compared to more than $9 billion in 2004. There are more than 800 products in the marketplace that have been developed using nanotechnology (Ehrmann, 2008). The launch

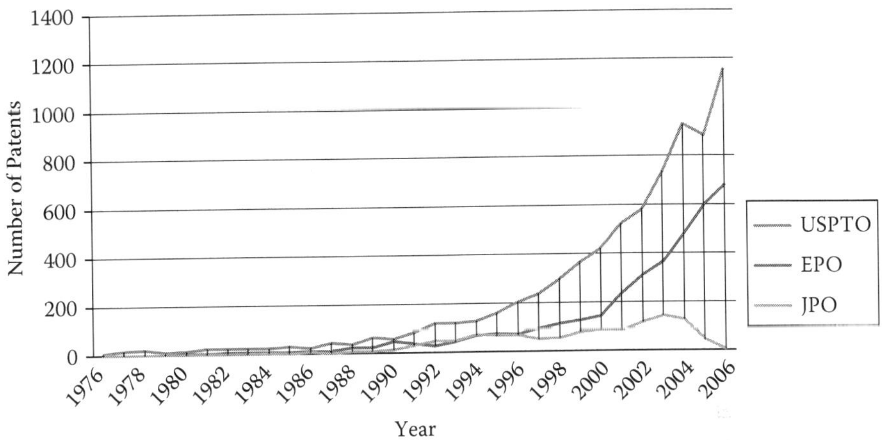

FIGURE 1.1 (See color insert)
Trends in nanotechnology patents filed globally, 1976–2006. (From "Trends in Nanotechnology Patents," by H. Chen, M. C. Roco, X. Li, and Y. Ling, 2008, *Nature Nanotechnology, 3,* p. 123. Copyright 2000 by Nature Nanotechnology. Reprinted with permission.)

of the U.S. National Nanotechnology Initiative (NNI) in 2000 inspired sustained research and development (R&D) programs in China, the European Community including Germany, Japan, Korea, and Taiwan. During 2001–2004, more than sixty countries established nanotechnology R&D programs (Roco, 2011). During 1976–2006, more than 12,000 patents were filed globally for nanotechnology (see Figure 1.1). There is also an increase in the number of scientific papers written every year on nanoscience and nanotechnology; the current world leaders for the period 1990–2006 are: European Union (14,700), United States (11,800), China (10,200), and Japan (4,800) (*Ten things*, n.d.). Global spending for nanotechnology research and development has increased exponentially. It is estimated that by the end of 2011, the total government funding for nanotechnology research worldwide will reach $65 billion and rise to $100 billion by 2014. If corporate and private funding are taken into account since 2004, then an estimate of nearly a quarter of a trillion dollars will have been invested into nanotechnology by 2015 (Harper, 2011).

Definitions

What is nanotechnology? The literature provides a number of definitions. According to the National Nanotechnology Initiative (NNI), nanotechnology is an area that encompasses the following traits/characteristics (Minoli, 2006):

TABLE 1.1
Matters of Scale

Item	Dimensions in nanometers (nm) [1 nm = 0.000000001 m]
Human hair	100,000
Diameter of bacterium cell	1,000–10,000
Range of visible light	400–700
Human immunodeficiency virus	90
Transistor dimensions on a CPU (2008)	43
Cell membrane	10
Drug molecule	≈5–10
Quantum dot	≈1–5
Diameter of DNA	2.5
Individual atom	0.1

1. Research and technology development at the atomic, molecular, or macromolecular levels, in the length scale of approximately 1–100 nm
2. Creating and using structures, devices, and systems that have novel properties and functions because of their small or intermediate size
3. Ability to control or manipulate matter on the atomic scale

According to Professor Stephen Fonash, nanotechnology involves manipulating matter at the atomic and molecular scale, seeing matter at the atomic and molecular scale, and exploiting the unique capabilities and properties of structures fabricated at the atomic and molecular scale (Fonash, 2008). In short, nanotechnology refers to the convergence of multiple disciplines and applied technologies dealing with particles and structures having dimensions in the range of a nanometer, one billionth of a meter. The diameter of a human hair is 100,000 nanometers. Presently, PC notebooks employ CPU chips that contain transistors having dimensions of 43 nanometers. It is said that today more nanoscale transistors are made in a year than grains of rice are grown in a year. (Figure 1.17 shows an engineered DNA strand.) Table 1.1 compares various entities and systems that occur in nature using a nanoscale.

Evolution of Nanotechnology

Nature is full of nanostructures (see Figures 1.2–1.7). Nanoparticles and structures have been used by humans for quite a long time. For example, in the fourth century, Romans used metal nanoparticles to make glass cups that changed their colors based on the transmission or reflection

FIGURE 1.2 (See color insert)
The iridescent colors of the Blue Morpho Butterfly's wings are produced by nanostructures that reflect different wavelengths of light. The wing span of a Blue Morpho Butterfly is about 10–15 cm. (*Source:* Wikimedia Commons, http://www.nisenet.org/viz_lab/image-collection)

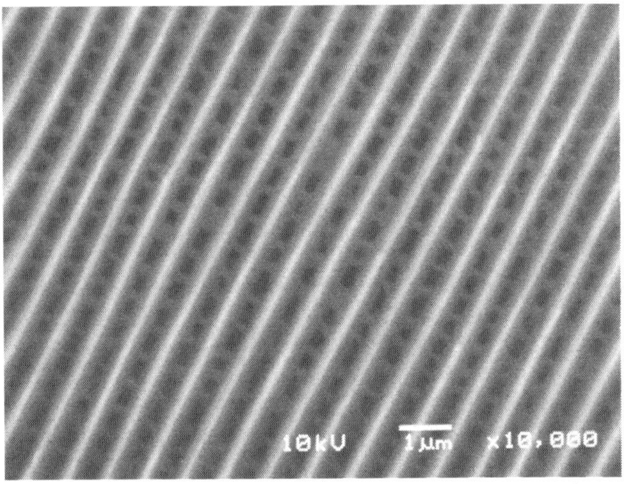

FIGURE 1.3
This scanning electron microscope image shows ridges on a Blue Morpho Butterfly wing scale. These ridges contain nanoscale structures that reflect light to create the Morpho's iridescent colors. Each ridge is about 800 nm wide. (*Source:* S. Yoshioka, Osaka University, http://www.nisenet.org/viz_lab/image-collection)

of light. An artifact from this period called the Lycurgus Cup resides in the British Museum in London. The cup, which depicts the death of King Lycurgus, is made from soda lime glass containing silver and gold nanoparticles (40-ppm gold nanoparticles and 300-ppm silver nanoparticles). The color of the cup changes from green to a deep red when a

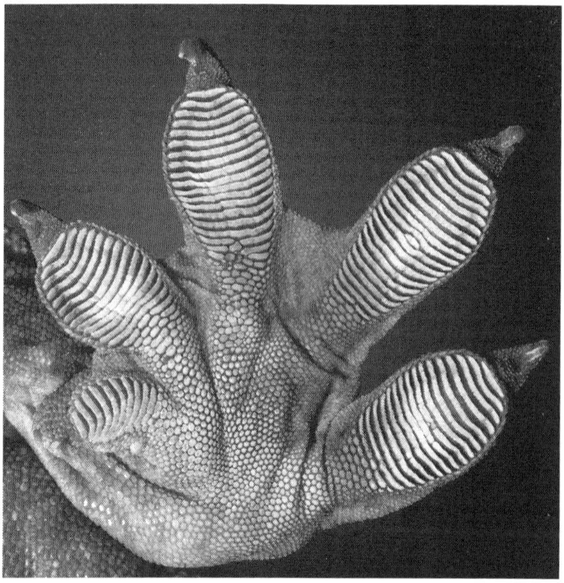

FIGURE 1.4 (See color insert)
The feet of the gecko cling to virtually any surface. This image shows the sole of a gecko's foot. The adhesive lamellae on the sole have millions of branching hairs that nestle into nanoscale niches on the contact surface. (*Source:* A. Dhinojwala, University of Akron, http://www.nisenet.org/viz_lab/image-collection)

light source is placed inside it (Barber & Freestone, 1990). The Irish manufactured stained-glass windows in AD 444 using nanoparticles. During the ninth–seventeenth centuries, glowing, glittering "luster" ceramic glazes used in the Islamic world, and later in Europe, contained silver or copper or other nanoparticles (National Nanotechnology Initiative, 2011). The Italians also employed nanoparticles in creating sixteenth-century Renaissance pottery (Poole & Owens, 2003). They were influenced by Ottoman techniques: During the thirteenth–eighteenth centuries, in the making of "Damascus" saber blades, carbon nanotube and cementite nanowires were used to provide strength, resilience, and the ability to hold a keen edge. In south Asia, traditional practitioners also employed fine particles in the composition of *Tib-e-Unani* and *Ayuvedic* medicines.

In 1959, prominent physicist Richard Feynman presented a visionary and prophetic lecture at the meeting of the American Physical Society entitled "There Is Plenty of Room at the Bottom," in which he speculated the possibility and potential of nanosized materials. In 1974, the term *nanotechnology* was used for the first time by Nori Taniguchi in Tokyo, Japan, at the International Conference on Production Engineering (Minoli, 2006). However, it was not until the 1980s, with the development of appropriate methods of fabrication of nanostructures, that a notable increase in research activity occurred and a number of significant developments materialized (Poole & Owens, 2003).

FIGURE 1.5
The feet of the gecko cling to virtually any surface. This scanning electron microscope image shows the branching hairs on the foot's adhesive lamellae. These hairs nestle into nanoscale niches on the contact surface. (*Source:* C. Mathisen, FEI Company, http://www.nisenet.org/viz_lab/image-collection)

Table 1.2 presents a summary of the modern-era discoveries and developments that led to the evolution of nanotechnology.

Nature at the nanoscale

- Blue Morpho butterfly (see Figure 1.2)
- Blue Morpho butterfly wing ridges (see Figure 1.3)
- Gecko foot (see Figure 1.4)
- Gecko foot (1660×) (see Figure 1.5)
- Nasturtium leaf (see Figure 1.6)
- Nasturtium leaf (2500×) (see Figure 1.7)

Nanoparticles have been used for ages, but the ability to understand the chemical and physical processes at the nano scale has been a recent development. Control and repeatability have been achieved; the knowledge base and applications have been refined. Table 1.3 presents a summary of the application of nanotechnology in various domains of science. Table 1.4 presents a summary of the equipment and processes used in nanotechnology research and development (R&D).

FIGURE 1.6 (See color insert)
The lotus effect describes water droplets rolling off leaf surfaces, removing dirt and contaminants. This phenomenon can also be seen in the more common nasturtium. These leaves are covered with wax nanocrystal bundles that trap air and force water to bead and roll off. (*Source:* A. Otten and S. Herminghaus, Göttingen, Germany, http://www.nisenet.org/viz_lab/image-collection)

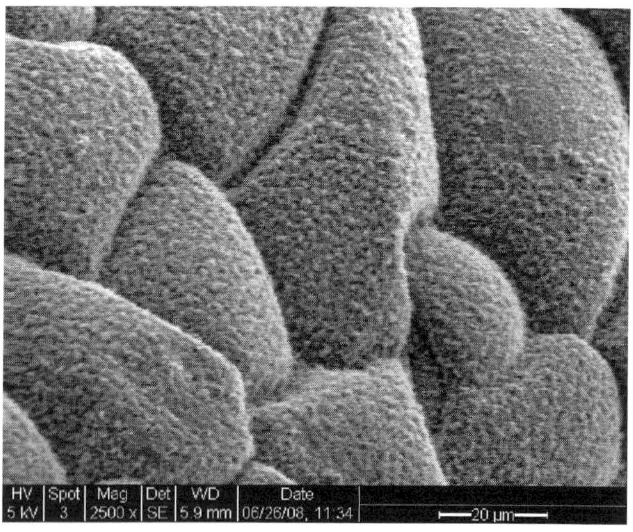

FIGURE 1.7
This image of the lotus effect shows leaf sections covered with wax nanocrystal bundles that trap air and force water to bead and roll off. (*Source:* A. Marshall, Stanford University, http://www.nisenet.org/viz_lab/image-collection)

TABLE 1.2

Timeline of Evolution of Nanotechnology

Year	Milestone
1857	Michael Faraday discovers colloidal "ruby" gold, demonstrating that nanostructured gold under certain illumination conditions produces different-colored solutions.
1936	Erwin Muller, at Siemens Research Laboratory, invents the field emission microscope, making it possible to achieve near-atomic resolution of imaged materials.
1947	John Bardeen, William Shockley, and Walter Brattain, at Bell Labs, create the first semiconductor transistor, ending the era of vacuum tubes and laying the foundation for solid-state electronics devices and the information era.
1950	Victor LaMer and Robert Dinegar develop the theory and a process for growing monodisperse colloidal materials. This controlled ability to fabricate colloids led to numerous industrial applications such as papers, paints, thin films, and dialysis treatment.
1951	Erwin Müller develops the field ion microscope, a way to image the arrangement of atoms at the surface of a sharp metal tip.
1956	Arthur von Hippel at MIT pioneered many concepts, and coins the term "molecular engineering" as applied to dielectrics, ferroelectrics, and piezoelectricity.
1958	Jack Kilby, at Texas Instruments, originates the concept of, designs, and develops first integrated circuit (IC).
1959	American Physical Society meeting on nanotechnology concepts is held. Richard Feynman describes foundational concepts in nanotechnology.
1965	Gordon Moore, Intel cofounder, forecasts trends in electronics; one trend known as "Moore's law" describes the density of transistors on an IC doubling every 12 months (later amended to every 2 years). Moore's law is still valid 50 years later, as nanotechnology replaces microtechnology in IC manufacturing.
1974	Norio Taniguchi gives a name to the new field in a scientific paper "On the Basic Concept of 'NanoTechnology.'"
1981	Gerd Binning and Heinrich Rohrer invent the scanning tunneling microscope at IBM's Zurich lab.
	Alexei Ekimov, in Russia, discovers nanocrystalline, semiconducting quantum dots in a glass matrix and conducts pioneering studies of their electronic and optical characteristics.
1985	Harold Kato, Sean O'Brien, Robert Curl, and Richard Smalley, at Rice University, discover the buckminsterfullerene (C_{60}), commonly known as a buckyball, which can be used to make carbon nanotubes (CNT).
1986	Invention of the atomic force microscope (AFM) makes it possible to view, manipulate, and measure matter at the nanoscale.
1989	IBM scientists spell IBM logo with 35 xenon atoms, demonstrating how nanoparticles can be manipulated.
1991	Carbon nanotube (CNT) is created by Sumio Iijima of NEC.
1992	C. T. Kresge and colleagues at Mobil Oil discover the nanostructured catalytic materials MCM-41 and MCM-48, presently used for refining crude oil, drug delivery, water treatment, etc.

TABLE 1.2 ***(Continued)***

Timeline of Evolution of Nanotechnology

Year	Milestone
1993	Moungi Bawendi, at MIT, invents a technique for controlled synthesis of nanocrystals (quantum dots).
1998	The White House forms the Interagency Working Group of Nanotechnology (IWGN) to investigate the state of the art in nanoscale and technology and to forecast possible future developments. The IWGN's study and report "Nanotechnology Research Directions: Vision for the Next Decade" (1999) defines the vision for and leads to formation of the U.S. National Nanotechnology Initiative (NNI) in 2000.
1999	Use of nanotechnology appears in the global marketplace.
2000	The National Nanotechnology Initiative (NNI) is launched to coordinate federal R&D efforts and promote U.S. competitiveness in nanotechnology.
2002	European Union launches gatherings called Nanoforums to educate the public about nanotechnology.
2003	Congress enacts the 21st-Century Nanotechnology Research and Development Act (Pub. L. No. 108-153).
	Naomi Halas, Jennifer West, Rebekah Drezek, and Renata Pasqualin at Rice University create gold nanoshells, which when "tuned" in size to absorb near infrared light, serve as a launching platform for integrated discovery, noninvasive diagnosis, and treatment of breast cancer.
2004	The European Commission adopts the communication "Toward a European Strategy for Nanotechnology," which proposes institutionalizing European nanoscience and nanotechnology R&D efforts with an integrated and responsible strategy.
2007	Angela Belcher and colleagues at MIT develop a lithium-ion battery with a common type of virus.
2008	The first NNI strategy for nanotechnology-related environmental, health, and safety (EHS) research is published.
2009–2010	Nadrian Seeman and colleagues at New York University create several DNA-like robotic nanoscale assembly devices.
2010	Toxic Substances Control Act requests regulation of commercial nanomaterial use.
	United Kingdom issues a lengthy report on nanotechnology and food, warning the country's food industry not to hide the use of nanotechnology
2011	The Senate Environment and Public Works Committee continues gathering evidence on revisions to the 30-year-old Toxic Substances Control Act to regulate the commercial use of nanomaterials.
	The NEST Subcommittee updates both the NNI Strategic Plan and NNI EHS research strategy based on inputs from stakeholders: government, academia, NGOs (nongovernmental organizations), public, etc.

Source: From "Nanotechnology Timeline," U.S. National Nanotechnology Initiative (http://www.nano.gov/timeline).

TABLE 1.3
Applications of Nanotechnology

Domain/area	Applications
Physics	Conductivity measurement of a single molecule
	Conduction through small junctions with few defects
	Observation of magnetic scattering of spin-polarized currents are possible
Optics	Fabrication of lasers and waveguides, optical switches, modulators, photonic crystals
Electronics	Fabrication of <10 nm by integrating electronics with MEMS and optics
MEMS/NEMS	Fabrication of micro- and nanomechanical systems-based sensors and actuators
Life sciences	Use of nanostructures to simulate biological structures, sort or detect cells or molecules, control cell growth
	Fabrication of nanoprobes for in vitro applications
Chemistry	Fabrication of monolayers, dendrimers, functionalized nanotube structures
	Development of chemical sensors and chemical mixing systems using microfluids
	Development of new materials and processes for nanostructure fabrication

Source: From National Nanotechnology Infrastructure Network (www.nnin.org).

TABLE 1.4
Nanotechnology Equipment and Processes

Equipment	Processes
Thin film deposition and growth	Evaporation, plasma chemical vapor deposition (CVD), sputtering, oxidation, molecular beam epitaxy (MBE), nanotube growth
Thin-film etching	Ion mill, reactive ion etching, plasma etching
Lithography	Electron-beam lithography, contact photolithography, projection photolithography
Synthesis, nanostructure coatings, and molecular assembly	Nanotube and nanowire growth, nanoparticle synthesis and characterization, molecular synthesis, electrophoretic particle deposition, porous materials, nanocomposites, self-assembly, cell culture, biological coatings, templates
Novel patterning, structuring, manipulation, and mesoscopic/microscopic assembly	Nanoimprint lithography, hot embossing, microcontact printing, nanoxerography, interferometric lithography, nanopore formation, scanned probe lithography, quantum dot growth/patterning, optical tweezers, nanosphere assembly, electrofluidic assembly of particles, chemically directed assembly of particles/devices, organomolecular coating
Inspection and characterization	Atomic force microscopy (AFM), fluorescence microscopy, confocal microscopy, ultrahigh-resolution electron microscopy, focused ion beam, spectroscopic ellipsometry

Source: From National Nanotechnology Infrastructure Network (NNIN) Capabilities (www.nnin.org). See Appendix A for photos of nanotechnology equipment and tools.

Characteristics of Bulk and Nanosize Materials

When the dimensions of solids are lowered to nanometers, the properties of these solids change: strength, melting temperature, color, electrical conductivity, thermal conductivity, reactivity, and magnetism. The magnitude of such change is size dependent; that is, the dimensions in nanometers determine the amount of change. When the dimensions of a material are reduced from large macroscopic values (meters or centimeters) to very small sizes, the properties remain the same at first, and then small changes start to take place until the dimensions fall below 100 nm, when significant changes in properties occur. Reduction in one dimension to nano size, but keeping the other two dimensions (2-D) large, results in a structure known as a *quantum well*. Reduction in two dimensions to nano size while one remains large results in a structure called a *quantum wire*, and the reduction of all three dimensions (3-D) to nano size results in the structure known as a quantum dot (QD).

At the nanoscale, the number of surface atoms is greater than that of the bulk material. The number of atoms in the nanocluster (surface-to-volume ratio) determines the change in electronic, structural, optical, and acoustic properties of the material. This surface-to-volume ratio relationship allows the possibility of using size in the nanometer regime to design and engineer materials with new and possibly technologically interesting properties (Poole and Owens, 2003). Also, between the gas and condensate phases of liquids and solids lies an intermediate state called *clusters*; these small agglomerations of atoms and molecules have properties not found in gases or condensed matter. This relatively new field of cluster science has the potential to make important advances in areas as far-reaching as material science, environmental chemistry, catalysis, and biochemistry (Castleman, 2007).

Applications of Nanotechnology

At the nanoscale, materials exhibit novel electronic, optical, and magnetic properties that have led to new applications of nanotechnology. Advances in nanoscience have enabled researchers to manipulate the behavior of a single cell, reverse disease, and repair and grow human tissues. Nanotechnology is supplying improved services in the areas of energy, lighting, computing, printing, and water filtration. Nanotechnology innovations such as quantum dots, semiconductor nanoparticles, carbon nanotubes, and nanoshells (see Table 1.5) have led to the fabrication of electronics hardware devices using the bottom-up approach in contrast to the present top-down approach.

TABLE 1.5

Potential Applications of Nanotechnology Innovations

Nanotechnology example	Potential applications
Buckyball: A soccer-ball-shaped molecule made of 60 carbon atoms 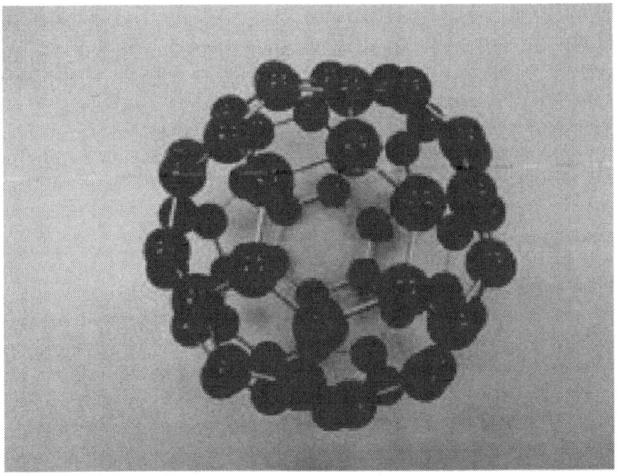	Composite reinforcement Drug delivery
Carbon nanotube: A sheet of graphite rolled into a tube (also see Appendix A) 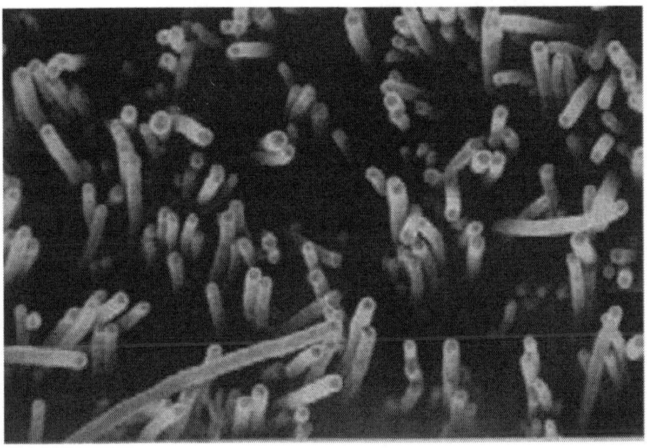	Fuel cells High-resolution displays Composite reinforcement
Quantum dot: A semiconductor nanocrystal whose electrons show discrete energy levels like an atom	Energy-efficient light bulbs Medical imaging
Nanoshell: A nanoparticle composed of silica surrounded by a gold coating	Medical imaging Cancer therapy

FIGURE 1.8 (See color insert)
The corral is an artificial structure created from 48 iron atoms (the sharp peaks) on a copper surface. The wave patterns in this scanning tunneling microscope image are formed by copper electrons confined by the iron atoms. The radius of the corral is about 7 nm. (*Source:* D. Eigler, IBM Almaden Research Center, http://www.nisenet.org/viz_lab/image-collection)

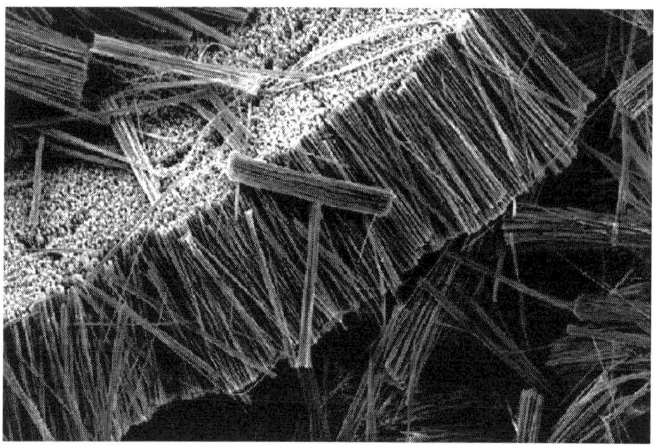

FIGURE 1.9
The orientation of the nickel nanowires shown in this scanning electron microscope can be changed by altering the direction of an applied magnetic field. The nanowires are 100–200 nm in diameter and about 20 μm in length. (*Source:* W. Crone, University of Wisconsin-Madison, http://www.nisenet.org/viz_lab/image-collection)

- Examples of nanotechnology
  - Quantum corral (see Figure 1.8)
  - Nickel nanowires (see Figure 1.9)
  - ZnO nanowires (see Figure 1.10)
  - A nanowire resting on a human hair (see Figure 1.11)

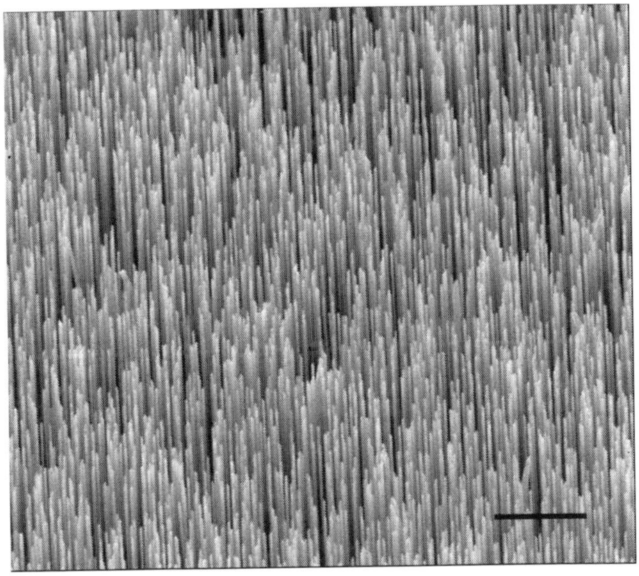

FIGURE 1.10
Vertical arrays of zinc oxide (ZnO) nanowires on a sapphire substrate. The sample displayed in the image is about 10 μm wide. (*Source:* S. Dayeh, University of California at San Diego, http://www.nisenet.org/viz_lab/image-collection)

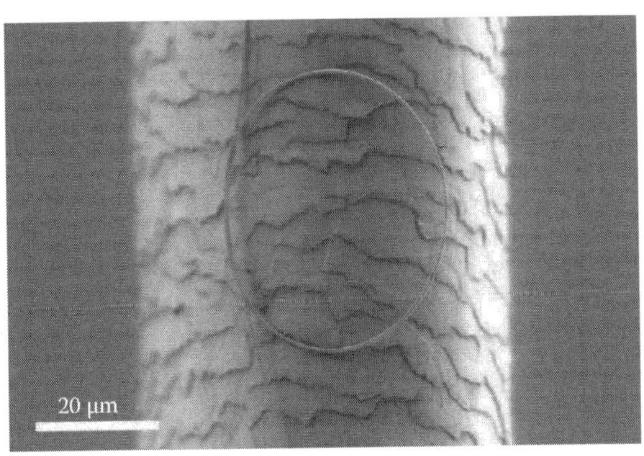

FIGURE 1.11
This scanning electron microscope (SEM) image compares the relative thickness of typical nanowire in comparison with a strand of human hair. The hair is about 100 μm in diameter; the nanowire's diameter is about 100 nm. (*Source:* S. Dayeh, University of California at San Diego, http://www.nisenet.org/viz_lab/image-collection)

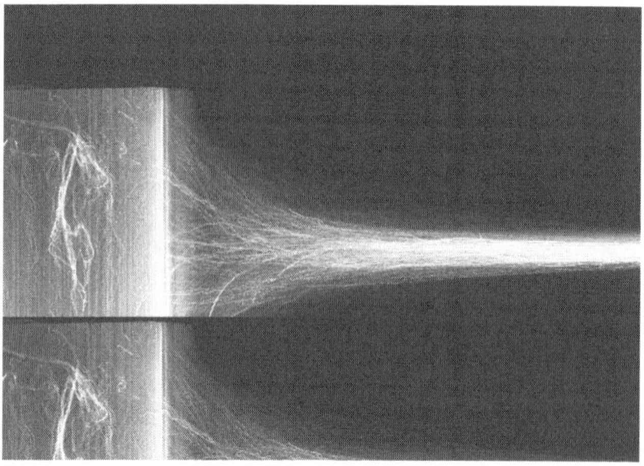

FIGURE 1.12
This scanning electron microscope (SEN) image shows nanotube yarn fibers drawn from a "nanotube forest." The yarn's diameter is about 1 μm. The nanotubes from which it is being drawn are each about 10 nm in diameter. (*Source:* Mei Zhang, UTD. Source: http://www.nisenet.org/viz_lab/image-collection)

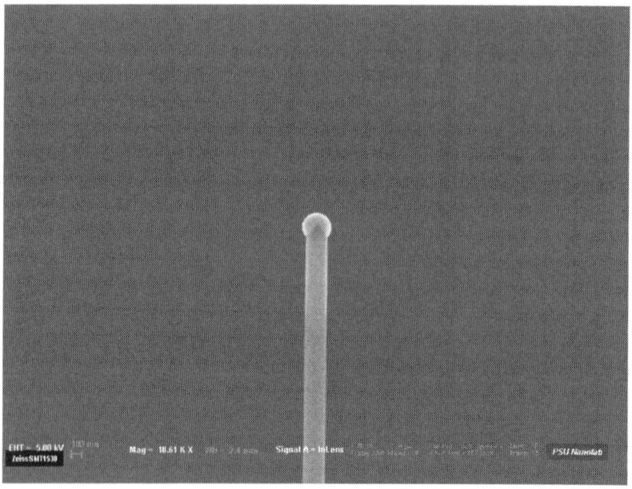

FIGURE 1.13
A silicon nanowire. (Courtesy Fonash Research Group, Penn State University.)

Nanotube yarn (see Figure 1.12)
Silicon nanowire (see Figure 1.13)
ZnO nanowire (see Figure 1.14)
ZnO nanowire growth (see Figure 1.15)

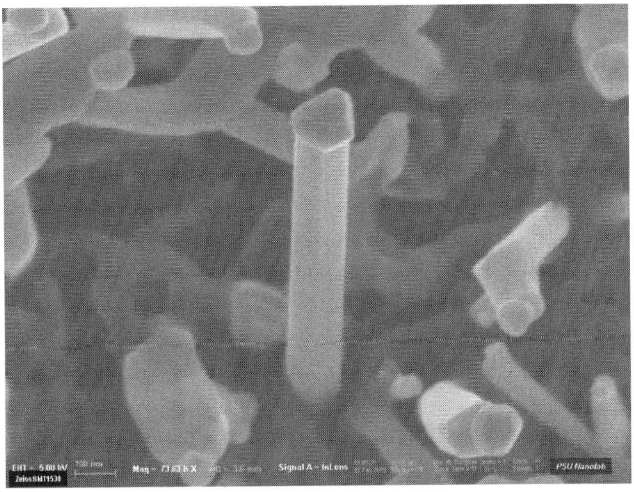

FIGURE 1.14
A ZnO nanowire. (Courtesy Fonash Research Group, Penn State University.)

FIGURE 1.15
ZnO nanowire growth. (Courtesy Fonash Research Group, Penn State University.)

Nanotechnology has numerous other applications (see Figure 1.16). In concrete manufacturing, the introduction of nanotubes increases the strength of cement. In plastics and polymers manufacturing, the use of nanoparticles allows precise control over color change and increases the strength of the materials. Nanocomposite materials are being used to manufacture various auto parts, including body panels, instrument panels, bumper fasciae,

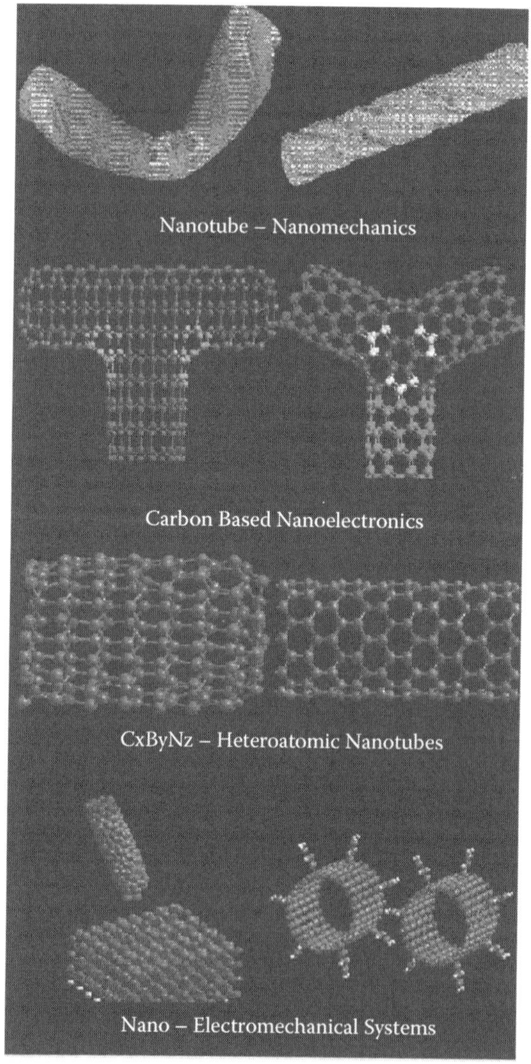

FIGURE 1.16 (See color insert)
Nanotube application. (Photo: NASA.)

front-end-module bolsters, etc. In the medical field, nanotechnology has been used to study cancer cell structure, and nanoparticles have been used for fluorescent imaging of tumors. Fluorescent nanoparticles have also been employed for the diagnosis and treatment of cancer. Nanoparticles of gold and palladium have been used to remove toxic chemicals from water. Nanoparticles have been used as catalysts, cancer therapies, sunscreens, antibacterial composites, and solar cells. More recently, advances in microfluidics and biosensors have made it possible to fabricate lab-on-a-chip devices.

Nanotechnology has enabled manufacturing processes to yield smaller, faster, and more energy efficient electronic, photonic, and optoelectronic devices. The first generation of nanotechnology (1990s–early 2000s) dealt with improving performance characteristics of existing micro materials; the second generation of nanomaterials (starting in 2006) is allowing the fabrication of devices that are cleaner, stronger, lighter, and more precise (Minoli, 2006).

Researchers at Penn State and the University of Southampton, U.K., have created a breakthrough in optical electronics by developing a technique to fill an optical fiber with micro and nano semiconductor structures. These structures enable the optical fiber to carry multiple wavelengths of light and thus tremendously increase data transmission capability.

New developments are making devices smaller and more efficient. About 25 nm was previously thought to be the limit for high-speed systems. At MIT, a 9-nm resolution of new high-speed electron-beam lithography was constructed. Scientists at Stanford University have created a nanoscale version of next-generation memory technologies using carbon nanotubes. Such nanotube-based memory will be much faster than the existing flash memory and can be manufactured at sizes less than 10 nm—a range where silicon-based memory cannot function (Courtland, 2011).

New developments in nanotechnology sensors have generated new interest in microelectromechanical (MEMS)-based systems. MEMS devices have applications in communications, medical diagnosis, commerce, the military, aerospace, and satellite systems. The extensive array of fabrication infrastructure developed for the manufacture of silicon integrated circuits has contributed to the development of systems having components of micrometer dimensions. Thin-film deposition processes coupled with advanced lithographic techniques are used to make MEMS devices. The major advantages of MEMS devices are miniaturization, multiplicity, and the ability to directly integrate the devices into microelectronics (Poole & Owens, 2003). The integration of MEMS and nanotechnology not only improves overall device performance such as reliability, but also reduces weight, size, power consumption, and production costs. Hybrid systems using MEMS and nanotechnology allow the design and development of novel electro-optic sensors, lasers, and radio frequency (RF)/mm-wave components such as phase shifters, switches, tunable filters, and micromechanical (MM) resonators, acoustic and infrared (IR) sensors, photonic devices, accelerometers, gyros, automobile-based control and safety devices, and unmanned aerial vehicles (UAVs) for battlefield applications (Jha, 2008).

When silver is transformed into very small particles, it exhibits antimicrobial properties that are not present in the bulk material. Following the recent discovery of these antimicrobial properties, silver structures and particles began to be used in the manufacture of various products. For example, a company manufactures socks with silver threads woven in with the cotton and other synthetic fibers to prevent odor. In the India-Pakistan subcontinent during the Mughal Period, it was a common practice in the households

TABLE 1.6

Nanotechnology Research and Applications in Food Science Areas

Food science area	Nanotechnology research/application
Food safety and quality	Development of sensors to detect a single molecule Research focus to improve the detection capability of biomolecules by exploiting nanomaterials such as carbon nanotubes, silicon wires, and zinc oxide nanorods Biosensors that can detect changes in the environmental conditions (temperature, oxygen, chemical contaminants, and microbial contaminants) Biosensors to allow detection of microorganisms and toxins
Ingredient technology and systems	Nanoparticle utilization for flavor, antioxidants, antimicrobials, bioactives, etc.
Food processing	Development of new membrane separation systems and catalysts
Food packaging	Development of low permeability, high-strength plastics High-performance edible packaging

to use metallic silver bowls for drinking water. Silver rings were used to treat various ailments, silver leaf was used to decorate desserts, and silver particles were mixed in various *Tibe-e-Unani* and *Ayuvedic* medicines.

Quantum dots (QD) have revolutionized the optoelectronics area. Quantum dots offer superior optical properties, high quantum efficiency (95%), and size-tunable emission. Quantum dots fabricated through the colloidal synthesis of semiconductors have many applications such as optical sources and in flat-panel displays (Xu et al., 2008).

Another major example of the application of nanotechnology is food science. Advances in nanotechnology have enabled food science to improve food safety and quality, food ingredient technologies, food processing, and food packaging. It is estimated that the nanofood market will be worth $20.4 billion by 2010. Table 1.6 presents a summary of nanotechnology research activities and applications in various food science areas (Floros, 2008).

Nanotechnology in the Military

Nanotechnology is also being employed for military applications. In the 1990s, the U.S. Department of Defense (DoD) identified nanotechnology as one of six "strategic research areas" (the other five being bioengineering sciences, human performance sciences, information dominance, multifunction materials, and propulsion and energetic sciences). The DoD nanotechnology program is classified into seven program component areas (PCAs), which emulate the PCAs of the U.S. National Nanotechnology Initiative (NNI). Table 1.7 lists the DoD PCAs (Berger, 2006).

In 2002, the U.S. Army established an advanced nanotechnology research center known as the Institute of Soldier Nanotechnologies (ISN) center at the

TABLE 1.7

DoD Nanotechnology Program Component Areas

	Description
PCA 1	Fundamental nanoscale phenomena and processes
PCA 2	Nanomaterials
PCA 3	Nanoscale devices and systems
PCA 4	Instrumentation research, meteorology, and standards for nanotechnology
PCA 5	Nanomanufacturing
PCA 6	Major research facilities and instrumentation acquisition
PCA 7	Societal dimensions

TABLE 1.8

Research Programs at Institute of Soldier Nanotechnologies (ISN)

Research program	Focus of research/themes
SRA 01	Development of lightweight, multifunctional nanostructured fibers and materials to serve as building blocks for clothing and other gear to provide soldier protection and survivability
SRA 02	Development of battle-suit medicine (nanostructured actuators, films and functional surfaces for battlefield medicine, noninvasive medical monitoring and drug delivery)
SRA 03	Development of ballistic and blast protection (lightweight nanoarchitectures for ultra-strong and energy absorbing materials, materials and structures for blast protection and injury mitigation, lightweight nanocrystalline alloy fibers for blast protection)
SRA 04	Development of chem/bio detectors for protection (functional, switchable, and microbicidal nanocoatings; ultrasensitive nanoengineered chemical detectors; and Nanostructured Origami™)
SRA 05	Development of nanoscale materials/components/devices that allow protection and survivability capabilities for the soldier (nanoelectronics, integrated fiber and fiber–fabric systems, non-RF fabric–enabled communications, and novel lightwave systems)

Source: From Institute for Soldier Nanotechnologies (web.mit.edu/isn/research/).

Massachusetts Institute of Technology (MIT). Initially, a $50-million contract was awarded to ISN to pursue a long-range vision of how technology can make soldiers less vulnerable to threats from the enemy and the environment. According to the ISN website (web.mit.edu/isn/) the ultimate goal is to create a twenty-first-century battlefield suit that combines high-tech capabilities with light weight and comfort. Table 1.8 summarizes research at ISN.

A significant amount of money is being spent on researching and developing these new nano weapons in the United States. The cost of going to nano research for the military has gone up dramatically over the past years. In 2000, the cost was only $70 million; in 2003 $243 million. In 2006, it was estimated that the Department of Defense had spent $436 million on nanotechnology, with only $1 million of that going to risk-related research (Berger, 2006).

In the international arena, the incorporation of nanotechnology in military hardware, like nuclear technology, could lead to a nano-arms race between nations. Countries with nanotechnology expertise will develop a new generation of chemical and biological weapons. This may lead to destabilization of existing arms control agreements such as the Biological Weapons Convention. A NATO (North Atlantic Treaty Organization) study group observed:

> The potential for nanotech-driven innovations in chemical and biological weapons are particularly disquieting as they can considerably enhance the delivery mechanisms of agents or toxic substances. The ability of nano particles to penetrate the human body and its cells could make biological and chemical warfare much more feasible, easier to manage and to direct against specific groups or individuals. (Berger, 2006)

Nanotechnology-based weapons applications will be invisible to the naked eye or satellites or chemical detection. It would be safe to say that a new arms race has quietly begun. Christine Peterson, vice president of public policy and cofounder of Foresight Institute, observed:

> The United States is to stay significantly ahead at all times, and work must begin now because other countries may have certain advantages. In China, for instance, nanotechnology research can be conducted at one-sixth of what it costs in the United States, and the number of engineering graduates exceeds that of the United States.... China is very technologically ambitious, and while the threat is not imminent, the United States could face stiff competition in 20 or 30 years. (Peterson, 2011)

These actions of countries competing to develop new nanoweapons are dangerous and can lead to unfortunate outcomes.

An example of a way nanotechnology research and development can be used in military applications is "smart dust," a sensor technology that uses tiny wireless microelectric mechanical sensors (MEMS). Smart dust is used to transmit information and track items or people. A group of these tiny remote sensors can form a two-way mesh radio network where devices communicate with each other, and thus could track the movements of people (Musthaler, 2004).

The application of nanotechnology in the military is a new area; there is a need at the international level to regulate the risks of and exposure to nanotechnology. Many ethical issues—ranging from security, privacy, environment, profit, and humanistic values—arise and must be examined. In war, the dilemma is that ethical principles are rarely followed. Each side may take unethical actions that give them an advantage. Globally, concern has been expressed about the use of depleted uranium in wars fought during the past two decades. The use of nanotechnology in chemical and biological weapons would be another prime example. The application of nanotechnology in military armament and explosives poses exposure risk to the environment and to humans. Battlefield blasts and explosions could release toxic nanoparticles in the air. The nanotechnology coating used on armor could also release toxic nanoparticles in the air when impacted.

Privacy is another very important issue. Presently, with everyone focused on terrorism in the United States and around the world, security is becoming more important, but it is also attenuating the domain of personal privacy. Nanotechnology can enable the use of miniature cameras, microphones, recording devices, and other such gadgets that would be so small that no one would notice them (Mnyusiwalla, Daar, & Singer, 2003). Will privacy become a relic of the pre–Information Revolution era?

Nanotechnology in Medicine

The application of nanotechnology in medicine will revolutionize health care for the next generation. The vision of the future is one of harnessing the qualities of nanotechnology to eventually provide health care that operates purely from a preventive state, identifying and stopping potential sources of disease and illness in the body before they even get started (Morrow, Bawa, & Wei, 2007). The nanomedicine market is expected to grow significantly in the coming years. It is estimated that in the United States, the demand for nanotechnology-related medical products (nanomedicine, nanodiagnostics, nanodevices, and nanotech-based medical supplies) will increase more than 17% per year to $53 billion in 2011 and $110 billion in 2016. The greatest short-term impact of nanomedicine will be in therapies and diagnostics for cancer and central nervous system disorders (Langwith, 2010).

Nanomedicine has the potential to revolutionize cancer treatment. Cancer has the distinctive challenges of being from the self, individually diverse, and heterogeneous; cancer is currently the top killer of persons younger than 85 years. New techniques are currently being developed at the National Cancer Institute (NCI) to improve early detection and surveillance of cancer. Novel cancer imaging and therapies using carbon nanotubes (CNT) are also under investigation to improve the imaging and treatment of cancer (Hinkel et al., 2011).

Nanotechnology and Education and the Workforce of the Future

What will be the impact of nanotechnology on education and the workforce of the future? It is estimated that by 2020, nanotechnology will be a $3-trillion-per-year global industry. In 1997 the investment in nanotechnology stood at $430 million, rising to more than $9 billion in 2004. The U.S. share of this is estimated at $1 trillion, which would correspond to about 5% of gross domestic product (GDP), compared to 0.8% of GDP in 2010. Presently, more than 800 products have been developed using nanotechnology (see Appendix B). Significant R&D investments were made by Brazil, India,

Russia, and other countries after the second generation of active nanotechnology products came to market in 2006 (Roco, 2011).

The development of nanotechnology requires multidisciplinary teams of highly trained researchers with backgrounds in biology, medicine, mathematics, physics, chemistry, material science, electrical engineering, and mechanical engineering. For innovative advances in nanotechnology, the key is researchers with expertise in multiple subsets of these disciplines, since so many implications and fields are linked to a nano "micro-revolution" (Khan & Qadeer, 2010).

Education and training in nanotechnology require special laboratory facilities that can be quite expensive (see Table 1.4). The cost of creating and maintaining nanotechnology facilities is a major challenge for educational institutions. But by using innovative approaches such as inter-university collaboration, academia-industry partnerships, and Web-based remote access to nanofabrication facilities, educational institutions can overcome the cost-related challenges and thus help students and faculty to become innovative nanotechnology researchers (Khan & Qadeer, 2010).

To address these new demands of the global marketplace, a skilled workforce is required that can move from industry to industry without retraining. The new workforce will consist of researchers, technicians, and educators. To develop this workforce, new interdisciplinary educational programs need to be developed and revised. To promote the conditions for the development of a nanotechnology workforce in the United States, the National Nanotechnology Initiative (NNI) program was established in 2001. The NNI (www.nni.org) provides a vision of the long-term opportunities and benefits of nanotechnology. The goals of the NNI are as follows:

- Advance a world-class nanotechnology R&D program
- Foster the transfer of new technologies into products for commercial and public benefit
- Develop and sustain educational resources, a skilled workforce, and the supporting infrastructure and tools to advance nanotechnology
- Support responsible development of nanotechnology

Ethical and Social Implications of Nanotechnology

> Addressing societal and ethical issues is the right thing to do and the necessary thing to do. It is the right thing to do because as ethically responsible leaders we must ensure that technology advances human well-being and does not detract from it. It is the necessary thing to do because it is essential for speeding technology adoption, broadening

> economic and societal benefits, and accelerating and increasing our return on investment.
>
> **Philip J. Bond (2003)**

Society is at the threshold of a revolution that will transform the ways in which materials and products are created. How will this revolution develop? The opportunities that will develop in the future will depend significantly upon the ways in which a number of challenges are met. As we design systems on a nanoscale, we develop the capability to redesign the structure of all materials—natural and synthetic—along with rethinking the new possibilities of the reconstruction of any and all materials. Such a change in our design power presents tremendous social and ethical questions. To enable our future leadership to make decisions for sustainable, ethical, economic nanotechnological development, it is imperative that we educate all nanotechnology stakeholders about the short-term and long-term benefits, limitations, and risks of nanotechnology. The social implications of nanotechnology encompass so many fundamental areas such as ethics, privacy, environment, and security (Gutierrez, 2004).

Nanotechnology, like its predecessor technologies, will have an impact on all areas. For example, in health care it is very likely that nanotechnology in the area of medicine will include automated diagnosis. This in turn will translate into fewer patients requiring physical evaluation, less time needed to make a diagnosis, less human error, and wider access to health-care facilities. If the average life span of humans increases with the use of nanomedicines, the large number of elderly persons requiring medical attention will result in increased health expenditures (Moore, 2004).

It is essential for the nanotechnology stakeholders to strive to achieve four social objectives (Khan, 2006):

1. Developing a strong understanding of local and global forces and issues that affect people and societies
2. Guiding local/global societies to appropriate uses of technology
3. Alerting societies to technological risks and failures
4. Developing informed and ethical personal decision making and leadership to solve problems in a technological world

In 2001, Roco and Bainbridge of the National Nanotechnology Initiative observed:

> Over the next 10–20 years, nanotechnology will fundamentally transform science, technology, and society. However, to take full advantage of opportunities, the entire scientific and technology community must set broad goals; creatively envision the possibilities for meeting societal needs; and involve all participants, including the general public, in exploiting them. (Roco & Bainbridge, 2001)

In 2001, the National Science Foundation (NSF) sponsored a workshop, and from it a report entitled *Societal Implications of Nanoscience and Nanotechnology* was published. The report is a collection of expert opinions and recommendations about how to: accelerate the advantages of nanotechnology while minimizing the risks; improve education and research; and provide guidance to the contributors of major institutions. The report makes a number of recommendations regarding how to inform, educate, and involve general public; how to create an infrastructure for interdisciplinary assessment of the scientific, technological, and social impacts in the short, medium, and long terms; and how to educate a workforce. One of their recommendations reads (Hornyak, Tibbals, Dutta, & Moore, 2009; Roco & Bainbridge, 2001):

> Make support for social and economic research studies on nanotechnology a high priority. Include social science research on the societal implications in the nanotechnology research centers, and consider creation of a distributed research center for social and economic research. Build openness, disclosures, and public participation into the process of developing nanotechnology research and development program direction.

Advances in nanotechnology also present numerous challenges and risks in health and environmental areas. Nanotechnology risk-assessment methods and protocols need to be developed and implemented by the regulatory bodies. Eric Drexler (1989), author of *Engines of Creation,* has identified four challenges in dealing with the development, impact, and effects of nanotechnology on society:

1. The Challenge of Technological Development (control over the structure of matter)
2. The Challenge of Technological Foresight (sense of the lower bounds of the future possibilities)
3. The Challenge of Credibility and Understanding (clearer understanding of these technological possibilities)
4. The Challenge of Formulating Public Policy (formulating polices based on understanding)

Table 1.9 presents a summary of the present and future impacts of nanotechnology. Table 1.10 lists nanotechnology application areas and potential benefits and risks. At the conclusion of the first decade of the twenty-first century, humankind faces tremendous challenges. Nobel laureate Richard E. Smalley has identified the top ten problems as (Smith & Granvist, 2010):

1. Energy
2. Water
3. Food
4. Environment

TABLE 1.9

Present and Future Impacts of Nanotechnology

Nanotechnology area	Present impact	Future impact
Dispersions and coatings	Thermal barriers Optical barriers Image enhancement Ink-jet materials Coated abrasive slurries Information-recording layers	Target drug delivery/gene therapy multifunctional nanocoatings
High-surface-area materials	Molecular sieves Drug delivery Tailored catalysts Absorption/desorption materials	Molecule-specific sensors Large hydrocarbon or bacterial filters Energy storage Gratzel-type solar cells
Consolidated materials	Low-loss soft magnetic materials High-hardness, tough WC/Co cutting tools Nanocomposite cements	Superplastic forming of ceramics Ultrahigh-strength, tough structural materials Magnetic refrigerants Nanofilled polymer composites Ductile cements
Nanodevices (See Figure 1.16)	GMR read heads	Terabit memory and microprocessing Single-molecule DNA sizing and sequencing Biomedical sensors Low-noise, low-threshold lasers Nanotubes for high-brightness displays
Biological (See Figures 1.17 and 1.18)	Biocatalysis	Bioelectronics Bioinspired prostheses Single-molecule-sensitive biosensors Designer molecules

Source: "Technological Impact: Present and Potential" by Siegel, Richard, Hu, Evelyn and Roco, M.C., Eds. (1999) WTEC Panel report Nanostructure Science and Technology, R & D Status and Trends in *Nanoparticles, Nanostructured materials, and Nanodevices*. Kluwer Academic Publishers, Norwell, MA 02061, available online at: http://www.wtec.org/loyola/IWG.Worldwide.study/execsum.pdf with permission.

5. Poverty
6. Terrorism and war
7. Disease
8. Education
9. Democracy
10. Population

How can technology in general and nanotechnology in particular help humankind deal with these problems? The following questions need to be answered by the all stakeholders of nanotechnology—consumers, businesses, industries, and national and international regulatory bodies:

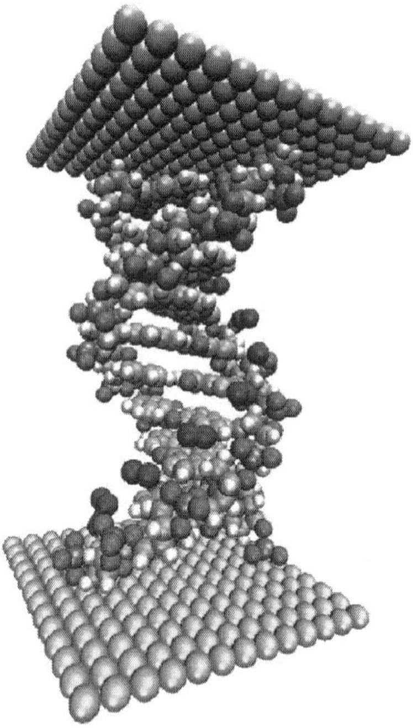

FIGURE 1.17 (See color insert)
Engineered DNA strand. (Photo: NASA.)

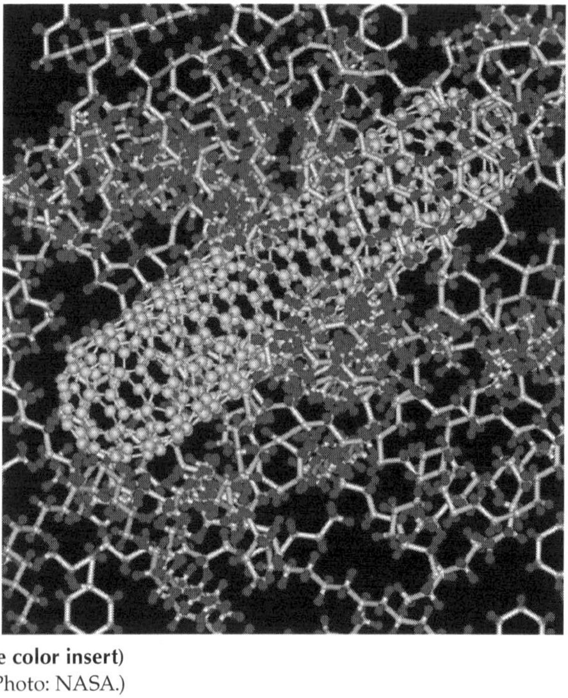

FIGURE 1.18 (See color insert)
CNT polymer. (Photo: NASA.)

TABLE 1.10

Nanotechnology: Benefits and Risks

Nanotechnology Application Area	Potential Benefits
Ecology	Nanoparticles have extremely high surface areas compared to their volume; this characteristic makes them ideal for the fabrication of: New catalysts Heat-reflection layers Aerogels for transparent damping layers in solar architectures Superthermal insulators Transparent layers showing resistance against wear and abrasion or antidamping properties
Energy	Nanodevices will allow cleaner energy production and improved storage. Small, compressed particles enable new photovoltaic cells, with simpler structure than conventional ones. Plastics are to be used as the electrode materials.
Dematerialization	Nanocrystalline particles, with a monodisperse size distribution, are to be formed into macroscopic parts with higher strength and resistance against mechanical and thermal load, despite the smaller amounts of material required. These parts can be hard and flexible in a unit and can replace scarce materials. New processing techniques using remarkably lower temperatures, offer possibilities for minimizing energy consumption during component fabrication.
Health	More effective pharmaceuticals with reduced secondary effects are due to improved basic understanding of the efficacy of natural human substances like insulin or hormones. New forms of localized drug delivery systems are based on the potential of water soluble, pharmacologically active substances when attached to nanometer-size particles. External control and incorporation of target information takes place by incorporating magnetic particles or antibodies into the drug delivery system.
Electronics/ telecommunications	Logical building blocks for digital electronics are based on particles or molecules. Nanosize electronic data storage and processing systems are possible.
Nanotechnology Application Area	**Potential Risks/Detrimental Effects**
Genetics/medicine/ health care	Artifacts based on nanotechnology incorporate genetic material or have genetic modification or repair as an objective. If the artifact incorporates some kind of computing and sensing element, say for the controlled delivery of a drug, additional risks arise for the patient if these elements should malfunction. Invasion of privacy and of the human body through the planting and implanting of computing-cum-communication devices without the knowledge of those affected has been done. The security and safety of a person is a problematic issue, since it will be very difficult initially to detect the presence of nanosize artifacts that are capable of breaching security and harming the individual. In warfare, controlled distribution of biological and nerve agents may become feasible.

TABLE 1.10 *(Continued)*

Nanotechnology: Benefits and Risks

Nanotechnology Application Area	Potential Risks/Detrimental Effects
Materials/ composites	The general problem with composite materials is that they are more difficult to recycle and consume more energy during recycling than pure materials. Wide-scale introduction of composite materials can increase environmental problems.
Self-assembling and self-replicating nature of nanotechnological processes	In the manufacturing area, many processes will need to be redesigned to embody new principles, particularly relating to containment of active or waste products.

Source: From *Socio-Economic Research and Nanoscale Science and Technology, Societal Implications of Nanoscience and Nanotechnology*, by M. Meyer, 2001, Arlington, VA: National Science Foundation, pp. 224–225. (http://www.wtec.org/loyola/nano/NSET.Societal.Implications/)

How should we manage exposure to nanowaste by humans and the environment?

How can we develop effective risk-management strategies dealing with the uncertainties of nanotechnologies?

How should people be educated about the potential benefits and risks of nanotechnology?

What will be the impact of nanotechnology on the privacy of the individual?

How can the proliferation of nanotechnology-based weapons be prevented?

Is humankind playing the role of God when it comes to applying nanotechnology in the areas of biotechnology and genetics?

Will nanotechnology be the catalyst for future military interventions and hegemony by the ruling world powers?

What will be the impact of human-machine union? How it will affect the ideology of existentialism relative to the "being" in general?

What will be the impact of nanotechnology on the developing countries?

What will be the governing moral principles for advancement of life by artificial means through nanotechnology?

What will be the social impact of implementing nanotechnology-based systems at workplaces, homes, industries, government, etc.?

Will nanotechnology intensify the divide between the haves and the have nots?

Law and Policy Issues

Nanotechnology is remaking the world at an alarmingly fast pace. Presently, nanoproducts are being developed in an environment of regulatory vacuum at national and international levels. The science of nanotechnology is quite different from the traditional microlevel knowledge base. As new understandings in chemistry, physics, and genetics are being reached, and nano products developed and marketed, it has become imperative that future policy makers and lawyers understand the complexities and implications of nanotechnology in order to serve the society with adequate insight vis à vis the issues of intellectual property, patents, and commercialization.

The biggest question is how to deal with uncertainty and risk assessment. In 2007, on July 31, Mike Honda, a U.S. congressional representative from California, summed up nanotechnology uncertainty in a press statement: "Uncertainty is one of the major obstacles to the commercialization of nanotechnology—uncertainty about what the risks might be and uncertainty about how the federal government might regulate the nanotechnology in the future," (http://www.nanwerk.com/news/newsid=2301.php). Risk assessment is defined as a process intended to calculate or estimate the risk to a given target organism, system, or (sub) population, including the identification of attendant uncertainties, following exposure to a particular agent, taking into account the inherent characteristics of the agent of concern as well as the characteristics of the specific target system (WHO, 2004).

All new and emerging technologies pose novel challenges and uncertainties; in the domains of science and technology these can be dealt with through additional research, but in the realm of law and regulation, immediate answers are sought. Uncertainty must be dealt with in law and regulation, and in the absence of straightforward regulations, methodologies are used to address uncertainty. One such methodology used for dealing with uncertainty in regulation is risk assessment (Sutton, 2011).

Conclusion

This chapter provided an introduction to the state of the art in nanotechnology by exploring the areas of evolution, characteristics, equipment, and processes. It also explored the applications, societal impact, ethical implications, benefits, and risks of nanotechnology. New demands on the curricula and workforce of the future were also explored. It is imperative for all nanotechnology stakeholders to guide society to the appropriate uses of nanotechnology, alert society to the risks and potential failures, and provide a vision to society in helping to solve societal problems that are related to nanotechnology. As the convergence of multiple disciplines occurs in the form of nanotechnology discoveries and developments, the time to remake the world will become shorter, and thus the nanotechnology stakeholders will be relied upon to guide society by making future decisions for humane, just, and responsible utilization of this "invisible" technology.

Nanotechnology has the potential to change society, positively or adversely. It will affect everyone. Therefore all members of society—all stakeholders—should have a voice in its development and commercialization phases. Presently, nanotechnology is in its infancy and, unfortunately, there is a lack of knowledge about its effects on humans and the environment vis-à-vis its application in the areas of food, agriculture, and medicine. As humankind marches forward, the key question is: How should we manage the risks and uncertainties? As one moves through the rest of this book, other chapters provide new insights and solutions to dealing with the uncertainties of nanotechnology.

Questions

1. What is nanotechnology? How does it differ from microtechnology?
2. How will nanotechnology impact society?
3. Do you think that the advantages offered by the nanotechnologies will outweigh their drawbacks?
4. Do you think that nanotechnology will enable humans to live longer and healthier? Yes or No. Explain your answer.
5. How will nanotechnology increase or decrease the stress level of humans?
6. What will be the impact of nanotechnology on privacy?
7. How do you think higher education will change with nanotechnologies developing in our society? What kind of courses will be even more urgent?
8. If you were to predict a nanotechnology future for humans in fifteen years, how would you describe it?
9. Discuss the risks and benefits of nanotechnology.
10. What kinds of risk-assessment methods and protocols need to be developed for nanotechnology?
11. What are the ramifications of the invisible nature of nanotechnology?
12. What is the uncertainty principle? Can we use it for dealing with risk assessment?
13. Using Internet resources, complete the following table:

Nanoparticle/structure	Applications	Ethical/social concerns
Quantum dots		
Silica		
Zinc oxide		
Iron oxide		
Liposomes		
Nano barcodes		
Alumino silicate		
Lithium iron phosphate		

14. Summarize your learning of new ideas and information presented in this chapter by completing the following flowchart:

Flow Chart

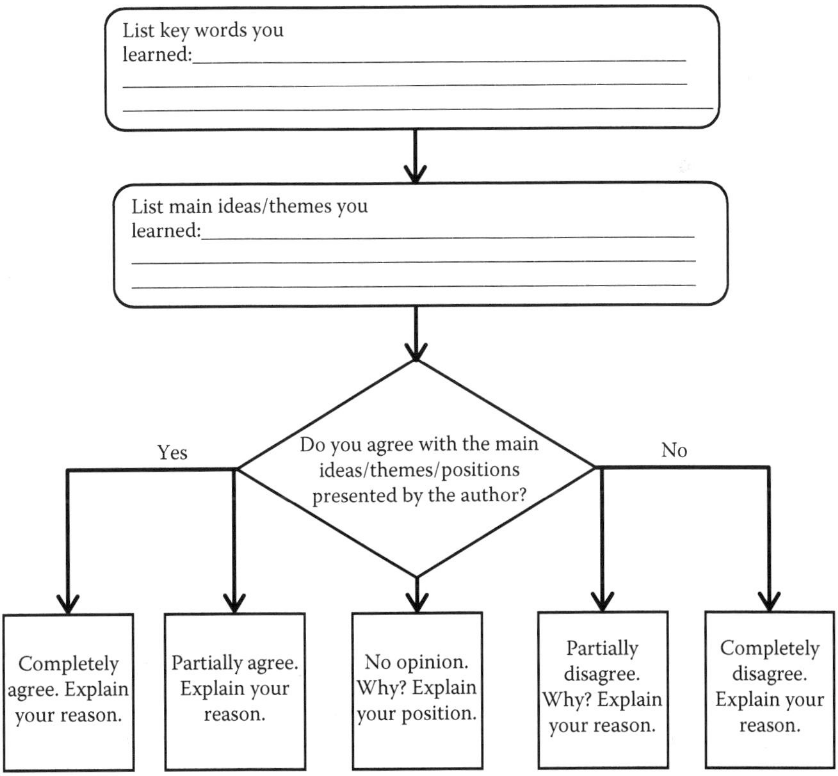

Scenario 1: Accident at a Nanoparticle Manufacturing Plant

Japan
December 1, 20XX

The Fujika industrial plant is located near the seashore in northern Japan. Fujika is a world leader in producing silver, gold, ZnO, and TiO_2 nanoparticles for commercial applications. Fujika employs around 800 people. At 9:30 p.m., a powerful earthquake shakes the plant and severely damages the nanoparticle manufacturing plant, causing the dispersion of nanoparticles in the air. Within the next 10 minutes, while employees are following the emergency procedure to evacuate the plant, a 32-ft-high tsunami wave comes in and floods the plant. All nanoparticles are carried away by the waters of the tsunami wave.

Response

1. Discuss the potential impact of nanoparticle dispersion in air and water on the local ecology.
2. What could have been done to prevent the dispersion of nanoparticles in water and air?
3. Do you think nanotechnology is a safe technology? Explain your answer.

Scenario 2: Release of Carbon Nanotubes into the Atmosphere

Chicago
October 4, 20XX, 10:19 a.m.

There is an accident on Lakeshore Drive. A truck carrying canisters of nanotubes collides with a school bus. The canisters are severely damaged; some fall on the road, and nanotubes are dispersed. As injured students are evacuated from the school bus, they are exposed to the nanotubes in the air. A strong wind blowing from Lake Michigan carries the nanotubes toward the nearby apartment buildings. The air intakes of the HVAC (heating, ventilation, and air conditioning) systems of some apartment buildings are in close proximity to the accident. It is suspected that some nanotubes will get into the residential units through the central air-conditioning system.

Response

1. Discuss the potential impact of the exposure of nanotubes to the students and residents of the apartment buildings.
2. Develop a strategy to prevent such accidents in the future.

Case Study 1: Language and Technology— Communication from the Past: Relics of Mohenjodaro

> Mohenjodaro or Moenjodaro, call it any way you wish, is an archaeological complex that is [a] landmark in the long journey that Man has so far completed and is fractional to what lies ahead of him. The galaxies after galaxies await the small feet of Man to trample them. Archaeology teaches us that if Man could survive with so small a number of "designs" as he has when he was organizing Mohenjodaro Society—civilizing himself—this same Man will reach stars known and unknown. The journey that Man has undertaken knows no end.

Haleem Sharar (Husain & Sharar, 1992)

> Mohenjodaro is a book, the opening chapter and the closing chapter of which are not available for study by the archaeologist—and [to] this add the element of ambiguity to fully understand the rise and fall of the most glorious part of the civilization of the south Asian Sub-Continent. (Husain & Sharar, 1992)

Dr. Ahmed Nabi Khan
Director General, Archeology and Museums, Pakistan

Mohenjodaro (2600–1800 BC), located in the province of Sind, Pakistan, was one of the largest urban settlements of the Indus Valley civilization. It was rediscovered in 1922 by British archeologists. Mohenjodaro has been designated a UNESCO World Heritage Site. Figure 1.19 contains two seals depicting the ancient language of Mohenjodaro.

Languages change over time because of different factors such as the evolution of human understating and knowledge, human needs to communicate clearly and efficiently, changing needs of society and business, and the impact of the advancement of technology. With the exponential growth of scientific and technological disciplines, humankind has benefited tremendously, but in the process has generated a vast amount of toxic waste. Present-day waste is stored and labeled using today's languages. The critical questions are: What will be the impact of technology on present-day languages? Will these languages survive over the next few centuries? Will the future generations be able to understand the "Cautionary labels" attached to the toxic waste left behind by the present generation?

Response

Use Internet and library resources to answer the following:

1. What are the top ten languages of the world today? What is their future? How will technological developments change these languages?

FIGURE 1.19 (See color insert)
Two seals depicting the language of Mohenjodaro (2600–1800 BC). (Photos: Ahmed S. Khan.)

2. What factors cause the death of languages?
3. Develop a strategy for labeling "toxic waste" so that future generations can comprehend the intended messages of danger depicted on these labels.

Case Study 2: Industrial Revolution, Information Revolution, and Nanotechnology Revolution

Toward the end of the eighteenth century and beginning of the nineteenth century, the Industrial Revolution started in England and swiftly spread through Europe and the United States (see Figure 1.20). The three primary characteristics of the period were substitution of (Landes, 1999):

1. Machines—rapid, regular, precise, tireless—for human skill and effort
2. Inanimate for animate (human and animal) sources of power, thus opening an almost unlimited supply of energy
3. New and far more abundant mineral materials for vegetables and animal materials

FIGURE 1.20 (See color insert)
During the Industrial Revolution, the demand for machine operators was high. The nanotechnology revolution is seeking graduates who have complex thinking skills and possess a multidisciplinary knowledge base.

The result was a substantial rise in productivity and in per capita income. The essence of the Industrial Revolution—"substitution of competition for the medieval regulation"—was made possible by (Doty, 1976; Toynbee, 1958):

1. The rapid increase in population, with a general decline of the agricultural population
2. Significant increase in the agricultural production by fewer farmers because of the end of the common-field system and the introduction of selective breeding, crop rotation, and improved implements

3. Ascendance of the factory system over the domestic system (cottage industry), especially in the cotton, textile, and iron industries, thanks to machines
4. Growth of trade due to improved communications
5. Redistribution of wealth into the hands of capitalists and farmers at the expense of laborers

Response

Using Internet resources, answer the following questions:

1. Complete the following table by highlighting the characteristics of the Information Revolution and the potential impact of the Nanotechnology Revolution on the nature of work, communication, economy, productivity, food, energy sources, trade, social etiquette, etc.

	Information revolution	**Nanotechnology revolution**
Nature of work		
Communication		
Productivity		
Food		
Energy sources		
Trade		
Social interaction		
Other		

2. How will nanotechnology shape the nature of jobs in the future?
3. What will be the impact of nanotechnology on the education of the future? How will it impact students and faculty?
4. How can nanotechnology provide a global energy system based only on wind, water, and solar power (WWS)? What is the global demand for energy? What is the global supply for energy? How many wind turbines, solar thermal plants, and hydroelectric plants will be needed to implement a global WWS system?

Case Study 3: Nanotechnology and Compressed Natural Gas (CNG)—A Win-Win Combination

Today the world is facing a growing problem of carbon emissions. A number of prominent scientists believe that carbon emissions are responsible for global warming. According to data from the U.S. Environmental Protection Agency (EPA), combustion of one gallon of petroleum in a vehicle produces nineteen pounds of carbon dioxide. There is an urgent global push toward alternative energy sources to reduce the carbon footprint at local and global levels. The use of compressed natural gas (CNG) in vehicles can reduce carbon dioxide emissions by 20%–30%; carbon monoxide emissions by 70%–90%; nitrogen oxide (NOx) emissions by 75%–95%; particle emissions by up to 90%; and volatile organic compound (VOCs) emissions by 89%. Thus natural gas vehicles (NGVs) yield a positive and direct impact on air quality and the environment (see Figure 1.21).

According to data provided by the International Association of Natural Gas Vehicles (www.iangv.org), the developing countries are leaders in the use of NGVs. With more than 2.7 million NGVs on the road in 2010, Pakistan has become a leader among the countries using CNG vehicles. Iran, with 1.95 million NGVs, is at second place, and Argentina, with 1.9 million NGVs,

FIGURE 1.21 (See color insert)
CNG-based public-transportation vehicles at Penn State University, State College, PA. (Photo: Ahmed S. Khan.)

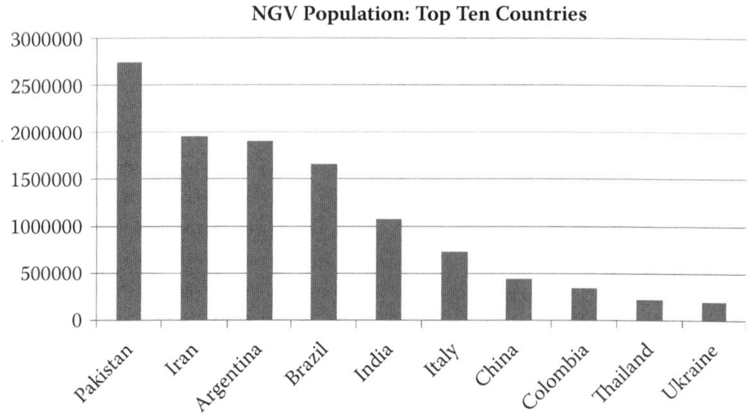

FIGURE 1.22
Global top 10 NGV countries.

holds third place in the list of top three countries using CNG vehicles (see Figure 1.22).

If nanotechnology-based exhaust systems could be developed to capture particulates and prevent their escape into the environment, the use of CNG coupled with nanotechnology has the potential to revolutionize transportation vehicles for promoting sustainable economic development.

Response

Using Internet resources, answer the following questions:

1. Why isn't the United States a leader in the use of CNG vehicles? What factors are vital in promoting CNG vehicles in the United States?
2. What is the state of the art in areas of emission-control technology? How is nanotechnology being used to design filters to capture pollutants from motorized vehicles and industrial applications? How will such systems help in reducing the carbon footprint at the local, national, and international levels?

Case Study 4: Brain Damage and Potential of Nanomedicine

Nadeem is a 45-year-old man. He suffers from dementia and requires help in managing his life. When he was three months old he was hospitalized for treatment of pneumonia. During his treatment at the hospital, one of the junior doctors adminstered an adult dose of antibiotics to three-month-old

Nadeem. As a result, Nadeem's nervous system was severely demaged, and he started to have seizures. His parents took special care of him. When Nadeem was fourteen years old, he was playing baseball with his friends when he suddenly had a seizure. He collapsed and hit his head on the ground and went into a coma. He remained in a coma for three weeks until doctors used electrical shock therapy to bring him out of the coma. After recovering, his seizures were gone, but he did not have any short-term memory. Neurologists believed that the front lobe of Nadeem's brain had been damaged during his coma. His parents are now in their 80s and are worried who will take care of their son when they are gone. They have started to look for some experimental therapy using nanomedicine that can restore Nadeem's short-term memory.

Response

Using Internet resources, answer the following questions:

1. What are some of the new nanomedcines and nanotechnology techniques that can be used to improve Nadeem's short-term memory? Does nanomedicine hold any potential in reversing brain injuries to restore brain memory functions?
2. What are some of the options for Nadeem's parents for taking care of him after they are gone?

Case Study 5: Nanotechnology and Globalization

Today, increasing globalization has become a controversial phenomenon. On one side of the debate are the proponents, who believe that globalization is the panacea for all global economic problems, and on the other side are the opponents, who claim that globalization is promoting economic inequity. On the pro side of the debate is Thomas Friedman, Pulitzer Prize–winning columnist of the *New York Times*. In his book *The World Is Flat: A Brief History of the Twenty-First Century* (2007), he argues that technological innovations and advances have led to a new form of globalization that is flattening the global playing field. Friedman argues that globalization has empowered millions of tech-savvy people all over the world, and enabled them to learn the new rules of the global economic system. On the opposite side is Joseph Stiglitz, winner of the 2001 Nobel Prize in economics. In his books *Globalization and Its Discontents* (2002) and *Making Globalization Work* (2006), he paints a less rosy picture of globalization, arguing that it produces winners and losers. Further, under the current rules of international trade, there are more losers than winners. He argues that a "one size fits all" solution does not work for everyone. He observes, "The hope of globalization was that it was like a rising tide lifting all boats, and so the poorest would see themselves go up. But the way it was managed, it might be more likened to riptide knocks over the weakest boat, and without life vests, without safety nets, a lot of the people in those weaker boats drowned." Why hasn't globalization worked? Stiglitz observes, "Economic globalization has outpaced political globalization."

Response

1. Which side of the globalization debate do you support? Explain your position.
2. What will be the impact on the gap between the rich and poor at the global level? Will the use and application of nanotechnology narrow or widen this gap? What is the Gini coefficient? (Use Internet resources to answer this question.)

Case Study 6: Nanotechnology Lab Environment and Safety

Johnny X is an international graduate student in the materials research program at a prestigious university. Johnny X arrived at the campus two years ago and in just one year completed his course work for the doctoral program. After passing the required comprehensive exam, he started his thesis work under the supervision of Professor Einstein Jr. During the last ten months, Johnny X has made excellent progress in his research work on the characterization of TiO_2 nanoparticles. Professor Einstein Jr. is very happy with the excellent pace of Johnny's research work. But during the past five months, Johnny has violated lab safety protocols. He has been warned four times by the lab safety officer. However, in his zeal to finish his research as soon as possible so that he can return to his native country, Johnny continues to violate lab safety protocols. Johnny's work involves the use of toxic gases and chemicals. Sometime he does not wear protective glasses, and sometime he does not follow procedures for using high-vacuum systems. As a result, toxic gases have been released in the lab environment. His peers believe that by not following the lab safety protocols, Johnny is endangering their safety.

The university's regulations state that after five violations, a student must be suspended from the program for one semester, and if the student still violates the policies, then the student should be expelled from the university. Johnny is not alone in violating lab safety policies: Large numbers of graduate and undergraduate students seem to not follow lab safety protocols. Johnny has been asked by the dean of his program to appear for a disciplinary hearing. Johnny's adviser is worried that if the disciplinary committee decides to expel Jonny from the university, he will return to his native country and will never be able to finish his doctoral program. Johnny is also afraid that if he returns to his country without finishing the doctoral degree, he will consider himself responsible for bringing shame to his family and country.

Response

If you were a member of a disciplinary committee:

1. What would be your recommendation? Should Johnny be expelled from the university? Or should he be put on probation for one term?
2. How can the university enforce lab safety policies?

Using Internet resources, answer the following questions:

3. How can schools promote best practices in lab safety protocols?
4. What are some design factors that must be considered in building laboratories that will provide a safe environment for researchers?

References

Barber, D. J., & Freestone, I. C. (1990). An investigation of the origin of the color of the Lycurgus Cup by analytical transmission electron microscopy. *Archaeometry, 32*(1), 33–45.

Berger, M. (2006). *Military nanotechnology: How worried should we be?* Retrieved November 25, 2011, from http://www.nanowerk.com/spotlight/spotid=1015.php

Bond, P. (2003, December). Preparing the path for nanotechnology. In M. C. Roco & W. S. Bainbridge (Eds.), *Nanotechnology implications: Maximizing benefits for humanity* (pp. 16–21). Report of the National Nanotechnology Initiative Workshop, Arlington, VA.

Castleman, A. W., Jr. (2007). Building nanomaterials one superatom at a time. *MRI, Focus on Materials,* (Summer), 10–13. Retrieved November 26, 2011, from http://www.mri.psu.edu/publications/focus_on_materials/summer-2007/

Courtland, R. (2011). Flash challengers work well at nanoscale. *IEEE Spectrum, 48*(8), 18.

Doty, C. S. (1976). *The industrial revolution* (p. 2). Huntington, NY: Robert E. Krieger Publishing.

Drexler, E. (1989). *The challenge of nanotechnology*. Retrieved November 25, 2011, from http://www.halcyon.com/nanojbl/NanoConProc/nanocon1.html#anchor528648

Ehrmann, Robert. (2008). Hands-on nanofabrication workshop for educators, Center for Nanotechnology Education and Utilization, Penn State University, State College, PA, December, 2–4.

Floros, J. (2008, December). *Nanoscale science and technology for food*. Paper presented at the Hands-on Nanofabrication Workshop for Educators, Center for Nanotechnology Education and Utilization, Penn State University, University Park, PA.

Fonash, S. (2008, December). *The overall picture: The world of nanotechnology*. Paper presented at the Hands-on Nanofabrication Workshop for Educators, Center for Nanotechnology Education and Utilization, Penn State University, University Park, PA.

Friedman, T. L. (2007). *The world is flat: A brief history of the twenty-first century*. New York, NY: Macmillan.

Gutierrez, E. (2004). *Privacy implications of nanotechnology*. Electronic Privacy Information Center. Retrieved November 26, 2011, from http://www.epic.org/privacy/nano/

Harper, T. (2011, July). Global funding of nanotechnologies and its impact. White paper, Cientifica. Retrieved November 25, 2011, from http://cientifica.eu/blog/research/market-reports/nanotech-funding-2011

Hinkel, G., Farrel, D., Hook, S., Panaro, N., Ptak, K., & Grodzinski, P. (2011). Cancer therapy through nanomedicine: The National Cancer Institute's plan to defeat cancer through engineered design. *IEEE Nanotechnology, 5*(2), 6–12.

Hjorth, L., Eichler, B. A., & Khan, A. S. (2007). *Technology and society: Issues for the 21st century and beyond*. Upper Saddle River, NJ: Prentice Hall.

Hornyak, G., Tibbals, H., Dutta, J., & Moore, J. (2009). *Introduction to nanoscience and nanotechnology*. Boca Raton, FL: CRC Press.

Husain, I, & Sharar, H. (1992). *Mohenjodaro, Moenjodaro: When human being said "go" for civilisation: A study of Indus Valley civilisation from Mehrgarh, Kotdiji, Harappa to Lakheenjodaro*. Karachi, Pakistan: International Press & Publications Bureau.

Jha, A. R. (2008). *MEMS and nanotechnology-based sensors and devices for communications, medical and aerospace applications*. New York, NY: CRC Press.

Khan, A. S. (2006). *Examining the impact of nanotechnologies for science, technology and society (STS) students*. Paper presented at 2006 ASEE Conference, Chicago, IL.

Khan, A. S., & Qadeer, A. (2010). *Incorporating Social and Ethical Implications of Nanotechnology in Science, Technology and Society (STS) Courses*. Paper presented at 2010 ASEE Conference, Louisville, KY. June 20–23.

Landes, D. S. (1999). In *The wealth and poverty of nations* (chap. 13). New York, NY: Norton.

Langwith, J. (2010). *Nanotechnology*. Farmington Hills, MI: Greenhaven Press.

Meyer, M. (2001). Socio-economic research and nanoscale science and technology: A European overview. In M. C. Roco & W. S. Bainbridge (Eds.), *Societal implications of nanoscience and nanotechnology* (pp. 217–240). Arlington, VA: National Science Foundation. Retrieved November 27, 2011, from http://www.wtec.org/loyola/nano/NSET.Societal.Implications/nanosi.pdf

Minoli, D. (2006). *Nanotechnology applications to telecommunications and networking*. Hoboken, NJ: John Wiley & Sons.

Mnyusiwalla, A., Daar, A., & Singer, P. (2003). "Mind the gap": Science and ethics in nanotechnology. *Nanotechnology, 14*(3), 9–13.

Moore, F. (2004). Implications of nanotechnology applications: Using genetics as a lesson. *Health Law Review, 10*(3), 9–15.

Morrow, K. J., Jr., Bawa, R., & Wei, C. (2007). Recent advances in basic and clinical nanomedicine. *Med. Clin. North Am., 91*, 805–843. Retrieved November 27, 2011, from http://www.rpitechnology.com/files/Recent_Advances_in_Basic_and_Clinical_Nanomedicine_%209_07.pdf

Musthaler, L. (2004, June 7). *Smart dust gets smaller and more intelligent.* Retrieved November 26, 2011, from http://www.networkworld.com/newsletters/2004/0607techexec1.html

National Nanotechnology Initiative. (2011). *Nanotechnology timeline: Premodern examples of nanotechnologies.* Retrieved November 26, 2011, from http://www.nano.gov/nanotech-101/timeline

Overby, D. (2011, September 23). Tiny neutrinos may have broken cosmic speed limit. New York Times, p. A8. Retrieved November 27, 2011, from http://www.nytimes.com/2011/09/23/science/23speed.html

Peterson, C. (2011). *Foresight Institute: Officers and staff.* Retrieved November 26, 2011, from http://www.foresight.org/about/Peterson.html

Poole, C., & Owens, F. (2003). *Introduction to nanotechnology.* Hoboken, NJ: John Wiley & Sons.

Roco, M. C. (2011, May). Nanotechnology: From discovery to innovation and socioeconomic projects. *CEP Magazine, 107*(5), 21–27. Retrieved November 26, 2011, from http://www.aiche.org/cep/issues/2011-05/index.aspx

Roco, M. C., & Bainbridge, W. (Eds.). (2001). *The societal implications of nanoscience and nanotechnology.* NSET Workshop Report. National Science Foundation. Retrieved November 26, 2011, from http://www.wtec.org/loyola/nano/NSET.Societal.Implications/nanosi.pdf

Schroeder, S. (2011, September 23). Scientists turn brain's visual memories into a mind-blowing video. *Mashable.* Retrieved November 27, 2011, from http://news.yahoo.com/scientists-turn-brains-visual-memories-mind-blowing-video-083203371.html

Smith, G. B., & Granvist, C-G. S. (2010). *Green nanotechnology: Solutions for sustainability and energy in the built environment.* Boca Raton, FL: CRC Press.

Stiglitz, J. E. (2002). *Globalization and its discontents.* New York, NY: W. W. Norton & Co.

Stiglitz, J. E. (2006). *Making globalization work.* New York, NY: W. W. Norton & Co.

Sutton, V. (2011). *Nanotechnology law and policy.* Durham, NC: Carolina Academic Press.

Ten things you should know about nanotechnology. (n.d.). Retrieved November 27, 2011, from http://www.nanowerk.com/nanotechnology/ten_things_you_should_know_6.php

Toynbee, A. (1958). *The industrial revolution.* Boston, MA: Boston Press.

WHO. (2004). *IPCS risk assessment terminology.* World Health Organization (WHO). Retrieved November 27, 2011, from http://www.who.int/ipcs/methods/harmonization/areas/ipcsterminologyparts1and2.pdf

Xu, J., Zhanao, T., Zhu, T., Zhang, F., Hedrick, B., Pickering, S., et al. (2008). *Developing bright and color-saturated quantum dot light emitting diodes: Towards next generation displays and solid state lighting.* Paper SaO4, presented at Asia Optical Fiber Communication and Optoelectronic Exposition and Conference, Center for Nanotechnology Education and Utilization, Penn State University, University Park, PA. Retrieved November 27, 2011, from http://www.opticsinfobase.org/abstract.cfm?uri=AOE-2008-SaO4

Appendix A: Nanotechnology Equipment and Tools

Modern vacuum systems are at the heart of nanotechnology equipment that is used for thin-film deposition, etching, lithography, patterning, inspection, and characterization. (Photos courtesy of Ahmed S. Khan in collaboration with Sebastian Meader, Terry Kuzma, and Dave Johnson, Center for Nanotechnology Education and Utilization, Penn State University, State College, PA.)

FIGURE 1.23 (See color insert)
Vacuum coating system: Chemical vapor-deposition system allows fabrication of thin films for various nanotechnology applications.

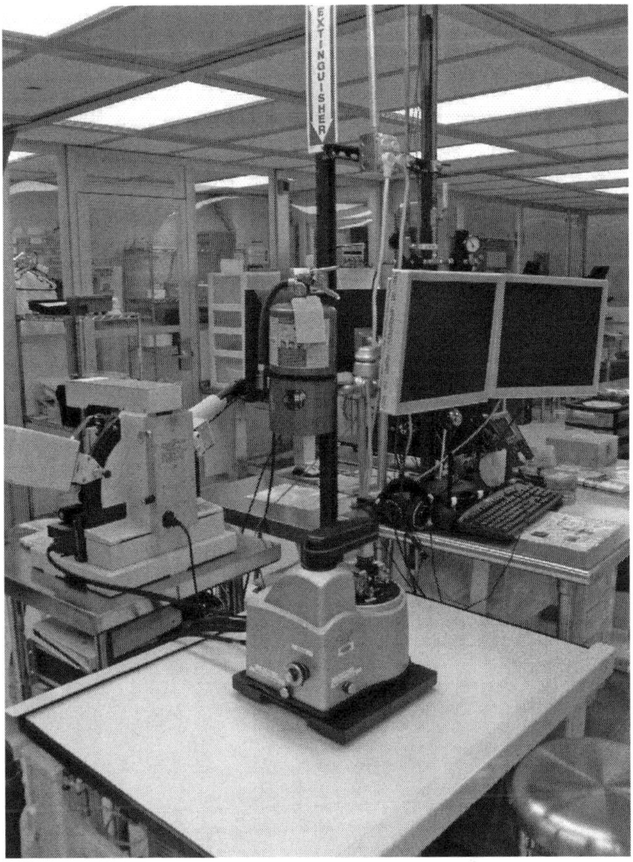

FIGURE 1.24
AFM system is used for characterization of nanotechnology samples.

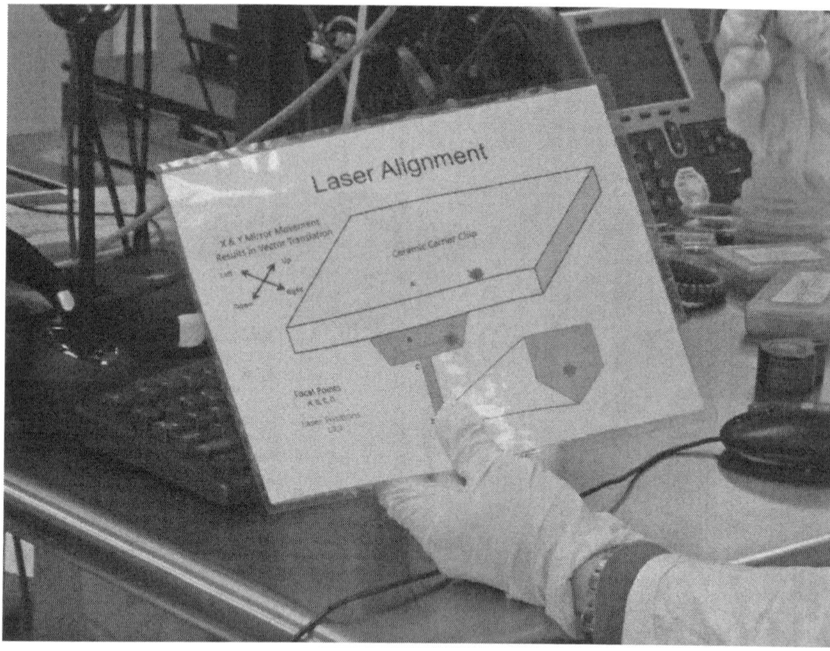

FIGURE 1.25
AFM system creates 3-D profiles of samples.

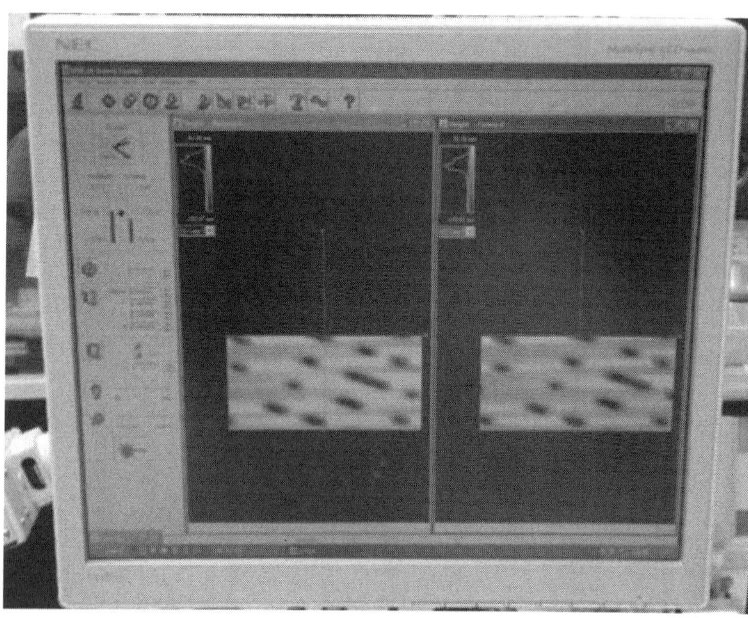

FIGURE 1.26 (See color insert)
AFM profile of a sample.

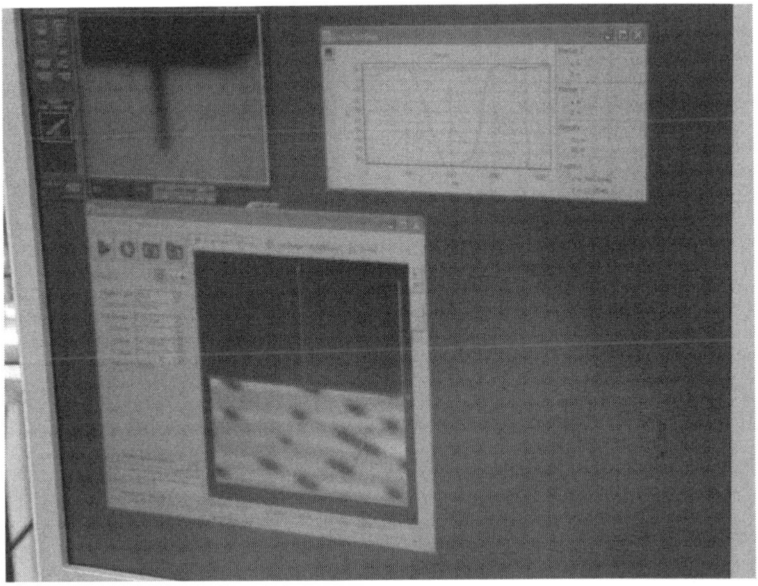

FIGURE 1.27 (See color insert)
Another view of AFM profile of a sample.

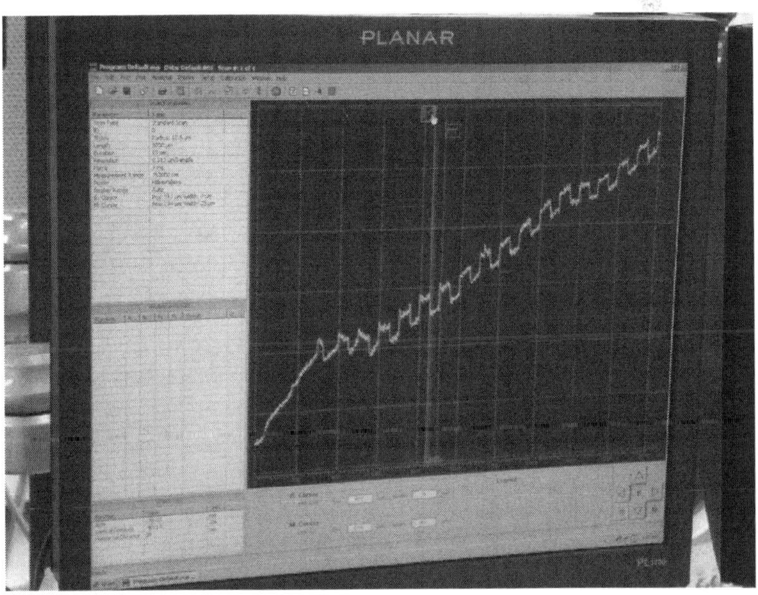

FIGURE 1.28
Another example of AFM profile of a sample.

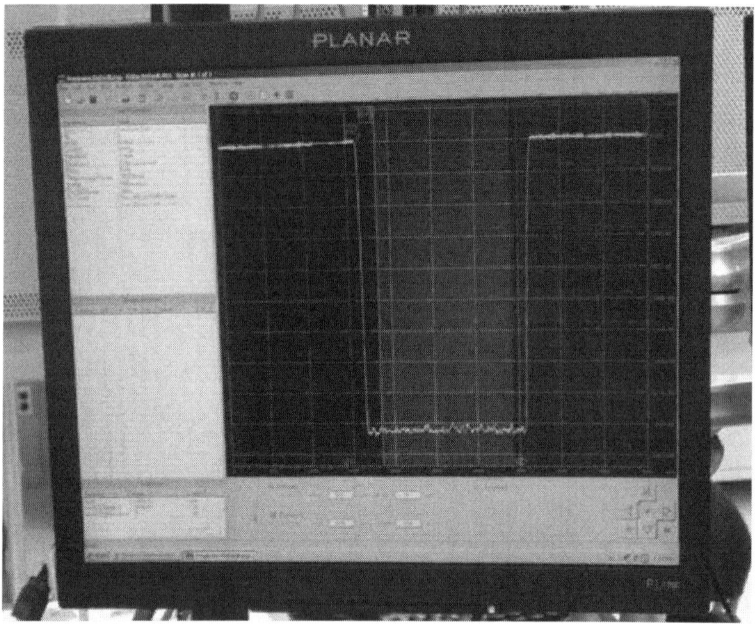

FIGURE 1.29
AFM software allows averaging of surface variations of a sample.

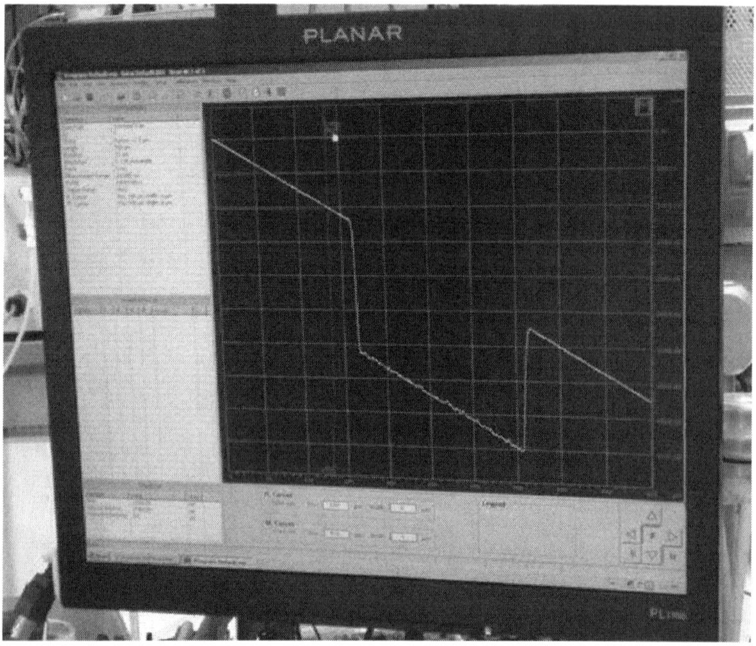

FIGURE 1.30
AFM software allows changing the orientation of an observed profile.

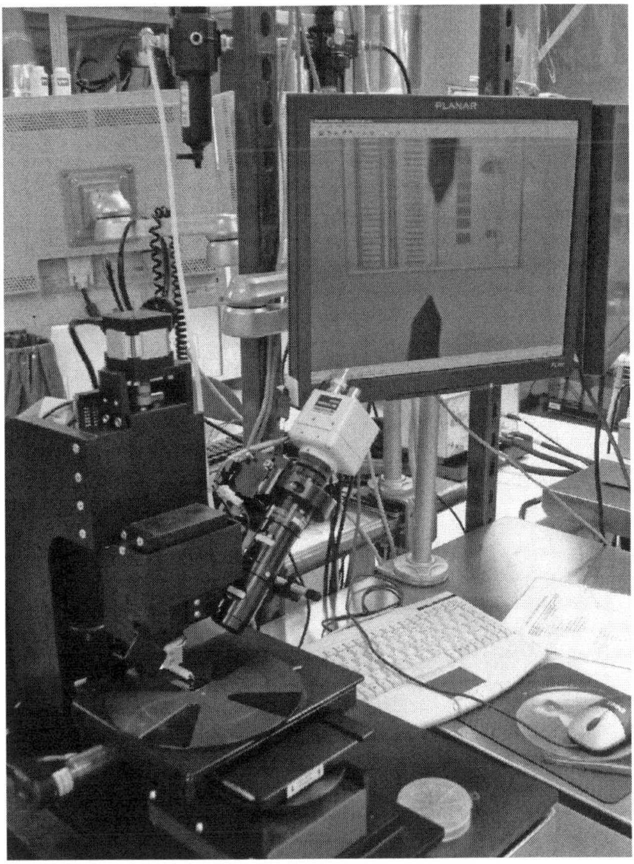

FIGURE 1.31
Optical profile meter provides a 2-D profile of samples.

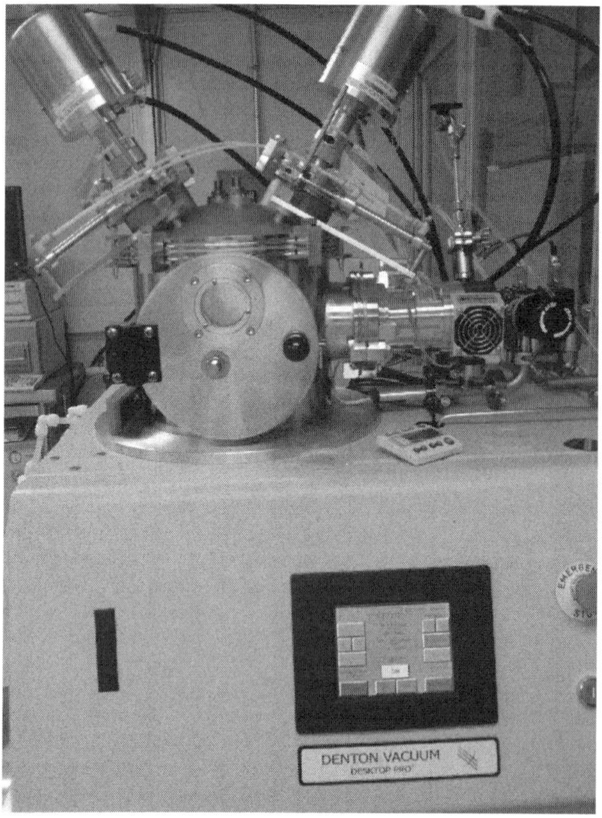

FIGURE 1.32
RF sputtering system is used for deposition of thin films.

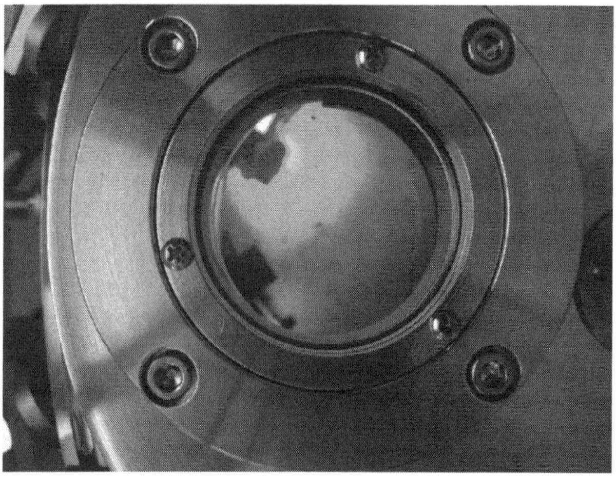

FIGURE 1.33
Plasma in an RF sputtering system allows atom-by-atom deposition of thin films.

FIGURE 1.34
Nanotechnology fabrication of components uses sophisticated vacuum systems.

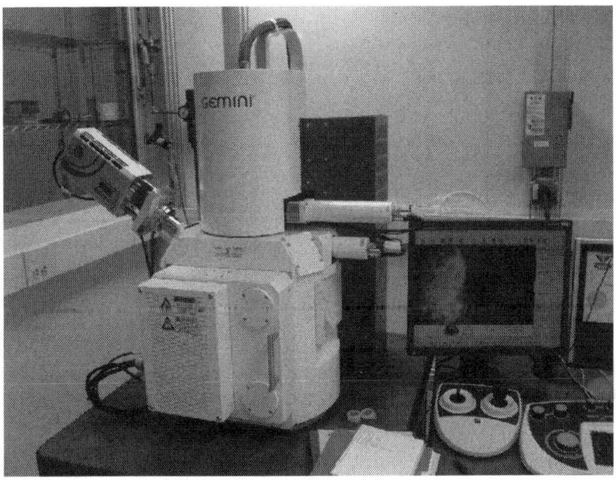

FIGURE 1.35
Scanning electron microscope (SEM) allows inspection of fabricated samples with a resolution of micro- to nanometer scale.

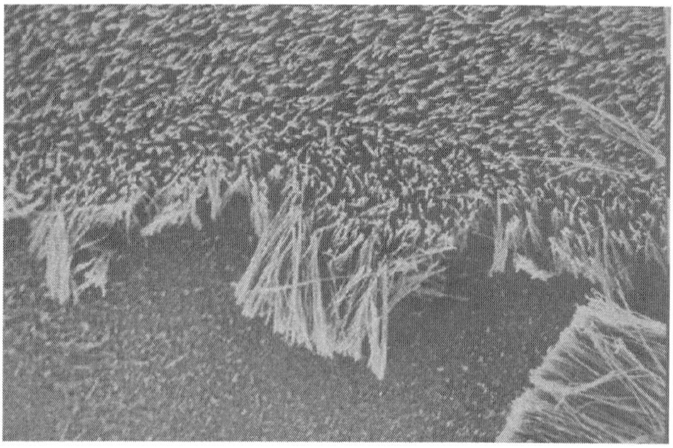

FIGURE 1.36
SEM image of Carbon nanotubes (CNTs).

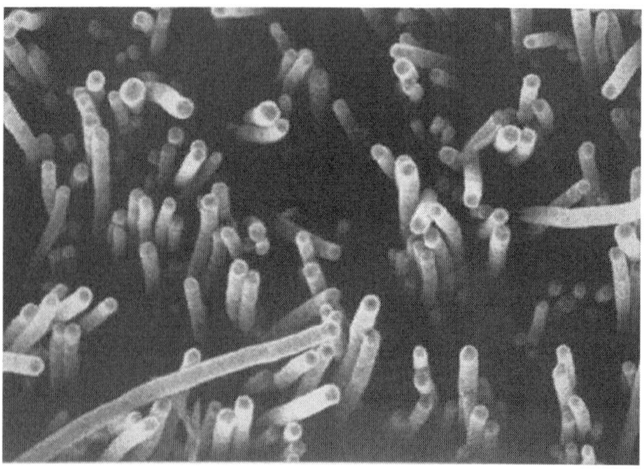

FIGURE 1.37
SEM image of CNTs.

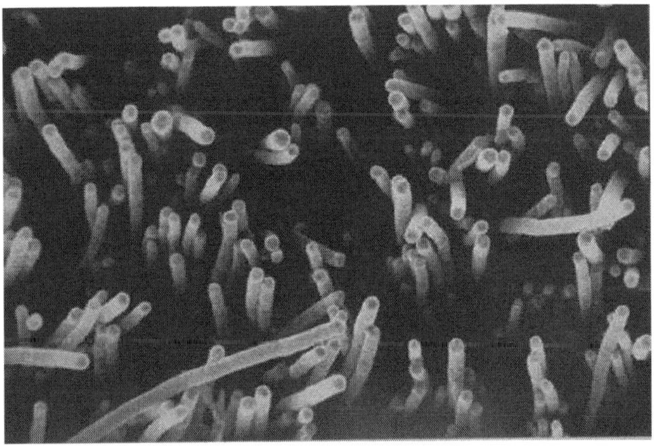

FIGURE 1.38
Another SEM orientation of CNTs.

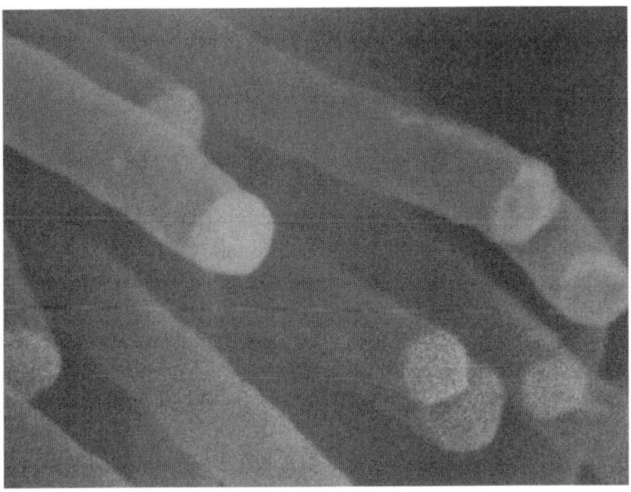

FIGURE 1.39
Magnified SEM image of CNTs.

FIGURE 1.40
Internal components of SEM.

FIGURE 1.41 (See color insert)
Internal view of SEM multiple specimen holder.

FIGURE 1.42 (See color insert)
SEM allows loading of multiple specimens for characterization.

Appendix B: Nanotechnology Products

The photos in Appendix B are courtesy of Ahmed S. Khan in collaboration with Robert Ehrmann, director of education and outreach, Penn State Center for Nanotechnology Education and Utilization, Penn State University, State College, PA.

FIGURE 1.43
More than 800 nanotechnology products are available in the market.

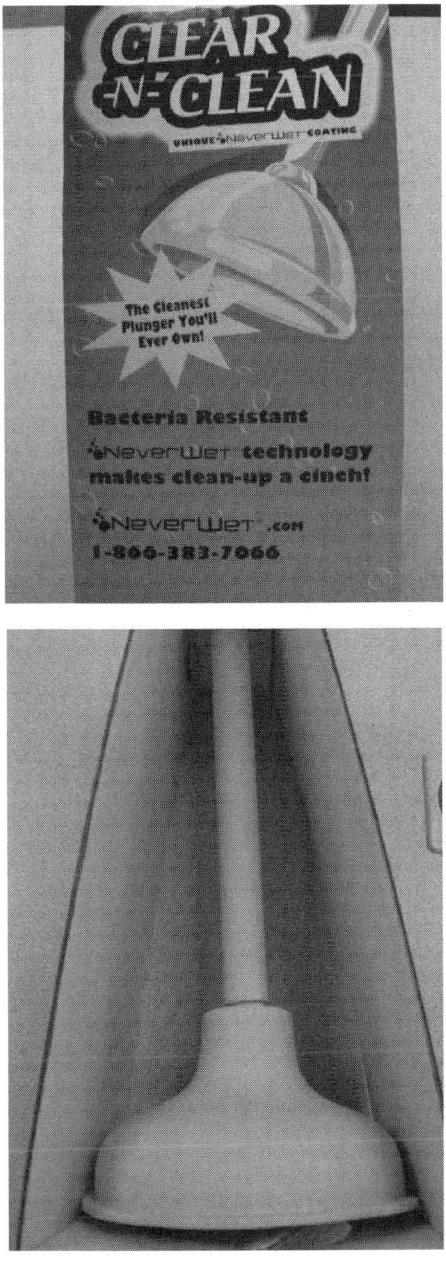

FIGURE 1.44
Use of nanotechnology creates hydrophobic surface on a plunger.

FIGURE 1.45
Use of carbon nanotubes (CNT) in tennis rackets and baseball bats provides extra strength.

FIGURE 1.46 (See color insert)
Use of nanoparticles in paint enhances smoothness and appearance.

FIGURE 1.47 (See color insert)
Use of nanoparticles in chrome polish enhances shine and covering of surface scratches and defects.

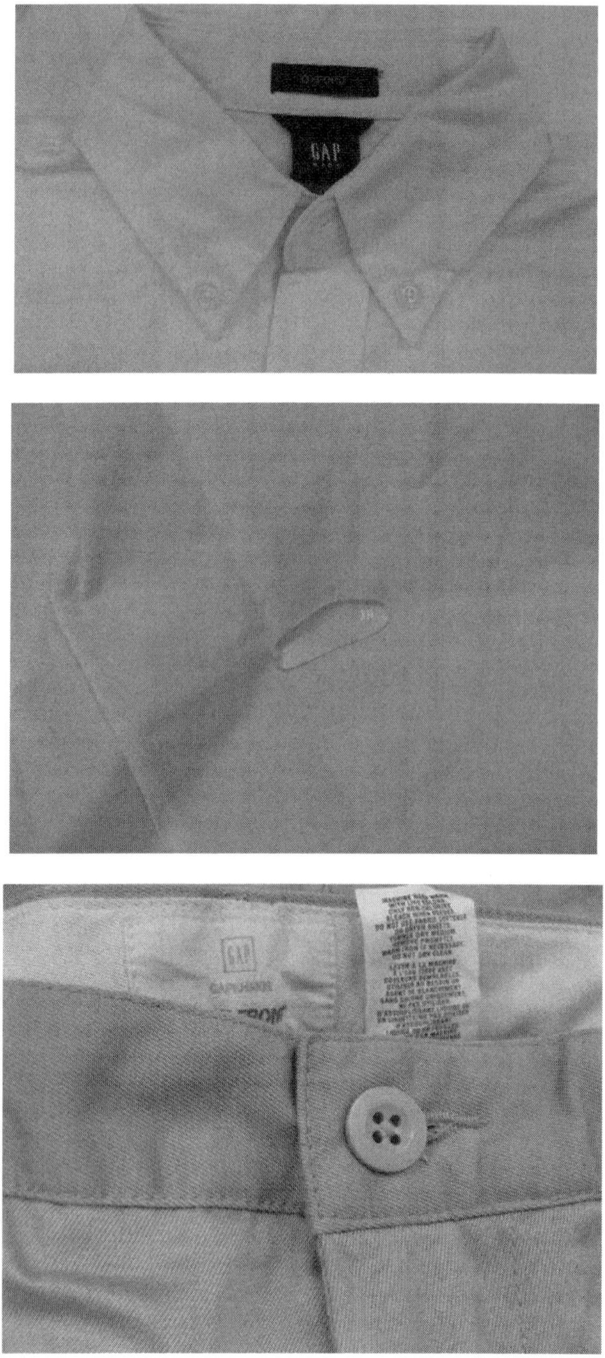

FIGURE 1.48
Use of nanotechnology in fabrics develops hydrophobic surfaces that repel water.

FIGURE 1.49
Use of silver nanowires in socks gives them an odor-fighting characteristic.

Nanotechnology National Initiative (NNI): List of Participating Departments and Agencies

Nanotechnology National Initiative (NNI)
http://www.nano.gov

Department of Agriculture
Cooperative State Research, Education, and Extension Service
www.csrees.usda.gov/nanotechnology.cfm
Forest Service
www.fpl.fs.fed.us

Department of Commerce
www.commerce.gov
Bureau of Industry and Security
www.bis.doc.gov
National Institute of Standards and Technology
http://www.nist.gov/nanotechnology-portal.cfm

Consumer Product Safety Commission
www.cpsc.gov

Department of Defense
www.defense.gov

Department of Education
www.ed.gov

Department of Energy
http://science.energy.gov/

Environmental Protection Agency
www.epa.gov/ncer/nano

Food and Drug Administration
www.fda.gov/ScienceResearch/SpecialTopics/Nanotechnology

U.S. Geological Survey
www.usgs.gov

Department of Homeland Security
www.dhs.gov
Intelligence.gov

International Trade Commission
www.usitc.gov

Department of Justice
www.usdoj.gov

Department of Labor
www.dol.gov

National Aeronautics and Space Administration
www.nasa.gov

National Institutes of Health
www.nih.gov

National Institute of Occupational Safety and Health
www.cdc.gov/niosh/topics/nanotech

National Science Foundation
www.nsf.gov/crssprgm/nano

Nuclear Regulatory Commission
www.nrc.gov

Patent and Trademark Office
www.uspto.gov

Department of State
www.state.gov

Department of Transportation
www.dot.gov

Department of Treasury
www.ustreas.gov

2

What's Different, Ethically, About Nanotechnology? Foundational Questions and Answers

Robert E. McGinn

CONTENTS

KEY WORDS: Relationship between ethics and nanotechnology; ethical issues related to nanotechnology in society; unique and new ethical issues raised by nanotechnology; nanoethics; criteria for a field of applied ethics to be a field in its own right; intrinsic features of a technical field; contingent features of a field's social contexts; autonomous technology; technological determinism; ways nanotechnology differs, ethically, from other fields of technical inquiry; paradigm shift; the ethical dimension of nanotechnology work

Introduction

In recent years, scholars have explored various ethical issues and implications of nanotechnology (Allhoff & Lin, 2008; Allhoff, Lin, Moor, & Weckert, 2007). One by-product of this work has been that the question of the extent to which nanotechnology differs ethically from other domains of technical activity has been posed and debated. This chapter focuses on several foundational questions relevant to that debate, such as:

1. Are there ethical issues related to nanotechnology?
2. Does nanotechnology raise unique ethical issues?
3. Should nanoethics be recognized as a distinctive subfield of applied ethics?
4. Are nanotechnology-related ethical issues engendered by intrinsic features of nanotechnology?

Hopefully, the discussion of these and related questions that follows will contribute to that debate, help clarify the general relationship of ethics and nanotechnology, and put subsequent consideration of specific nanotechnology-related ethical issues on a firmer footing, such that they can be more clearly articulated and effectively analyzed.

Are There Ethical Issues Related to Nanotechnology?

Many, if not most, of the scholars who have pondered the question, "Are there ethical issues related to nanotechnology?" have answered it in the affirmative. However, to answer that question unambiguously requires clarity about a related meta-ethical question: "What is an ethical issue?" Before answering the latter question, let us get clear about what is meant by an *issue*. According to one source, an *issue* is "a point or matter of discussion, debate, or dispute" (Editors, 1992, p. 957); for another, it is "a point or matter in contention" (Friedrichsen, 1973, p. 1119); for a third, it is "a point in question or a matter that is in dispute; a point, matter, or dispute, the decision of which is of special or public importance" (Stein & Urdang, 1983, p. 757). In the following discussion, an *issue* means a contested matter about which disputants differ in judgment or belief.

An *ethical issue* is an issue on which disputants' differences in judgment or belief stem from differing assessments of the matter made from an *ethical perspective*.[1] An ethical perspective is one in which action assessment hinges on the assessor's beliefs about its likely consequences for the well-being of parties affected or likely to be affected by it, or on belief that the action is intrinsically good or right, evil or wrong, or obligatory.[2] Thus, "the infliction of physical suffering on a prisoner of war in order to extract information from him or her" and "physician-assisted suicide" are ethical issues. To designate them thus is not, however, to judge them "ethical" in the positive sense that is the opposite of the negative sense connoted by "unethical." Rather, it is to claim that matters thus designated fall within the domain, purview, or scope of ethics, as opposed to falling outside that domain, as most aesthetic issues and some legal issues do. Ethical issues fall within the domain of ethics, so it is appropriate to bring to bear the concepts and principles of ethics to make judgments about the moral acceptability of related actions.[3] Ethical issues related to nanotechnology are those associated with nanotechnology research and development (R&D) or manufacturing activity, or to the diffusion, regulation, or use of nanotechnology materials and final products. For example, consider the matter of nanotechnology researcher conduct when working with a material newly available at the nanoscale, and the matter of regulatory administrator conduct when a new nanotechnology product, believed by experts to pose a nontrivial but publicly unrecognized or underappreciated risk of harm to humans, is under consideration for release to the marketplace. Both matters qualify as ethical issues related to nanotechnology and are appropriate targets of ethical scrutiny.

The community of nanotechnology practitioners is neither oblivious nor indifferent to ethical issues related to nanotechnology. In fact, there is evidence that many more nanotechnology practitioners believe there are significant ethical issues related to nanotechnology than do not. A 2005–2006 survey of researchers working at thirteen nanotechnology research laboratories belonging to the U.S. National Nanotechnology Infrastructure Network[4] revealed that 50.1% of the respondents somewhat or strongly agreed that there are "significant ethical issues related to nanotechnology," compared with 26.9% who somewhat or strongly disagreed with that proposition (McGinn, 2008, p. 104). The same survey yielded another important finding: that among researchers who agree that such issues exist, there is little indifference to ethical issues related to nanotechnology.[5] In a number of scholarly articles about ethics and nanotechnology, authors have listed or alluded to various ethical issues related to nanotechnology, or have argued that there must be such issues. For example, in an article section entitled "Issues in Nanoethics," Allhoff wrote, "I want to present briefly some of the ethical issues that nanotechnology allegedly raises." He proceeds to list a wide array of issues, under the following headings: legal and regulatory issues, research funding and priorities, equity and distributive justice, environmental health and safety, toxicity, therapy/enhancement, and privacy (2007, pp. 189–193). While some of these items are reasonably regarded as ethical issues, for example, the risk of harm to researchers and to the environment posed by some nanomaterials or nanotechnology, others seem more accurately termed ethical concerns that are related to ethical issues, for example, privacy. Yet others seem more like nonethical social issues than ethical issues per se, for example, patent law (Allhoff, 2007, p. 189). It was to avoid vagueness and conflation that care was taken to indicate clearly what an *ethical issue* means in this chapter, and to characterize referenced ethical issues as neutrally as possible.

Instead of writing about ethical *issues* related to nanotechnology, Grunwald references various ethical *questions* posed by developments in nanotechnology, including "questions of the acceptability and comparability of risks," "questions of the distribution of the opportunities for the use of as well as the spatial and temporal distribution of the opportunities and risks of nanotechnology," and "questions of monitoring and of data protection" (2005, pp. 192, 194). What Grunwald calls *ethical questions* posed by developments in nanotechnology are closely related if not identical to what I termed *ethical issues related to nanotechnology*.

Swierstra and Rip refer to certain "well-known ethical issues, say privacy and new information and communication technologies (ICT), or point-of-care diagnostics and professional-medical responsibilities," as ones that, while not new to nanotechnology, "can become more pressing" with the rise and progress of nanotechnology (2007, p. 3). Clearly they believe that there are ethical issues related to nanotechnology.

Van de Poel distinguishes between "types of ethical issues, like privacy and enhancement," and "the concrete ethical issues as they arise during the further development of nanotechnologies" (2008, p. 26). Thus, for him, the existence of specific ethical issues is not in question, only at what points and in what forms they manifest themselves.

Clarity would be well served by distinguishing ethical concerns, protectable interests of stakeholders, and issues. Thus, privacy is an ethical concern; privacy preservation is a protectable ethical interest; and the violation of human privacy with nanodevices is a specific (nanotechnology-related) ethical issue. That said, it is clear that many scholars and nanotechnology practitioners believe that there are nanotechnology-related ethical issues, be they abstract or concrete, and actual or potential.

Does Nanotechnology Raise Unique Ethical Issues?

Having (a) established what is meant by "ethical issues related to nanotechnology," (b) cited evidence that a number of scholars and many nanotechnology researchers believe that there are significant ethical issues related to nanotechnology, and (c) listed a few issues identified as such by scholars, the following question naturally arises: Does nanotechnology raise unique ethical issues?

This question is tantamount to asking whether there are qualitatively new kinds of ethical issues that only nanotechnology raises. Some scholars ask a closely related question: Does nanotechnology raise new ethical issues?

Van de Poel notes that this question "has already raised a lot of discussion" (2008, p. 31). The latter question is a slightly weaker but more determinable question than: Does nanotechnology raise unique ethical issues? Granted, if nanotechnology raised an ethical issue of a kind not previously raised by any other technology, that issue would be (qualitatively) new (at least in relation to technology) at that point in time. But it could turn out that the same kind of ethical issue was also raised at a later time by another, yet-to-be-developed technology. If so, then looking back on the situation at a point in time after the second technology had emerged, one could say that at the point when nanotechnology first raised the issue, it was new, but that it eventually proved not to be unique, since nanotechnology turned out not to be the only technology that raised it. Thus, one can know when a new-technology-related ethical issue first appears whether it is qualitatively new, but can never know at any given point in time that a then-new ethical issue will prove to be raised only by the technology that first raised it, and hence be unique to that first technology. For this reason, in what follows, the question to be considered is the determinable one: Does nanotechnology raise qualitatively new ethical issues?[6]

Biotechnology, encompassing both the activity and its products, appears to have raised some qualitatively new ethical issues, ones that had not arisen previously in any other field of technical inquiry. Examples of qualitatively new ethical issues arguably raised for the first time by biotechnology include: predetermination of the sex of human offspring via various technical means; creation of new forms of plant and animal life via r-DNA splicing techniques; and, eventually, human reproductive cloning via somatic cell nuclear transfer. There are vigorous disputes over the permissibility of such undertakings, stemming from differing assessments of related actions and practices from an ethics perspective, whether it be consequentialist, deontological, or mixed in nature. Does nanotechnology raise any qualitatively new ethical issues? Barring a valid deductive argument demonstrating that a particular technology (or family of technologies) cannot raise any qualitatively new ethical issues, it is impossible to know a priori that there is no matter related to that technology that can engender an unprecedented kind of ethical issue. Nevertheless, in the case of nanotechnology, based on the examples given and arguments made to date in support of the claim that there are such issues, the answer appears to be that there are no grounds for thinking that such issues exist.

Thus far, the ethical issues cited by some scholars and activists as newly raised by nanotechnology all seem to be instances of "the usual generic suspects," such as risk of privacy violation, creation of a nontrivial risk of harm to humans or the natural environment, preservation of a safe workplace, and access to the fruits of publicly funded R&D activity. None of these issues, or any other with which the writer is familiar, is qualitatively new and raised by nanotechnology. The same generic issues emerged in earlier technological contexts having nothing to do with nanotechnology.

But might nanotechnology raise qualitatively new ethical issues when some of its more exotic projected developments emerge? Arguably, the most promising domain within which to look for qualitatively new kinds of ethical issues raised by nanotechnology is the area encompassing NBIC convergence work, i.e., work building on the projected convergence of research in four technical fields: nanotechnology, biotechnology, information sciences, and cognitive science.[7] For example, one can envision, through eventual application of the fruits of NBIC R&D activity, the emergence of what might be termed a new, hybrid form of human existence. This would be one in which humans have a range of nanodevices with monitoring and communication capabilities, including ones having to do with human cognitive functioning, permanently implanted in them as neonates. This would ostensibly be done for purposes of health preservation, illness prevention, or performance enhancement. Suppose that this imaginable practice became the norm in a society, rather than being a rare exception or something reserved for seriously ill or economically affluent patients.

This imaginable development might seem to engender one or more novel ethical issues. However, upon further reflection it appears that none is qualitatively new. For example, the arguably ethical issue of cognitive and physical enhancement by technical means has already been raised in the context of biotechnology, through genetic engineering and pharmacological means. If the implantation of a range of nanodevices into neonates for health-monitoring purposes is reserved for beings whose families can tap sufficient economic resources, two familiar ethical issues are implicated: the distribution of access to medical resources, and whether it makes sense to speak of a moral right to health care, here preventive rather than therapeutic in nature. The ethical issue of whether parents have an ethical responsibility to effect such implantation in their newborns is a special case of the general ethical issue of parental ethical responsibility to spare their (and other parents') children preventable harms. Deciding to effect such implantation in one's newborn in order to prevent medical harm is at bottom no different than deciding to have one's child vaccinated to protect her/him (and her/his schoolmates) against diseases like diphtheria and pertussis. That miniature technological devices are involved in one case and drugs in the other does not alter the kind or type of ethical issue involved.

What do the authors cited previously, who believe that there are ethical issues related to nanotechnology, say about whether nanotechnology raises qualitatively new ethical issues?

Swierstra and Rip are somewhat elusive on this question. On the one hand, they state, seemingly in a deflationary vein, that "there might not be a specific nano-ethics" (2007, p. 3), a claim that seems to express doubt that there are qualitatively new ethical issues raised by nanotechnology and/or that novel ethical concepts or principles are required to deal with whatever ethical issues nanotechnology raises. On the other hand, after allowing that "there might be ethical issues somewhat specific to nanotechnology," they immediately add, "We think there are" (2007, p. 16).

But what are these "ethical issues somewhat specific to nanotechnology"? Swierstra and Rip do not explicitly designate any specific issues as qualitatively new ethical issues raised by nanotechnology. Rather, they argue that there must be some without listing any. They contend that there must be such issues because "nanotechnology introduces new ambivalences and enhances existing ones" (2007, p. 16). One of these ambivalences is the fact that "at the nanoscale, the small size (of particles) creates unexpected new properties," ones that "can create new and unexpected benefits as well as potential problems." This is apparently one of the "nano-specific issues" (2007, p. 3) that the authors recognize, although do they not term it a *nano-specific ethical issue*. They characterize this nanospecific issue thus: "how size matters" (2007, p. 3). But while this ambivalence is a fact about the behavior of elements at the nanoscale, perhaps even only at the nanoscale, the present writer fails to discern any ethical issue related to this ambivalence that is qualitatively new and raised by nanotechnology. That an ambivalence may be new and perhaps unique to a technology does not by itself show or imply that that technology raises any qualitatively new ethical issue.

The second ambivalence Swierstra and Rip identify is "the delegation of agency to nano-enabled technology like smart dust and active systems in general," more generally to "smart devices" (2007, pp. 3, 17). While this is a seductive *façon de parler,* no genuine delegation of agency from the human being to a smart nanodevice would occur. The latter would not "do something…on its own accord" (2007, p. 17). For example, if there is any delegation of agency on the part of a person who is voluntarily implanted with smart nanodevices or whose blood is injected with tumor-targeting nanoworms, the delegation is not to the reactive nanoproducts but to the designer and/or doctor who prescribed them. Such devices only do as programmed by their designers, perhaps in collaboration with medical researchers. Similarly, smart dust—extensive arrays of information-gathering nanosensors—could eventually make torrents of valuable local or global information of various sorts available, leading lay individuals or professional experts to make or recommend decisions to alter individual or societal behavior to realize desired outcomes. While the imagined applications of various kinds of smart nanodevices are sometimes stunning, and while the familiar ethical issues of informed consent, risk of privacy violation, and equity of access are likely to rear their heads once again, no qualitatively new ethical issues allegedly raised by such devices or arrays have been identified to date.

In short, while Swierstra and Rip (2007) believe that there are "ethical issues somewhat specific to nanotechnology" and claim "that there are some nano-specific issues: in how size matters, and when agency is delegated to smart devices," they do not identify any specific, qualitatively new ethical issues (in either the normative or meta-ethical branches of ethics) raised by nanotechnology.

Allhoff and Lin argue that "some [ethical] issues are emerging that appear unique to nanotechnology, namely the new environmental, health and safety (EHS) risks arising [from] nanomaterials" (2006, p. 183). This claim is unpersuasive. These may be new instances of risks of the same sort raised by other technological materials, but they are not a new kind of ethical issue. They are new examples of a well-known category of risks that various scientific and technological materials, products, and processes have posed, are posing, and will pose in the future: namely, risks to environmental safety and human health. Hence, these issues are not qualitatively new issues raised only by nanotechnology.

Grunwald takes a fairly strong position on the newness question, stating that "technical enhancement of human beings themselves—if this would be possible at all—would in any case pose a series of new ethical questions" (2007, p. 196). He goes on to claim that "nanotechnology, in combination with biotechnology and medicine, opens perspectives for fundamentally altering and rebuilding the human body" (2007, p. 197). However, the only allegedly qualitatively new ethical questions he goes on to identify are these: (a) "how far human beings...should...go in remodeling the human body, and to what end(s) this should or could be done" and (b) whether aging and death should be "acknowledged as a predetermined initial condition of human existence" or be "seen as conditions which are, whenever possible, to be abolished" (2007, p. 197). While efforts will probably be made to enlist nanotechnology to remodel the human body and combat aging, and while it is clear that, as Grunwald states, "ethical reflection is needed in this field" (2007, p. 198), these are not qualitatively new questions first raised by nanotechnology. After all, the ethical issues of how far humans should go in remodeling the human body and of whether aging and death should be accepted or combated were previously raised by other technological developments, such as cryogenic preservation, organ transplantation, genetic engineering, cosmetic surgery, and nutritional supplements.

Godman adopts a mixed position about the question of whether there are qualitatively new ethical issues raised by nanotechnology. On the one hand, she seems to contemplate the possibility that there might be such issues: "If nanotechnology truly affords all the new and intriguing technological possibilities as it promises to do, one would suspect an arrival of new challenges to our society and its values" (2008, p. 392). On the other hand, she devotes a good part of her essay to showing that taking what she calls "the uniqueness approach"—that is, trying to identify qualitatively new ethical issues raised only by nanotechnology—is problematic. Among the problems Godman sees with taking this approach is that "people thinking in the pattern of the uniqueness approach" are apt to "[zero] in on unlikely scenarios such as a Drexlerian 'grey goo scenario'" (2008, p. 395). The problem with this claimed tendency, she contends, is that preoccupation with "these scenarios [will] over-shadow other ethical issues that may be more important but not as unique to nanotechnology." The phrase *not as unique* implies that Godman believes that there are ethical issues unique to nanotechnology and that they are unique to differing degrees.[8] However, she does not identify any specific qualitatively new nanotechnology-related ethical issues of any degree of uniqueness.

The upshot is that none of the authors discussed here has identified a qualitatively new ethical issue raised by nanotechnology or made a compelling argument that any such issue exists. Therefore, the present writer's response to the question "Does nanotechnology raise qualitatively new ethical issues?" is skepticism based on available evidence and argument coupled with adoption of a defeasible presumption: that the nanotechnology-related ethical issues claimed to be new (and sometimes unique) amount to old ethical wine in new technological bottles. To the best of my knowledge, no scholar has yet identified any specific qualitatively new ethical issue raised by nanotechnology, or by nanotechnology taken in conjunction with other "technologies that converge at the nanoscale," to borrow a phrase from the subtitle of the journal *NanoEthics*. Of course, this presumption can be invalidated at any time and would be abandoned if any persuasive new example or compelling argument were to be put forward.

Should Nanoethics Be Designated a Distinctive Subfield of Applied Ethics?

In light of the answer just given to the "qualitatively new ethical issue" question, a third question can now be addressed: Should nanoethics be regarded as a distinctive field of applied ethics or, in the words of Allhoff and Lin, as "a field in its own right" (2006, p. 182)? Besides providing an example of an allegedly new ethical issue raised by nanotechnology (2006, p. 183), Allhoff and Lin offer four "good reasons for believing nanoethics to be a distinct field" (2006, p. 183). Let us examine each.

Their first reason might be called the *argument from widespread substantial investment*:

> Nanoethics...commands a significant amount of attention and money, though far less than the amount poured into nanotechnology. In the U.S. the NNI [National Nanotechnology Initiative] currently sets aside approximately $43 million for the "identification and quantification of the broad implications of nanotechnology for society, including social, economic, workforce, educational, ethical, and legal implications." So, it would certainly be strange that there would be so much invested by various government agencies, universities, publishers and other organizations globally, if nanoethics were not a distinct or intelligible field. (Allhoff & Lin, 2006, p. 183)

This reason is not persuasive. The amount of money invested in an intellectual endeavor in no way justifies the belief that it is a "distinct or intelligible field." The feeling that a substantial level of investment in nanoethics would be "strange" if it wasn't such a field is equally beside the point. After all, such investment might be motivated by prudential or political-economic considerations, not by intellectual conviction about the distinctness of nanoethics as a field of inquiry. Whether something is a distinct field depends on properties of the field itself and the kinds of issues it raises compared with those of other fields of applied ethics, not on how much various organizations invest in it or how that investment is perceived by observers. At most, substantial investment in a new area of inquiry might show that the investing organizations perceive or *believe* it to be a distinct field, but it cannot show that it actually *is* a distinct field.

Second, Allhoff and Lin claim that it is "unclear why we should accept the litmus test that, to be a true discipline, nanoethics must either raise new ethical issues or larger ethical issues than other technologies" (2006, p. 183). The disjunctive litmus test that the authors reject is flawed. While its first element has merit—if, that is, *new* here means *qualitatively new kind of*—the second is problematic. Its meaning is unclear. Does "larger ethical issues" mean that they have a larger domain of impact, or an impact that is quantitatively larger in magnitude, or that these issues are in some qualitative sense more important? Given this flaw, the disjunctive litmus test referenced by the authors does not compel assent. But the fact that nanoethics does not qualify as a distinctive field under a flawed litmus test is not a compelling positive reason for regarding nanoethics as a distinct field or a field in its own right. Another defeasible presumption needs underscoring: that a new area of ethics inquiry is not to be designated or treated as a distinct field in its own right unless and until there is a compelling reason for doing so. Thus far, no such reason has been adduced for that condition.

Third, Allhoff and Lin argue that

> to the extent that nanotechnology is a convergence of many disciplines in the first place, it should be no surprise that nanoethics is a convergence of many ethical issues as well. So, even if a new area of ethics requires raising new or larger issues, that standard may no longer apply with the discovery or creation of nanotechnology, which uniquely draws from other disciplines like no other discipline before it. (2006, 183–184)

The fact that a new field (at a new scale) may arguably "draw from" other long-established fields, even possibly "uniquely"—the latter as stated, not shown—does not provide a good reason why a set of criteria for a field of ethics to be distinctive "no longer apply" here to nanotechnology. Hence, the arguably amalgam-like nature of the nanotechnology field does not by itself warrant treating nanoethics as a field or area in its own right, or as a distinctive subfield of applied ethics.

Fourth, Allhoff and Lin argue that

> Nanoethics does seem to raise new ethical issues insofar as it adds a new dimension or "flavor" to current ethical debates. For instance, though privacy may be a relatively old debate, the possibility of creating near-invisible and undetectable devices did not meaningfully exist prior to nanotechnology; so nanotechnology may help shift the privacy debate in an entirely new direction: whereas worries about unauthorized or unwanted surveillance have traditionally focused on a few agencies, notably governmental organizations, the possibility of cheap, ubiquitous tracking devices—emerging now with radio frequency identity chips (RFID) technology...and later to a greater extent with nanotechnology—"decentralizes" surveillance and changes the terms of the debate. (2006, p. 184)

This argument, too, is invalid. That nanotechnology R&D may eventually enable "cheap, ubiquitous tracking devices" could result in concerns over privacy violation becoming more salient and urgent, hence could give ethical debates over privacy in relation to nanotechnology a more urgent "flavor." However, that is not to be confused with that prospective development's raising a qualitatively new kind of ethical issue. This is not the case any more than the fact that certain technologies enabled portable, affordable, and widely diffused glucose test devices entitles one to say that this phenomenon raises one or more qualitatively new ethical issues not present before such devices existed, a time when comparative access to medical treatment was already widely recognized as an ethical issue.

The present writer's view about the criteriological question is that nanoethics would deserve to be regarded as a distinctive subfield of applied ethics or a field in its own right—one on a par with bioethics and environmental ethics—only if (a) there is at least some qualitatively new ethical issue raised by nanotechnology activity and its products, and (b) this qualitatively new ethical issue is due solely or predominantly to the nature of the nanotechnology field.[9] Having argued that, to date, no qualitatively new kinds of nanotechnology-related ethical issues have been identified, let us consider their genesis.

Are Ethical Issues Related to Nanotechnology Engendered by Features of Nanotechnology?

Some of the nonunique ethical issues related to nanotechnology identified to date are partly attributable to the characteristic features or aspects of nanotechnology phenomena and products, namely, the fact that nanomaterials and nanoproducts are extraordinarily small in physical scale, and the fact that, due to quantum effects, elemental materials often exhibit quite different physical properties at the nanoscale than they do at larger scales. However, all of the currently identified or projected ethical issues related to nanotechnology are no less attributable to the contingent features of the existing societal contexts in which nanotechnology R&D work is done and in which its fruits are exploited. Consider the following features of the societal context of contemporary nanotechnological activity:

- To date, U.S. federal regulations designed specifically for nanotechnology materials, processes, and products are virtually nonexistent.
- Many national governments have made huge investments in nanotechnology R&D activity.
- National governments that have made such investments have earmarked relatively very small amounts of money for safety research.
- Many companies active in commercial nanotechnology work believe the potential financial gains are substantial.
- Nanotechnology researchers believe they have a chance to make significant contributions to the development of a new and exciting field and that the rewards for doing so could be substantial.
- Some nanotechnology laboratories, staffed with research personnel drawn from countries around the world, have lab cultures that are substantially laissez-faire in character.
- In contemporary society, much or most new scientific knowledge can be turned into technological products and introduced into the marketplace with unprecedented rapidity.
- Nanotechnology R&D is being pursued by many nations with a clear eye to enhancing their international economic and military competitiveness.

Thus, for example, the ethical issue of nanotechnology-related risk to privacy is attributable not only to the extremely small scale of nanotechnology phenomena and products, an aspect clearly conducive to surreptitious monitoring. It is also due to the contingent politics, priorities, and preferences of the governmental institutions and organizations that will shape the outcomes of nanotechnology R&D into products with certain specifications and deploy and use them for certain purposes, all that occurring under particular legislative, regulatory, legal, political-economic, and cultural conditions that also help set the stage for such ethical issues to emerge.

Similarly, the nanotechnology-related ethical issue of researcher behavior vis-à-vis a material newly available at the nanoscale is engendered not just by the cardinal fact that elemental materials often exhibit radically different properties at the nanoscale. It also arises from extrinsic contingent factors, such as the specific social conditions that prevail in particular nanotechnology laboratories (e.g., their respective cultures); the time-dependent natures of the job markets in which nanotechnology researchers shape their research and seek jobs; the amounts, directions, and priorities of government funding for nanotechnology research; and the natures of the reward and punishment systems that prevail in nanotechnology research communities. Such considerations show that ethical issues related to nanotechnology are engendered not solely or predominantly by the nature of nanotechnology phenomena and products. The factors that engender nanotechnology-related ethical issues are sociotechnical, not purely technical, in nature.[10] Therefore, in light of the two necessary conditions, stipulated previously, for a new area of ethics inquiry to be regarded as a distinctive field in its own right, the conclusion here is that nanoethics does not deserve to be so regarded.

Are There Any Ways in Which Nanotechnology Is Different, Ethically, From Other Fields of Technical Inquiry?

Despite the deflationary character of some of the answers given thus far, nanotechnology is different, ethically, from other fields of contemporary technical inquiry in at least two ways. However, both depend on viewing nanotechnology not as an isolated domain or field of technical inquiry, but as a total sociotechnical R&D, manufacturing, and regulatory enterprise.

The history of public resistance in the United States to biotechnology R&D and some of its fruits—from contested genetic engineering experiments in the mid 1970s to the continuing controversy over embryonic stem cell research—is well known. Mindful of that history, the nanotechnology community is anxious over the possibility that widespread public resistance to continued major funding for nanotechnology research might emerge in the future. These (and perhaps other) characteristics of the historical and contemporary social contexts of nanotechnology have engendered an unprecedented situation. Some of nanotechnology's key institutional promoters, for example, the U.S. National Science Foundation (NSF), have supported the upstream exploration of nanotechnology's social and ethical implications in parallel with the pursuit of basic nanotechnology research.

The rationale appears to be to protect the field against the possibility of strong negative public reaction downstream if nanotechnology were to be implicated in serious social harm through negligent or irresponsible practitioner action, regulatory oversight, or manufacturing practices. Put differently, upstream study of ethical (and social) issues related to nanotechnology is viewed by some forces that support it as an investment in stable public funding support for nanotechnology R&D work in the future.[11]

A second way in which nanotechnology is different, ethically, from other existing technical fields of inquiry has to do with the apparent emergence in the nanotechnology community of a challenge to a long-dominant paradigmatic belief in the scientific research community regarding the ethical responsibilities of researchers. A central tenet of the scientific community's dominant belief system has been that the only ethical responsibilities of researchers revolve around the Holy Trinity of Research Ethics: lab safety, data integrity, and respect for intellectual property (e.g., not plagiarizing the ideas or words of others and giving all contributors credit in publications in proportion to their respective contributions to the achievements in question). An important complementary dual tenet has been that society alone is responsible for whatever is done with the fruits of basic research and that researchers cannot be held responsible for that which they do not control, namely, the end-use applications of their work. In the words of Nobel physics laureate, Leon Lederman,

> Our lame but perhaps time-honored response is that scientific knowledge is not good or evil; it is enabling. Modern science, however abstract, is never safe. It can be used to raise mankind to new heights or literally to destroy the planet. As democratic government spreads, it is the people and their representatives who must use the power provided by science. We give you a powerful engine. You steer the ship. (Lederman, 1999)

One finding from the National Nanotechnology Infrastructure Network (NNIN) ethics survey shows unambiguously that the great majority of responding NNIN researchers reject this conventional position. Respondents were asked about their degree of agreement with the following claim: "If a nanotech researcher has reason to believe that her/his work will be applied in society so as to pose a significant risk of harming human beings, s/he has an ethical responsibility to alert appropriate parties to the potential dangers" (McGinn, 2008, p. 126). While 4.0% of respondents strongly disagreed, 0.8% somewhat disagreed, and 13.4% somewhat agreed, a remarkable 76.3% of the respondents strongly agreed with the claim (McGinn, 2008, p. 115).

This response profile suggests that in the nanotechnology community a new paradigm of ethical responsibility in technoscientific research may be emerging to challenge the one reflected in Lederman's conventional views. Nanotechnology appears to be one of the first fields, if not the very first field, of contemporary technoscientific inquiry in which this challenge is being posed and played out. What is reasonably clear is that most NNIN respondents do not exclude nanotechnology researchers from the list of groups whose members they believe have ethical responsibilities toward society at large.[12]

In short, nanotechnology as a research enterprise is being required by its institutional patrons to address upstream the ethical (and social) implications of work in the field, and many if not most nanotechnology practitioners seem to be departing from the established paradigm of researchers' ethical responsibilities by acknowledging that even researchers can have ethical responsibilities to society at large, not just to other researchers in their labs, those with whom they publish, and those working in their fields. These are two ways that nanotechnology is different, ethically, from other fields of contemporary technical inquiry. However, neither is due to any intrinsic technical features of the field; rather, both stem from contingent aspects of its (historical and contemporary) social contexts, such as those sketched in this discussion.

Is It Desirable to Think, Write, and Speak of Nanoethics?

Despite the apparent absence of any qualitatively new ethical issues related to nanotechnology, and the absence of any compelling rationale for regarding nanoethics as a distinctive field of applied ethics or a field in its own right, the fact that there are ways in which nanotechnology differs, ethically, from other fields of technology might make it seem desirable or even valid to continue to think, write, and speak about nanoethics. However, rather than continuing to refer to *nanoethics,* the present writer's position is that, all things considered, it is preferable to refer to *ethical issues related to nanotechnology.*[13] If it were clear that use of the term *nanoethics* was intended only to focus attention on the nanotechnology field so that it does not escape ethical scrutiny, then one could have no quarrel with such usage. But casual talk of nanoethics may leave listeners and readers with the impression that it is a distinctive subfield of ethics, on a par with bioethics and environmental ethics, and this could backfire. How so?

Suppose that qualitatively new ethical issues related to nanotechnology developments are not identified and that it is not shown that at least some ethical issues related to nanotechnology are solely or predominantly attributable to intrinsic features of the nanotechnology field. In that case, expectations, fueled by use of the term *nanoethics,* that there are such distinctive issues and intrinsic grounds for attribution will go unmet and nanotechnology practice and products might come to be regarded as ethically unimportant or irrelevant and become the object of benign neglect. For those who care about protecting and enhancing societal well-being and environmental quality in the context of rapid sociotechnical innovation involving nanotechnology or other emerging technologies, that would be an unfortunate and counterproductive outcome.[14]

The term *nanoethics* invites those who encounter and use it to think that the ethical issues involved are attributable, solely or predominantly, to the nature of the nanotechnology field rather than, as argued previously, to variable combinations of the nanotechnology field's intrinsic features and contingent extrinsic aspects of the societal contexts in which nanotechnology activities are carried out. Although less compact and elegant, the expression *ethical issues related to nanotechnology in society* is less likely to be interpreted so narrowly and, hence, is more conducive to analysts giving more accurate genetic accounts of these issues. The use of that expression would help counter (rather than reinforce) the tendency to attribute potential ethical issues and societal problems to nanotechnology, per se, rather than to the dynamic interaction of nanotechnology and its contingent social situation and related socially shaped practices (e.g., funding, public understanding, and regulation). It has taken major intellectual effort to discredit the seductive notions of autonomous technology and technological determinism; care should be taken to avoid inadvertently rehabilitating them.

Should Nanotechnology Practitioners Study the Ethical Dimension of Their Work?

Notwithstanding the absence of compelling arguments that qualitatively new nanotechnology-in-society-related ethical issues exist, or that nanoethics should be treated as a distinctive field of applied ethics, pressure to familiarize nanotechnology practitioners with the ethical dimension of their work is increasing. In 2004, after extensive study, the United Kingdom's Royal Society and Royal Academy of Engineering published a joint report on nanotechnology. The document recommended that

> consideration of ethical and social implications of advanced technologies (such as nanotechnologies) should form part of the formal training of all research students and staff working in these areas and, specifically, that this type of formal training should be listed in the Joint Statement of the Research Councils'/AHRD's Skills Training Requirements for Research Students.[15]

More recently, the America Creating Opportunities to Meaningfully Promote Excellence in Technology, Education, and Science Act—the America COMPETES Act—became law. Section 7009 of this law stipulates that

> the Director [of NSF] shall require that each institution that applies for financial assistance from the Foundation for science and engineering research or education describe in its grant proposal a plan to provide appropriate training and oversight in the responsible and ethical conduct of research to undergraduate students, graduate students, and postdoctoral researchers participating in the proposed research project.[16]

The increasing pressure to do so aside, there are several reasons why providing for such study makes excellent sense in the case of nanotechnology. First, only 36% of 933 researcher respondents in the NNIN ethics survey indicated that they had taken "a course in which ethical issues closely related to science, technology and engineering were discussed." Indeed, only 16.7% had taken "an ethics course focused on ethical issues closely related to science, technology, or engineering" (McGinn, 2008). Not surprisingly, the majority of respondents rated themselves not at all, slightly, or moderately well informed about ethical issues related to nanotechnology.[17] Happily, most respondents supported the idea of making study of ethical issues related to science and engineering—not just to nanotechnology—"a standard part of the education of future engineers and scientists."[18] In short, many nanotechnology researchers recognize a tension between their thin educational backgrounds and modest levels of understanding about nanotechnology-related ethical issues on the one hand, and their belief that such issues are important, likely to be encountered in their careers, and worth considering on the other. This situation suggests that it makes sense to provide such education to nanotechnology (and other) researchers, just as it has proven useful to provide intending researchers with opportunities to improve other underdeveloped work-related skills, for example, technical communications skills.

Second, such study is likely to lead to greater attentiveness to the presence of such issues in the practice of nanoscale science and engineering. In turn, to the extent that the NNIN nanotechnology community is typical, that will probably mean less indifference to such issues,[19] something that should contribute to diminishing the chances of harmful outcomes through a lack of awareness or negligence.

Third, the workforce in nanotechnology labs in more-developed countries is becoming increasingly global in character, with researchers in such labs originating from many countries around the world, including China, Russia, India, and Turkey. Consequently, it would be naïve to assume that all researchers in a given nanotechnology facility have the same understanding of the basic ethical responsibilities of laboratory researchers. Requiring suitable ethics study for nanotechnology researchers could help ensure that they have a shared notion of the most basic such responsibilities, something that should promote more responsible professional behavior in settings in which pressures to engage or acquiesce in one or another form of misconduct, such as shortcutting that violates published lab safety procedures, can be intense.

Fourth, such education is worthwhile because if nanotechnology practitioners, acting ethically irresponsibly, do or fail to do something that results in significant harm to the public, the natural environment, or public welfare, the contemporary research enterprise of nanotechnology could suffer a funding backlash from the public. This could preclude or delay the realization of important individual and societal benefits and prevent the elimination or mitigation of existing harms.

But even if there are good reasons for providing such education, is it feasible? Some technical professionals deny or doubt that it is; others believe that even if it is, it is not an appropriate subject for researcher education.

But such views often stem from the fact that those holding them have deep misconceptions about the teaching of ethics. For example, some technical professionals believe that teaching ethics is like teaching religion or engaging in political indoctrination.

Others see teaching ethics as having the misguided goal of trying to cultivate or create virtuous individuals, something they believe doomed to fail since, as they see it, virtuous behavior is at bottom a matter of character, will, and intent. In refuting such misconceptions, it should be pointed out that the purpose of nonpartisan education about ethical issues related to science and engineering in society is threefold: to alert present and future practitioners in these fields to the full range of harms that irresponsible technical practices, decisions, and actions can cause; to provide tools for thinking systematically, nonsuperficially, and critically analytically about ethical issues related to science and engineering; and to eliminate the classic excuse of those previously denied such exposure: that they were unaware that there were important ethical issues in a particular domain of technical practice.

Conclusion

While significant ethical issues related to nanotechnology have been identified, none appears to be qualitatively new in kind. Having examined alleged examples of such issues and assessed arguments that there must be such issues, adoption of a defeasible presumption seems justified: that the nanotechnology-related ethical issues claimed to be new or unique amount to old ethical wine in new technological bottles. Moreover, the nanotechnology-related ethical issues identified to date have not been engendered solely or predominantly by intrinsic features of the nanotechnology field. Rather, their genesis is also due to contingent extrinsic features of nanotechnology's micro-, meso-, and macrosocial contexts.[20] All known ethical issues related to nanotechnology have roots that are sociotechnical, not purely technical, in nature. Taken together, the qualitative non-novelty of nanotechnology-related ethical issues and the fact that such issues are engendered by combinations of intrinsically technical and extrinsically social-contextual factors strongly suggest that nanoethics does not merit being regarded or treated as a distinctive subfield of applied ethics or as a field in its own right.

Despite the qualitative non-novelty of ethical issues related to nanotechnology, there are ways in which the nanotechnology field is different, ethically, from other fields of technical inquiry. Two salient differences are institutional support for upstream study of nanotechnology-related ethical issues as part of the nanotechnology research mission, and the emergence in the nanotechnology community of a challenge to the long-established paradigmatic belief that researchers have no ethical responsibilities to society as a whole based on the end uses made of the results of their work. These differences, too, stem from contingent features of the nanotechnology enterprise's historical and contemporary social contexts, not solely or predominantly from the field's intrinsic nature or features.

All things considered, it is preferable to think, write, and speak about *ethical issues related to nanotechnology in society* rather than about *nanoethics*. An important reason why is to counter persistent temptations, posed by use of the expression *nanoethics*, to believe that intrinsic features of nanotechnology phenomena and products alone engender nanotechnology-related ethical issues.

Extrinsic contingent features of nanotechnology's multiple societal contexts, features subject to alteration given sufficient social pressure and suitable educational initiatives, play at least as important a genetic role. Finally, for various reasons, not least to ensure that nanotechnology practitioners of diverse national origins understand the general ethical responsibilities of scientists and engineers and the specific ethical responsibilities to which they give rise in contemporary nanotechnology workplaces, nanotechnology practitioners should be required to study the ethical dimension of their work as a part of their formal education.

To reiterate what was stated at the outset: Hopefully, the questions, arguments, and answers provided in this chapter will contribute, if only through identification, elaboration, and rectification of their shortcomings, to putting subsequent discussion of specific nanotechnology-in-society-related ethical issues on a firmer footing.

Questions

1. What is the relationship between ethics and nanotechnology? What are ethical issues related to nanotechnology?
2. Does nanotechnology raise unique ethical question?
3. Should nanoethics be designated a distinctive subfield of applied ethics? Yes or no. Explain why.
4. Are ethical questions related to nanotechnology engendered by features of nanotechnology?
5. Are there any ways in which nanotechnology is different, ethically, than other fields of technical inquiry?
6. Should nanotechnology practitioners study the ethical dimensions of their work?
7. Use Internet resources to compare the ethics system of the following religions:

Religion	Source of ethics	Main principles of ethical system
Christianity		
Judaism		
Islam		
Shintoism		
Buddhism		
Hinduism		
Sikhism		

8. Discuss the common themes in their ethical systems. What are the differences?
9. What is the difference between a maxim and a universal law? Can a universal maxim be a universal law?
10. Summarize your learning of new ideas and information presented in this article by completing the following flowchart:

Flow Chart

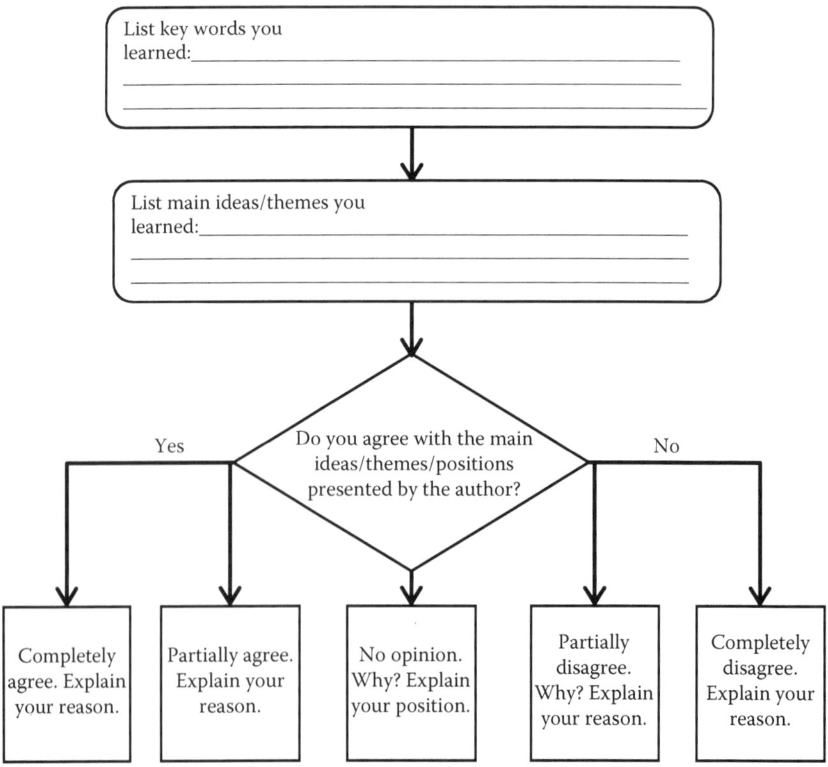

Acknowledgment

I wish to acknowledge the valuable criticisms made by an anonymous referee and the editor's helpful suggestions regarding the scholarly literature. The wording of the title question is due to Keith Baker. This chapter was originally published in *Nano Ethics,* Vol. 4, No. 2, 115–128. Reprinted with permission of Springer Science & Business Media B.V.

Notes

1. An *ethical perspective* is to be understood here as a "perspective of ethics" and to be contrasted with, say, a "legal perspective" or an "aesthetic perspective."
2. In some cases the assessment might be of a closely related practice or policy instead of an action.
3. For the present writer, whether an issue is properly termed an ethical issue, i.e., one that falls within the domain of ethics, hinges ultimately on whether the playing out of related actions or practices bears significantly on the well-being and protectable interests of affected parties.
4. The National Nanotechnology Infrastructure Network (NNIN) is a network of U.S. university-based nanotechnology research laboratories that is supported in part by the U.S. National Science Foundation (NSF). Hereinafter, this survey, whose focus was the views of nanotechnology researchers about ethical issues related to their work, will be referred to as the NNIN ethics survey. See McGinn (2008).
5. Among respondents who somewhat or strongly agreed that significant ethical issues related to nanotechnology exist, 61.5% were quite or very interested in them, while only 7.8% were slightly interested in them and 0.8% were not at all interested (McGinn, 2008, p. 106).
6. Van de Poel distinguishes "different notions of 'newness'" of technology-related ethical issues. "[O]ne is that an ethical issue is new if it is not raised by an existing technology or not dealt with in another field of applied ethics" (2008, p. 31). (Note: it is likely that he meant to write "...if it is not raised by an existing technology *and* not dealt with in *any other* field of applied ethics.") His second notion is that "an ethical issue is new if we (still) lack adequate normative standards to deal with it" (2008, p. 31). This is a more problematic sense of "new," since, under it, whether or not a technology-related ethical issue qualifies as new depends on judgments being made about whether existing normative standards are "adequate...to deal with it." The vagueness of this criterion invites the proliferation of nanotechnology-related ethical issues deemed "new," something not in its favor.
7. According to Grunwald, "New questions are also posed by the fact that previously separate lines of ethical reflection converge in the field of nanotechnology" (2005, p. 187). However, he does not appear to identify any such specific new questions, referring only to "the integrative and cross-sectional nature of

many ethical challenges" posed by nanotechnology. He also writes of "new topics and questions" being "concentrated in nanotechnology," but with no greater specificity (2005, p. 198).

8. Interestingly, just as Godman speaks of ethical issues related to nanotechnology as being unique to different degrees, Swierstra and Rip refer to ethical issues as being "specific to nanotechnology" to different degrees, e.g., "somewhat specific to nanotechnology" (2007, p. 16).
9. The expression *nanotechnology field* is intended to encompass elemental nanotechnology phenomena and products.
10. The degree to which a given nanotechnology-related ethical issue is attributable to intrinsic features of the nanotechnology field and to contingent extrinsic features of the societal contexts involved will typically vary from issue to issue.
11. In the 1990s, the U.S. Department of Energy and the National Institutes of Health devoted 3%–5% of their annual Human Genome Project (HGP) budgets to supporting study of the ethical, legal, and social issues raised by the availability of new human genetic information. This came to be known as the ELSI program. It might seem, therefore, that NSF's stipulation regarding study of social and ethical implications (SEI) is not unprecedented, hence that this is not a respect in which nanotechnology differs, ethically, from all other technologies. But the difference is that, in the case of nanotechnology, study of its social and ethical implications is to be done as a constituent part of the mission of the consortium doing the research, not outsourced to scholars outside the enterprise, as with the HGP.
12. Asked about their degree of agreement with the proposition that "academic researchers in nanotechnology have an ethical responsibility for the effects that new nanomaterials and nanodevices have on society," 65.3% of respondents somewhat or strongly agreed, 18% somewhat or strongly disagreed, and 5% agreed as much as disagreed.
13. Examination of ethical issues related to nanotechnology falls within the scope of the cross-cutting subfield of applied ethics that Harris calls "the ethics of science and technology" (2004, p. 9). Interestingly, Grunwald uses the identical expression (2005, p. 192).
14. As discussed previously in this chapter (section entitled "Should Nanoethics Be Designated a Distinctive Subfield of Applied Ethics?"), Allhoff and Lin (2006) defended the view that nanoethics is a distinctive field in its own right. However, a year later, Allhoff espoused a different position. Having identified a number of ethical issues related to nanotechnology, he concluded that "none of them is new or novel in any substantial way" (2007, p. 193). Moreover, he argued against the notion that although "the issues themselves might not be unique, they nevertheless are instantiated to such a degree that extant moral frameworks will be ill-equipped to handle them" (p. 185). Instead, he concluded that while there is no "metaphysical" justification for regarding nanoethics as a distinctive field, viewing nanoethics as such can be justified "pragmatically" (p. 206). For Allhoff, this means that in light of "the impacts that nanotechnology will have on society..., nanotechnology deserves ethical attention," and that nanotechnologies "must be evaluated along whatever ethical dimensions they manifest effects, whether well-being, rights and liberties, fairness, or whatever" (p. 207). Although the present writer's views are considerably closer to Allhoff's 2007 position than to his (and Lin's) 2006 position, I submit that, as argued in this section, the

goal of coming seriously to grips with ethical issues engendered jointly by nanotechnology and its societal contexts is best served by jettisoning talk of "nanoethics" in favor of talk about "ethical issues related to nanotechnology in society."

15. AHRD is the Academy of Human Resource Development. See the Royal Society and the Royal Academy of Engineering, "Nanoscience and Nanotechnologies: Opportunities and Uncertainties," RS Policy Document 19/04, July 2004, p. 87, Recommendation 17. See also http://www.royalsoc.ac.uk and http://www.raeng.org.uk.
16. 42 U.S.C. 18620-1. See, for example, http://www.nist.gov/mep/upload/PL110-69_8907.pdf. Pursuant to Section 7009, in January 2010 NSF began requiring that applications for research funding include research-ethics training plans. See http://edocket.access.gpo.gov/2009/E9-19930.htm.
17. Only about a fifth of the respondents believed themselves to be quite (15.9%) or very well (5.6%) informed about ethical issues related to nanotechnology, whereas almost four-fifths rated themselves not at all (10.5%), slightly (31.8%), or moderately (34.6%) well informed about these issues (McGinn, 2008, p. 115).
18. Almost two-thirds (63.1%) of the NNIN ethics survey respondents agreed quite a bit or very much that study of such issues should become a standard part of future engineering and science education, while only 14.9% indicated that they supported this change slightly or not at all (McGinn, 2008, p. 117).
19. In the NNIN ethics survey, of the 516 respondents who somewhat or strongly agreed that significant ethical issues related to nanotechnology exist and who answered the ensuing "interest" item, 61.5% were quite or very interested in such issues, whereas only 7.8% were slightly interested in them, and 0.8% were not at all interested. Thus, among respondents who agreed that significant ethical issues related to nanotechnology exist, there was little indifference to such issues (McGinn, 2008, pp. 104–105).
20. The microsocial domain is the everyday work world of the nanotechnology practitioners: R&D laboratories and manufacturing facilities. The mesosocial domain is the domain that encompasses interactions between nanotechnology practitioners from the microsocial domain and representatives of mediating societal institutions and organizations, such as funding and media organizations, and policy-making institutions (executive, legislative, regulatory, and legal). The macrosocial domain is the domain of society at large, at either the local, regional, national, or global level.

References

Allhoff, F. (2007). On the autonomy and justification of nanoethics. *NanoEthics, 1*(3), 185–210.

Allhoff, F., & Lin, P. (2006). What's so special about nanotechnology and nanoethics? *Int. J. Appl. Philos., 20*(2), 179–190.

Allhoff, F., & Lin, P. (Eds.). (2008). *Nanotechnology and society: Current and emerging ethical issues*. Berlin, Germany: Springer.

Allhoff, F., Lin, P., Moor, J., & Weckert, J. (Eds.). (2007). *Nanoethics: The ethical and social implications of nanotechnology*. Hoboken, NJ: Wiley.

Editors of the American Heritage Dictionaries. (1992). *The American Heritage dictionary of the English language* (3rd ed.). Boston, MA: Houghton Mifflin.

Friedrichsen, G. W. S. (1973). *The shorter Oxford English dictionary on historical principles* (rev. 3rd ed.). Oxford, UK: Clarendon.

Godman, M. (2008). But is it unique to nanotechnology? Reframing nanoethics. *Sci. Eng. Ethics, 14*, 391–403.

Grunwald, A. (2005). Nanotechnology: A new field of ethical inquiry? *Sci. Eng. Ethics, 11*,187–201.

Harris, J. (2004). *On cloning*. London, England: Routledge.

Lederman, L. M. (1999, July 24). The responsibility of the scientist. *NY Times,* p. A15.

McGinn, R. (2008). Ethics and nanotechnology: Views of nanotechnology researchers. *NanoEthics, 2*(2), 101–131.

Stein, J., & Urdang, L. (1983). *The Random House dictionary of the English language* (unabridged ed.). New York, NY: Random House.

Swierstra, S., & Rip, A. (2007). Nano-ethics as NEST-ethics: Patterns of moral argumentation about new and emerging science and technology. *NanoEthics, 1*(1), 3–20.

Van de Poel, I. (2008). How should we do nanoethics? A network approach for discerning ethical issues in nanotechnology. *NanoEthics, 2*(1), 25–38.

3

The Beginning of Ethics: Confucius and Socrates

Jiyuan Yu

CONTENTS

This chapter is an effort to better understand, through a comparison, how Confucius and Socrates initiate their ethical inquiries that have laid down, respectively, the foundations of Chinese and Western ethics. Since both Confucius and Socrates claim to have a divine mission to undertake their investigations, the chapter focuses on the issue of how religion and rational philosophy are related when ethics begins. It shows that both have serious religious belief, yet each has secular rational grounds for doing what he is doing. Finally, each philosopher has a different view about how human beings are related to the divine being, and the difference determines their different approaches to ethics.

Confucius (551–479 BC) and Socrates (469–399 BC) have been regarded as the founders, respectively, of Chinese and Greek ethics. This does not mean that they themselves knew that they were starting a philosophical discipline called *ethics*, and indeed neither of them knew the term *ethics*.[1] Each of them, however, is well aware that he is doing something that is tremendously significant, and each claims that his undertaking is a divine mission. At *Analects* 3:24, a visitor says to the disciples of Confucius: "The world has long been without the way [*dao*]. Heaven is about to use your Master as the wooden tongue for a bell."[2]

Dao, usually translated as way, refers to a thing's characteristic mode of existence or action, and also to theory or discourse. In ancient Chinese belief, Heaven (*tian*, meaning literally sky) has an impersonal ordering force. Heaven is said to have its dao, and each thing has its own dao as the individualization of the dao of Heaven in that thing. Whether each thing gets its dao depends on whether it exists or acts in accordance with the dao of Heaven. According to the quoted passage from the Analects, the dao of Heaven does not prevail in Confucius's time, and Heaven has Confucius wake people up and restore the dao. Confucius himself also claims that "there is no one who understands me. It is only Heaven that understands me."[3]

With striking similarity, Socrates also claims to have a divinely ordered mission to investigate ethical issues. He declares in his trial:

> I was attached to this city by the god—though it seems a ridiculous thing to say—as upon a great and noble horse which was somewhat sluggish because of its size and needed to be stirred up by a kind of gadfly. It is to fulfill some such function that I believe the god has placed me in the city. (*Apology*, 30e)[4]

This proclaimed divine mission matches well with Confucius's in *Analects*, 3:24. As Heaven uses Confucius as the wooden clapper for a bell, the Delphic god uses Socrates as a gadfly.

Confucius's divine mission leads us to question whether he is Heaven's missionary or a rational philosopher. Socrates' divine mission also makes us question the rationality of his ethics. Since both are founders of their respective ethical traditions, the drawing of them together raises a more general and significant question: Is the beginning of ethics based on a rational ground, or is it a result of divine command? In other words, how are religion and rational philosophy related when ethics begins?

Ironically, neither of these two founders of ethics wrote anything down. Socrates' philosophy is mainly preserved in the dialogues of Plato (427–347 BC).[5] Plato, however, uses Socrates as the speaker for almost all of the major doctrines in his dialogues. It becomes the case that Plato keeps writing, but Socrates keeps speaking.

This creates the notoriously difficult "Socrates problem," that is, how to determine to what extent Socrates in Plato's dialogues represents the historical Socrates and to what extent he just serves as the mouthpiece of Plato. Needless to say, different answers to the Socrates problem result in different pictures of Socrates. The *Analects*, credited to Confucius, is a collection of his sayings and anecdotes, compiled and edited by his disciples and the disciples of disciples over several centuries.

There is also a "Confucius problem," that is, how to distinguish between the authentic Confucius dicta and the late interpolations. However, although this Confucius problem has been a long-standing scholarly issue,[6] it does not prevent commentators from reconstructing Confucius out of the *Analects* as a whole. This is, I think, partly because the ideas included in the *Analects* are not as systematically different as the ideas found in Plato's dialogues, and partly because none of the editors of the *Analects* assumes the towering status that Plato has. In contrast, the Socrates problem has made a profound impact on our reading of Plato's dialogues.

It is not necessary for me to get into the details of this issue here, but I should briefly state where I stand. In the spectrum of various solutions to this problem in our time, there is the maximal view, represented by Gregory Vlastos (1991), and the minimal view, represented by Charles Kahn (1996), and there are numerous others that fall somewhere in between. The maximal view holds that the early and pre-middle Platonic dialogues are reproductions of the thought of the historical Socrates,[7] and the minimal view claims that there is a crucial distinction between the *Apology* (which is not a dialogue) and the dialogues proper.[8] The common point that is accepted across the board is the high degree of historical accuracy of the *Apology*.[9] The *Apology* is, of course, Plato's writing; yet, Socrates' trial is a historical event that takes place in public. This puts constraints on Plato's invention. With the hope that the Socrates who is compared with Confucius in this chapter is a widely acceptable one, I concentrate on the Socrates depicted in the *Apology*, and use materials from other dialogues when I think they are consistent with the views of the *Apology* and help illuminate these views.

The Divine Mission

Socrates' announcement of his divine mission seems to have presented a sharp conflict with his well-known statement that "the unexamined life is not worth living for men" (*Apology*, 38a). In the history of Western philosophy, the image of Socrates is the ideal representative of rigorous rational inquiry, and he is thought to have laid down the basis for Western rationalism. Indeed, he himself issues a manifesto for rationalism:

> We must therefore examine whether we should act in this way or not, as not only now but at all times I am the kind of man who listens only to the argument that on reflection seems best to me. (*Crito*, 46b)

Such a rational Socrates is apparently at odds with a religious Socrates who claims to do philosophy by the god's command.

How to deal with the tension between Socrates' divine mission and his rationalism has been a topic of intense debate among interpreters of Socrates. A vast literature has been produced. Of various interpretations, some suggest that Socrates' belief in the divine command should not be taken seriously and that there is "not a trace of voluntarism in Socrates' 'obedience' to the god" (Nehamas, 1999, p. 46). Others hold that Socrates is serious about the god's command, for it assures him of the moral certainty of his undertaking (Brickhouse & Smith, 1988, p. 107). Still others seek to show that Socrates' commitments to rationalism and his commitments of obedience to the divine command can somehow be reconciled, although they differ as to how it should be done.[10]

It is interesting to note that Confucius's divine mission has never been an issue for scholarly dispute among Confucian commentators.[11] It has never been thought of as a disturbing problem, and hence there has been little attempt to explain it away. The scholarly indifference in this regard can be explained, I think, partly because, unlike Socrates, Confucius does not make an explicit announcement of rationalism and hence does not present a sharp conflict between religion and rationalism in his own thought, and partly because of the traditional assumption that, whereas Western philosophy is rational, Chinese (or Oriental) philosophy is a sort of religion, or at most a mixture of religion and philosophy. Hence, the divine mission is not surprising. Now, if we bring it together with Socrates' divine mission and relate the issue to the beginning of ethics, Confucius's mission becomes an interesting issue.

Let us first have a look at how Socrates gets his command from the god to do ethics, and how Confucius receives his mission from Heaven. Beginning with Socrates, according to the *Apology*, Socrates' friend Chaerephon once visited Delphi to ask the oracle whether there was anyone who was wiser than Socrates. The Delphic oracle replied that there was no one wiser (*Apology*, 21a).[12] This puzzles Socrates. On the one hand, he is well aware of his own lack of wisdom; on the other, he knows that the god does not lie. To solve the puzzle, he sets out to question his fellow Athenians who enjoy the reputation of being wise, in order to find someone who is wiser than himself. The reasoning is that, if he can find a counterexample to the oracular pronouncement, he refutes it. In the course of his questioning, however, Socrates finds that people do not really know what they claim to know, whereas he himself at least knows that he does not know. Hence, instead of a refutation of the oracle, he comes to an understanding of it: The god, in saying that Socrates is the wisest, uses him as an example to show that "human wisdom is of little or no worth" (23a).

Socrates, however, does not stop his examination when he has found the meaning of the oracle. On the contrary, he says in the trial: "Even now I continue this investigation." Moreover, the scope of his examination is broadened from those who have a reputation for being wise to "anyone I happen to meet, young and old, citizen and stranger" (*Apology*, 30a). Why does he do this? The answer is: "The god has commanded me that I should live philosophizing, examining myself and others" (28e). "To do this has been commanded [*prostetaktai*] me, as I maintain, by the god through divinations and through dreams and every other means through which divine apportionment has ever commanded [*prostetaktai*] anyone to do anything" (33c).

Here a problem arises. The oracle says only that no one is wiser than Socrates. There is no explicit command about what he should do or how he should do it. Clearly, it is Socrates himself who chooses to continue his examination beyond the test of the meaning of the oracle and turns it into his full-time occupation. In his own words, he *oiethen* (supposes or thinks) and *hupelabon* (believes) that the god had ordered him "to live the life of a philosopher, to examine myself and others" (*Apology*, 28e). In other words, he assigns to himself a divine mission.

Similarly, Confucius's Heaven never directly orders him to discover and restore the dao. Heaven does not really say anything, let alone give explicit commands. Socrates' god has at least the oracle and other channels such as dreams and *daimonion* (a divine voice) to communicate with him. Confucius's Heaven does not even have that. How, then, does he get the order from Heaven? In one place, Confucius contends that, although Heaven does not talk, it can still convey its message. "The Master said: 'What does Heaven ever say? Yet there are the four seasons going round and there are the hundred things coming into being. What does Heaven ever say?'" (*Analects*, 17:19). On the strength of this passage, Heaven's will is known through observation and understanding. We infer, then, that this is also how Confucius derives his sense of the divine mission.

It turns out that the divine mission is a direct prescriptive neither for Confucius nor for Socrates. Each derives a belief from his own understanding that his mission is divine. What, then, are these reasons that make Confucius and Socrates believe in this way? If they have independent reasons for doing ethics, why do they claim that it is a divine mission?

Socrates: Eudaimonia and Virtue

What is the justification for Socrates' status as "the inventor of ethics"? The following reasons must be the most important.[13] First, he shifts the attention of philosophy away from natural knowledge and makes the discussion of ethical issues the central area of philosophizing. According to Aristotle's report, Socrates ignores the world of nature as a whole and occupies himself with ethical matters.[14] This accords with Cicero's well-known remark that Socrates is "the first to call philosophy down from the heavens and set her in the cities of men and bring her also into their homes and compel her to ask questions about life and morality and things good and evil" (Cicero, 1971, p. 435). Of course, moral reflections existed before Socrates, but they were associated with natural philosophy and were often taken as implications of the latter. His belief that moral knowledge is what we need in order to live well effectively promotes ethics as a separate field.

Second, Socrates makes happiness the starting point of ethical reflection. *Happiness* is the translation of the Greek term *eudaimonia*. Eudaimonia, etymologically associated with being "favored by the gods," or "good fortune,"[15] refers to overall well-being, achievement, or flourishing.[16] From Socrates on, it becomes the universal premise for Greek ethics that happiness is the supreme end of life. A history of Greek ethics is a history of disputes about what happiness is and how to achieve it.

Third, Socrates approaches eudaimonia by an appeal to virtue. The notion of virtue itself is not new,[17] yet Socrates associates virtue with the state of the soul and places virtue at the central position in a happy life. He raises numerous questions and also holds a set of interrelated beliefs about happiness and virtue, virtue and the soul, virtue and knowledge, virtue and other goods in life, and so on. These issues set the agenda for subsequent Greek ethics.

None of these reasons seems to be directly related to the divine being. The god does not order him to abandon natural philosophy. In *Phaedo* (96a–99d), a text usually regarded as Socrates' intellectual autobiography, Socrates tells us that, when he was young, he was enthusiastic in learning natural philosophy. But he came to be dissatisfied with this kind of study and finally gave up, for the reason that it cannot really explain the cause of things. In the *Apology*, he denies that he has conversed with people on these topics.

The idea of eudaimonia is a traditional belief.[18] Socrates provides reasons why happiness is the starting point of ethical reflection. It is the only end that we pursue with no further reason and hence the only one that we do not pursue for the sake of a further end. We do every action and every thing for an end, and the end is the good we desire. While we can ask about everything else, what it is for, it would not make sense to ask why one wants happiness. "There is no need to ask further, 'What's the point of wanting happiness?'" (*Symposium*, 205a2–3). Socrates connects happiness with the soul and virtue. This is because of his idea that virtue has sovereign value in human life. He declares: "Wealth does not bring about excellence, but excellence makes wealth and everything else good for men, both individually and collectively" (*Apology*, 30b).

Nevertheless, it is Socrates' reaction to the oracle that starts his typical philosophical activity. Having heard the oracle, he seeks to find a counterexample in order to *elegxein* it (*Apology*, 21c). *Elegxein* means *refute*, *cross-examine*, or *censure*; and from here Socrates' examination is called *elenchus* or *elenctic examination*. The process always ends in the interlocutor abandoning his original position, reduced to perplexity about what to believe. For Socrates, the *elenctic examination* not only reveals what the god means by his oracle, but also demonstrates that the god is right about human arrogance. It also turns out to be the most effective way to make people aware of their self-conceit. Hence, the examination or refutation of the oracle develops into Socrates' characteristic way of doing ethics, that is, the examination and refutation of people's moral beliefs.

Socrates then relates his examination with happiness by claiming that the goal of his examination is to make Athenians happy: "The Olympian victor makes you think yourself happy; I make you be happy" (*Apology*, 36e). This does not mean that he makes people feel pleased, as we know well that people indeed hate him for this. Given the objective meaning of happiness, to make you happy means to promote your well-being. In Greek, eudaimonia is synonymous with living well or doing well.[19] This is why Socrates also claims that the goal of his examination is to make people "live in the right way" (*Apology*, 39d).

It is from his examination that Socrates connects virtue with happiness. What he examines are the beliefs in people's souls. In the *Apology*, virtue is taken to be "the best possible state of the soul" (30b1–4). "To care for virtue" and "to care for the soul" are used interchangeably (31b1–4; 41e3). Moral virtue constitutes a good condition of the soul, whereas vice corrupts the soul.[20] Virtue is the only thing worth having and is all that is relevant to happiness, even at the risk of death (28).

Apparently, Socrates' examination is inspired by the god, but the god says nothing about his examination and nothing about eudaimonia and virtue. Yet it is through the examination that the relation between happiness and virtue becomes the chief theme of Greek ethics. What determines the nature of Socrates' ethical study is not the oracle but his rational reaction to the oracle.

Confucius: Dao and Ren

The reasons for Confucius to be called the founder of (Chinese) ethics are as follows: First, he raises the question of "Where is the *dao* of being a human?" (that is, the dao of Heaven as individualized in a human life). By asking this question, he introduces the concept of dao (way) as the starting point of his ethical reflection. From Confucius on, it becomes the common starting point and goal for the classic Chinese philosophers to find and establish the dao. Just as Greek ethicists are all concerned with eudaimonia, Chinese philosophical schools in the classical period (which roughly corresponds to the flourishing period of Greek philosophy) offer competing accounts of what dao is.[21]

Second, Confucius introduces his key notion: *ren*. Ren has various translations. Along with *benevolence*, frequent choices also include *love*, *good*, *humanity*, *humanness*, *virtue*, and *complete virtue*. The word *ren* appeared in pre-Confucian texts, referring to *manhood*, or *kindness*, or *the aristocratic clan*. In employing this term as the central notion of his teaching, Confucius distinguishes between ren as a general disposition and ren as a particular trait of character. As a particular trait of character, ren means love or benevolence (*Analects*, 12:22), and is contrasted with other particular virtues such as courage and wisdom. However, ren in Confucius is more often described as a general quality that embraces particular character traits, a quality that makes a man a *junzi* (the noble man, the gentleman, the exemplary man).[22] This quality is even more important than one's life: "It may happen that they [the men of ren] have to accept death in order to have ren accomplished" (*Analects*, 15:9). Many translators use *humanity* or *humanness* to translate this general sense of ren. They are on the right track, because ren is intrinsically related to what is human. According to both *Mencius* and *The Doctrines of the Mean*, "To be ren is to be man."[23] I would like to point out, however, that, whereas *humanity* or *humanness* can refer to the inborn characteristic of a person, ren is a cultivated disposition based on the human characteristic. Hence, *virtue* or *human excellence* might be better terms.[24] In any case, ren as a general virtue means the quality that makes a person truly human. A person of ren is one who has fulfilled and manifested what is genuinely human.

Third, he works out a theory of what ren is and of how we achieve ren (human excellence). Confucius approaches human excellence first by an appeal to traditional values. He claims that to achieve ren (human excellence) is "to return to li" (12:1). *Li* refers to the traditional ritual and cultural practices, fully developed in the early Zhou dynasty. Its core is a humane social hierarchy modeled on family relationships. "Let the ruler be a ruler, the subject a subject, the father a father, the son a son" (*Analects*, 12:11). "To return to li" means to be shaped or transformed by traditional values. Since traditional culture is recorded in the ancient classics, it is necessary to turn to them to find the dao. In Confucius's time, the classics that recorded ancient civilization were not in good condition. It is Confucius who starts the serious and systematic investigation of the traditional texts. He collects, studies, and edits them, and then uses them for the literary curriculum of his own school.[25]

He even takes it that his divine mission of restoring the dao is associated with the preservation of culture. Once, in traveling to advocate his teaching, Confucius was under siege in a place called Kuang, where

> The Master said: "With King Wen dead, is not culture invested here in me? If Heaven intends culture to be destroyed, those who come after me will not be able to have any part of it. If Heaven does not intend this culture to be destroyed, then what can the men of Kuang do to me?"

In the *Analects*, Confucius describes himself as a transmitter: "The Master said, 'I transmit but do not innovate; I am truthful in what I say and devoted to antiquity'" (7:1). To take seriously Confucius's claim that he does nothing more than hand down the old, however, is as naïve as taking seriously Socrates' claim that he knows nothing. Confucius transmits these classics to explicate the dao that he believes to be embedded in them. For this purpose, he has to single out traditional norms and values, select them according to certain standards, explain them in some refined notions, justify their significance, and arrange them in some kind of order. To explicate the embedded dao in the traditional classics is a rational process of creative inquiry and reflection.

In summary, Confucius raises the question about where the human dao is and makes it the subject matter of systematic ethical reflection. In his pursuit of human dao, he appeals to ren (human excellence). His philosophy is a ren (human excellence)-centered dao. In explicating what ren as human excellence is, he develops, through his critical editing of traditional classics, a virtue theory: "There is one single thread binding my dao together" (*Analects*, 4:15). According to this unified theory, to be a person of ren (human excellence) involves the internalization of traditional values. But it is not just that. It also involves an emotional aspect. Ren also requires forming an emotional pattern. "The Master says that to be ren you need to 'love your fellow men'" (*Analects*, 12:22). Furthermore, it involves an intellectual aspect. "The Master said, 'In his dealings with the world the gentleman is not invariably for or against anything. He is on the side of what is appropriate (*yi*)'" (4:10). This intellectual aspect enables a Confucian moral agent not to follow traditional values blindly. Indeed, as mentioned previously, Confucius's own transmission of tradition is selective and critical. Confucius's vision of the dao shapes the subsequent development of Chinese philosophy. Whereas there is a pre-Socratic period in Greek philosophy, there is no pre-Confucian period in Chinese philosophy. Confucius's beginning of ethics is also the beginning of Chinese philosophy in general. His dao is the target that all rival schools have to respond to and attack, and it is also what his followers defend and develop in response to the attacks from rival schools.

From our foregoing discussion, we can see that the role of Heaven is very limited in Confucius's approach to ethics. Surely, the dao is Heaven's dao, and the question of dao is raised because of traditional beliefs. However, the question is prompted by the social reality that Confucius faces. In Confucius's time, the Zhou dynasty, which had ruled China since the eleventh century BC, declined. The Zhou rulers used to claim to have possessed the dao of Heaven. When the Zhou family overthrew the Shang dynasty (ca. 1766–1050 BC), the act was justified on the grounds that the Shang dynasty, because of its misrule, had forfeited the mandate of Heaven (*Tian-ming*), a different expression of the dao of Heaven. The decline of the Zhou rule entailed the breakdown of the original moral and political order, and that is why the divine mission passage at *Analects* (3:24) asserts that "The world has long been without the way [dao]." It is the chaotic social reality that leads Confucius to think about the following question: Where is the dao that the Zhou once claimed to possess but is now lost?

Furthermore, Heaven says nothing about ren (human excellence). It is Confucius's own innovation that to grasp the way of being a human, one has to pay attention to what makes a true human being. Heaven also does not say what is the dao that needs to be restored, and it does not say how to restore it; it suggests no connection between Heaven's dao and the traditional culture. It is his own understanding that restoring the dao amounts to the task of regaining the spirit of traditional culture.

Doing Ethics and Being Pious

We have explained the secular rational grounds for both Socrates and Confucius to do what they are doing. Indeed, there are several significant similarities shared by these two beginnings of ethics. First, both are concerned with the life as a whole rather than particular moral acts. The goal of Socrates' examination is eudaimonia, the flourishing life; and the goal of Confucius's investigation is human dao, the way of being a human. Both address the common question of how one ought to live, although they formulate the same question in different ways. Second, both philosophers' ethics are virtue-centered. Socrates specifies eudaimonia in terms of virtue, and Confucius explicates human dao in terms of his theory of ren. For Socrates, virtue is said to be the best state of the soul; for Confucius, ren (human excellence) means what makes a human being a true human being. The content of what virtue is, however, is different, as we shall see in the next section.

Our examination of their respective rational grounds, however, has not yet made sense of why they each claim to practice ethics out of the divine command. Given the strong secular motivation to do philosophy, as we have seen on both sides, it seems tempting to think that the divine mission is a fabrication by each of them to facilitate their own causes. But this is hardly convincing. For both, the belief that their respective missions are divinely charged is so sincere and serious that neither is willing to give in, even in the face of grave danger. In Confucius's case, when Huantui, the minister of war in Sung, threatens to kill him, "The Master said, 'Heaven is the author of the virtue that is in me. What can Huantui do to me?'" (*Analects*, 7:23). When Yen Hui, Confucius's favorite disciple and most promising successor, suffers a premature death, Confucius takes it as an issue between himself and Heaven: "Alas! Heaven has bereft me! Heaven has bereft me!" (11:9). On Socrates' side, even if he has the opportunity to be acquitted on the condition that he stops practicing philosophy, his response is: "Gentlemen of the jury, I am grateful and I am your friend, but I will obey the god rather than you, and as long as I draw breath and am able, I shall not cease to practice philosophy" (*Apology*, 29d).

There is no question that their religious belief is serious. Indeed, although examining people's beliefs is Socrates' daily occupation, he never applies his elenctic examination to the issue about the existence of the gods. He makes one significant change in the traditional notion of gods. Contrary to the traditional picture of Greek gods, as depicted in the works of Homer and Hesiod—that they do all sorts of immoral things such as deception, adultery, quarreling, etc.—Socrates insists that gods must be morally good. Indeed, the major reason for him to take the Delphic oracle so seriously is that "for surely he [the Delphic god] does not lie; it is not legitimate for him to do so."[26] Other than this, Socrates, being consistent with his disavowal of knowledge and his lack of interest in natural philosophy, maintains that an adequate knowledge of the gods goes beyond human wisdom, and he disavows any positive knowledge about life after death (*Apology*, 29a–b, 37b).

Equally, Confucius never questions the ordering force of Heaven, and, like Socrates, he has little desire to theorize about the nature of Heaven and the afterlife.[27] More significant, both Socrates and Confucius believe that to conduct their respective missions is what the virtue of piety means. Impiety is one major accusation against Socrates, but it is also the major task of Socrates' defense in the *Apology* to reject it. He insists that his philosophical activity is his service (*latreia, huperesia*) to the god.[28] This claim suggests a clue that would link the *Apology* to another dialogue, the *Euthyphro*. The subject matter of the *Euthyphro* is the definition of piety. After a series of answers and refutations, Socrates' interlocutor Euthyphro finally comes up with the answer that "piety is that part of justice that is a service [*huperetike*] of humans to the gods" (*Euthyphro*, 13d7). But when Socrates presses Euthyphro to specify what service human beings can provide to the god, Euthryphro fails to do so and loses his patience. Socrates then says to him:

> You could tell me in far fewer words, if you were willing, the sum of what I asked, Euthyphro, but you are not keen to teach me, that is clear. You were on the point of doing so, but you turned away. If you had given that answer, I should now have acquired from you sufficient knowledge of the nature of piety. (14b8–c4)

Euthyphro turns away at the very moment he is close to giving an answer that satisfies Socrates. Since the point at which he turns away is Socrates' question about what kind of service we human beings can provide to the god, if we are allowed to relate this to Socrates' claim in the *Apology* that his elenctic examination serves the god by improving the moral state of the human soul, we have an answer as to what service human beings can provide. According to this reading, Socrates, in saying in the *Apology* that his examination is a service to the god, means that his philosophical activity is not impious, but represents what piety really is.

Confucius, as we recall, maintains that he is the transmitter of the ancient tradition in *Analects* (7:1). In *The Doctrines of the Mean* (chap. 19), the term *transmitter* is related to the virtue of piety. It states that what makes a son filial is his ability "to continue (*ji*) the will (*zhih*) and to transmit (*shu*) the work of his father."[29] If we relate this statement to Confucius's self-description as a transmitter, the following two points can be made: first, that Confucius transmitting the tradition is analogous to what a filial son does to his father's work. What is emphasized in the passage just quoted is not the son's personal affection toward his parents or his duty to biologically continue the family line. Rather, what a filial son transmits should be the aspirations, causes, and ideals of his father. Analogically, what Confucius transmits is the spirit, value, and ideals of the tradition. Furthermore, because a filial son does not just hand down his father's dreams, but seeks to actualize them, Confucius's transmission can also be taken to mean to bring out and actualize the underlying traditional values and ideals. In short, the purpose of transmission is to carry on and fulfill the dao embodied in the tradition.

Second, since a son's transmission of his father's work shows his virtue of piety, Confucius's philosophical activity of transmitting traditional value can also be regarded as his piety to the authentic tradition. Because tradition is where Confucius believes the dao of Heaven is embedded, we can say that to be pious to tradition amounts to being pious to Heaven.

The link with piety that can be seen in both Confucius and Socrates provides a perspective for understanding the nature of their respective divine missions. In the *Euthyphro*, at one stage of the elenctic examination about what piety is, Euthyphro proposes a definition that piety is what all the gods love (9e). In response to this definition, Socrates raises his influential question that is often used to criticize the Divine Command Theories of morality: "Is the pious being loved by the gods because it is pious, or is it pious because it is being loved by the gods?" (10a).

We can ask the same question of Socrates' own divine mission and Confucius's. Given our discussion of their respective rational grounds, the answer should be clear. Neither of them simply holds that their undertaking is noble because it is what the god or Heaven commands. Indeed, although the mission is divinely charged for both, neither Socrates' god nor Confucius's Heaven dictates it. For each of them, it is because his mission itself is meaningful that he believes that it is divine. On the one hand, both of them subscribe to the traditional belief in the existence of the divine being. Yet, neither Greek nor Chinese religion was supported by documents with the status of the Bible. What the divine being wants human beings to do or what is the best way to serve the divine being have not been seriously reflected upon.[30] On the other hand, when Confucius and Socrates think about the issue of how one ought to live, each must have a clear awareness that his undertaking is something fundamentally new and important. What Confucius and Socrates do, independently but commonly, is to combine the unique sense of the significance of their work to the traditional faith in the divine being, to the effect that what the divine being really wants from human beings is that we lead a meaningful and fulfilled life, and that the best way to be pious is to think about and find out the right way to conduct our lives. This, I submit, is the basis for the claim of divine mission. In founding their respective ethical traditions, each of them is the first, in their own tradition, to give a systematic reflection upon what the divine being intends human beings to achieve and how we human beings should relate to the divine being.

We usually think that philosophy develops by breaking away from traditional religious belief and by replacing the mythological account of the world with a rational explanation of nature. This appears to be true for the beginning of natural philosophy in Thales. Natural philosophy does not leave room for supernatural gods in its cosmology. Yet, if our comparison of Confucius and Socrates makes sense, the beginning of ethics seems to be a different story. Instead of breaking away from religious beliefs, the beginning of rational ethics is a divine mission. Ethics begins by giving new meaning to religious beliefs and a new interpretation of the relation between the divine and the human.

Different Paths of Ethics

Although both Socrates and Confucius link doing ethics with the virtue of piety, they apparently have different notions of being pious. For Socrates, to be pious is to examine people's moral beliefs, whereas for Confucius, to be pious is to transmit and develop wisdom embedded in the tradition. Different forms of piety entail that they have different ways of carrying out their respective divine missions. As a result, although Socrates' inquiry into eudaimonia and Confucius's inquiry into dao both focus on virtue, they differ about what virtue is and how to pursue it.

In closing this chapter, I would like to list some of the most important differences between Confucius and Socrates that have shaped the subsequent development of Chinese and Greek ethics in significant ways. Space does not allow me to present in detail these differences and spell out the implications, but I hope that pointing out these differences is helpful in promoting further thinking.

First, they have different attitudes toward traditional and social values. Socrates is critical of traditional values. He claims that he is the "gadfly" to "rouse each and every one" (*Apology*, 30e), and his mission is to examine people's moral values to show that their priorities in life are wrong (30a). In his effort to expose the inconsistencies of people's moral beliefs, Socrates embarrasses all sorts of respected people by showing that they know little or nothing (23c) and are "not living in the right way" (39d). These remarks leave the impression that the Athenian ethos is problematic, and all other Athenians are not living the right way. Socrates intends to reexamine and reevaluate these values.

For Confucius, the tradition is where the dao is embedded. One major dimension of being virtuous is to return to *li*, that is, traditional ritual and cultural practices. *Li* would be precisely what Socrates wants to examine and refute. In contrast to Socrates' hostility toward tradition, Confucius is characterized by his deep respect and affection for the rich cultural past. Whereas Socrates spent his days in the marketplace examining people's beliefs and values, Confucius spent most of his time systematically transmitting ancient classics that record traditional values and practices.

Second, they have different conceptions of virtue. For Socrates, to pursue virtue is an intellectual process. Virtue is the best state of the soul (*Apology*,

30b). He identifies the soul with reason only and treats the appetites and emotions as properties of the body.[31] The perfection of the soul means, then, to have wisdom or truth, and thus, "Virtue is knowledge."[32] Since virtue is sufficient for happiness, it follows that knowledge is sufficient for happiness.[33] The virtuous differ from the vicious not in their motivations, but in their intellects; nobody does wrong willingly, and weakness of the will is impossible.[34] In contrast, Confucius does not regard virtue as being purely an intellectual matter. For him, the pursuit of ren (human excellence) involves a full-fledged development of moral character. It requires one to cultivate traditional ritual practices (*Analects*, 12:1), to love his fellow men (12:22) that is rooted in filial piety and brotherly love (1:2), and also to develop an intellectual aspect of knowing what is appropriate in applying the general requirement of traditional values in particular circumstances (4:10).

Third, they have different views about the relation between politics and virtue. Socrates sets up a contrast between political life and the pursuit of justice. He claims that he hears a divine voice (*daimonion*) that has forbidden him to take part in public affairs:

> If I had long ago attempted to take part in politics, I should have died long ago, and benefited neither you nor myself.... A man who really fights for justice must lead a private, not a public, life if he is to survive for even a short time. (*Apology*, 31d–32a)

He thus denies the positive value of politics in the cultivation of virtue, and he is also deeply contemptuous of politicians in general (*Apology*, 21c–d). In contrast, for Confucius, ethics and politics are inseparable. Politics, in his understanding, means *rectification* (*Analects*, 12:17), and hence the most important function of politics is for politicians to set good examples to guide people in their cultivation of virtue. To engage in political activity is also an indispensable dimension in each individual's cultivation. According to the *Great Learning*, the cultivation of character is not separable from the regulation of one's family and from the ordering of one's state.[35] Contrary to Socrates' contrast between ethical undertaking and political engagement, Confucius, in addition to the compilation of the classics and teaching, takes to the road and travels with his disciples from state to state to persuade the rulers to accept and practice his theory.

Fourth, both Confucius and Socrates are the greatest teachers that human beings have ever had, but they are different types of teachers. Socrates actually denies himself being a teacher to answer the charge that he corrupts the youth. His reason is that he does not charge fees. People freely listen to him and join in his conversation as they wish (*Apology*, 33a). However, Socrates does not deny that he has attracted young followers, and he himself describes his influence in this way:

> The young men who followed me around…take pleasure in hearing people questioned; they themselves often imitate me and try to question others. I think they find an abundance of men who believe they have some knowledge but know little or nothing. (23c)

The youth are amused to watch Socrates refute those who make false claims of wisdom (23e2–e4, 33b9–c4), and they then imitate Socrates and undertake to examine their elders on their most vital and cherished beliefs. Hence, Socrates is a teacher. However, what he teaches is not a positive doctrine, but the spirit that in moral areas one should not take anything for granted and should always be critical.

Confucius teaches his students to think and reflect. As he says, "When I have pointed out one corner of a square to anyone and he does not come back with the other three, I will not point it out to him a second time" (*Analects*, 7:8). Nevertheless, he puts emphasis on learning positive things and teaching positive things. As he says, "To study without thinking is worthless, to think without study is dangerous" (*Analects*, 2:15). Describing his self-image, he says, "Quietly to store up knowledge in my mind, to learn without flagging, to teach without going weary—Is this not me?" (7:2).

Last but not the least, they have different views regarding the end of ethics. For Socrates, the most important thing is to examine, for "the unexamined life is not worth living" (*Apology*, 38a). Out of his understanding of what piety is, he spends most of his time examining people in the street. As described in the *Apology* (29d–e), to practice ethics means "to exhort you [the Athenians] and in my usual way to point out to any one of you whom I happen to meet" that one should seek to possess the best possible state of the soul. While examination can certainly discover the inconsistencies in one's beliefs, and help illuminate the puzzles in one's life, it is unclear whether it can establish and defend positive moral truth and hence provide us with a constructive understanding of how to live. Socrates seems to believe that his elenchus can lead to positive truth. "I prefer nothing, unless it is true" (*Euthyphro*, 14e). But he never explains how his elenchus can reach this goal.[36] Given his repeated disavowal of knowledge, and given that many earlier dialogues in which Socrates performs elenchus thoroughly are aporetic, it should be safe to say that, for Socrates, ethics is more a matter of critical examination.

In contrast, Confucius believes that ethics must provide positive guidance and offer a vision of how to live properly. Whereas Socrates claims that he knows nothing, Confucius declares that he has a unified theory. Whereas Socrates claims that the unexamined life is not worth living, Confucius maintains that to have a positive knowledge of the dao is the most important thing in human life. "The Master said, 'He has not lived in vain who dies the day he is told about the dao'" (*Analects*, 4:8).

Questions

1. Our knowledge of the teachings of Socrates and Confucius have not come to us in the form of any documents they themselves wrote. How were Socrates' teachings handed down to posterity? How were Confucius's teachings communicated to later generations?
2. The terms Socrates uses to refer to happiness or blessedness (eudaimonia) and how it is attained (virtue) relate most closely to what two Chinese words that Confucius used?
3. Assume that Confucius was correct, and that it would be possible to apply his methods to a Western culture. In an effort to apply his methodology in the West, what would be the *li*, the traditional ritual and cultural practices that one would aim to draw upon in order to direct society back to the dao?
4. In the cases of both Socrates and Confucius, divine involvement in the development of ethics is not present throughout the entire process. These two philosophers would credit god or Heaven for what stages of that development?
5. Suppose you had to rely on one of two philosophers' methods to come to a conclusion about moral truth. Do you think you would be more likely to find that truth by analyzing the cultural beliefs and practices of the past and their effects like Confucius did, or do you think you would be more likely to discover moral truth by the Socratic method of rational inquiry unanchored to tradition? Explain the reasons for your choice.
6. Explain what suggestions Confucius might have for Socrates to help him bring about more fulfillment of eudaimonia in ancient Greek society. Explain what suggestions Socrates might have to assist Confucius in his attempt to bring society back to the dao.

Using Internet sources, answer the following questions:

7. Compare the ethical systems of Christianity, Judaism, and Islam with the ethical teachings of Confucius.
8. How can the teachings of Confucius be used in the development and applications of nanotechnology?
9. Summarize your learning of new ideas and information presented in this chapter by completing the following flowchart:

Flow Chart

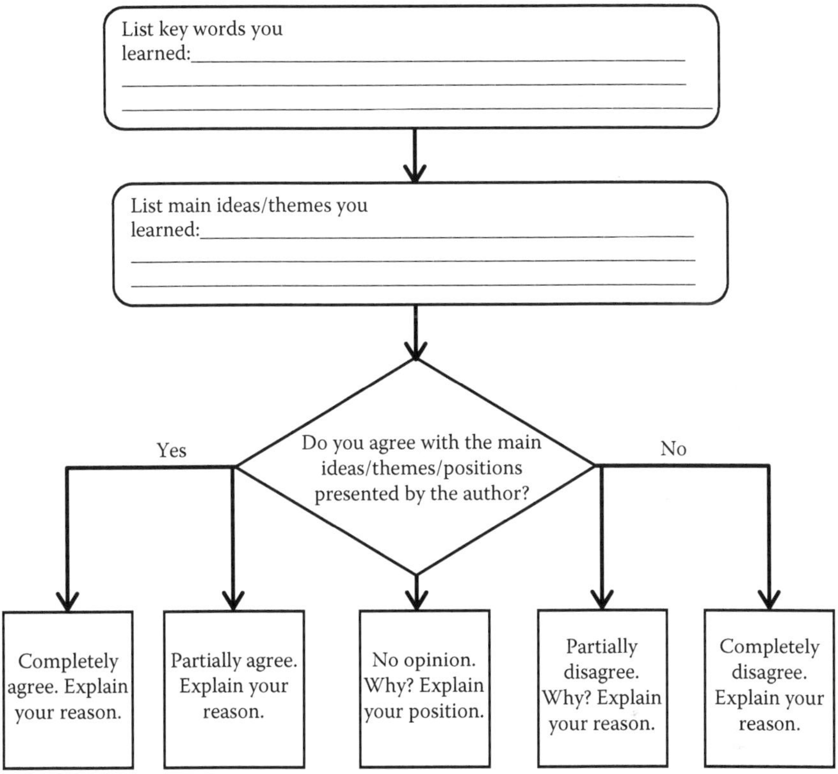

Acknowledgment

This chapter was originally published in *Philosophy, 15*(2), 173–189. Reproduced with permission.

Notes

1. It appears that Aristotle (384–322 BC) was the first to introduce the term *ethics* to refer to a particular branch of philosophy; see *Nicomachean Ethics* (ii.1, 1103a16–18).
2. Unless otherwise indicated, the translation of the *Analects* is from Lau (1979).
3. *Analects* (14:35), translation modified. Cf. also 7:23, 9:5.

4. Cf. also *Apology* (30a, 33c, 40b–c). Unless otherwise indicated, all translations of Plato are from Cooper (1997).
5. Other sources include Aristophanes' *Clouds*, the writings of Xenophon (ca. 428–354 BC), and Aristotle's testimonies. They give very different portraits of Socrates. The *Clouds* is a comedy, and its credibility as historical evidence has been seriously questioned. Xenophon is not a philosopher. His portrait of Socrates lacks philosophical interest and can hardly justify Socrates' status as the founder of ethics. Finally, Aristotle's report is secondhand.
6. For a recent major work in this regard, see Brooks and Brooks (1998).
7. For Vlastos, these dialogues include the elenctic dialogues (*Apology, Charmides, Crito, Euthyphro, Gorgias, Hippias Minor, Ion, Laches, Protagoras, Republic I*) and transitional dialogues (*Euthydemus, Hippias Major, Lysias, Menexenus, Meno*) (Vlastos, 1991, pp. 46–47).
8. In contrast, the dialogues belong to an established literary genre called *Sokratikoi logoi* (*Socratic Discourses* or *Conversations with Socrates*). Many of Socrates' disciples wrote dialogues, and Plato is only one of them. As literary works, dialogues are fictions designed to convey the author's own messages rather than to give historical representations of Socrates. For a summary of his minimal view, see Kahn (1996, pp. 88–95).
9. For Kahn, the *Apology* reports an historical event and "can properly be regarded as a quasihistorical document" (Kahn, 1996, p. 88). Vlastos also regards the *Apology* "as a touchstone of the like veracity of the thought and character of Socrates depicted in Plato's other early dialogues" (Vlastos, 1971, p. iv).
10. See, for instance, Reeve (1989, pp. 62–74), Vlastos (1991, p. 171), and McPherran (1996, pt. 4).
11. This phenomenon is also noticed by Ivanhoe (2000, p. 9, no. 7): "Confucius' sense of mission is an often overlooked and under appreciated aspect of his thought. But he clearly saw himself in such terms."
12. At *Apology*, 21b5–6, a stronger form of the oracle's saying is given: "What does he mean by saying that I am the wisest?"
13. Diogenes Laertius, *Lives and Doctrines of the Philosophers*, II, 16. Democritus (ca. 460–350 BC), a contemporary of Socrates, left us with numerous fragments on ethics and expressed many views similar to Socrates. Julia Annas (2000, p. 304) thus legitimately suspects that to say Socrates is the first Greek ethical philosopher "may be doing an injustice to Democritus." But she admits that "as with all philosophy we can read only in the form of isolated sayings, Democritus' ethics poses a problem of seeing it in a general framework" (p. 307). Kahn also noticed similarities between Socrates and Democritus, but believes that "from the point of view of the history of ethics, Democritus stands in the shadow of Socrates" (Kahn, 1998, p. 42). In addition, it is also important that Democritus's ethical thinking is the implication of his natural philosophy.
14. *Metaphysics*, i.6, 987b1–3.
15. For Greeks, fortune is distributed by gods. See, e.g., Homer, *Iliad*, 24, 525–531.
16. In Vlastos's remark, "Socrates' true place in the development of Greek thought" is that "he is the first to establish the eudaimonist foundation of ethical theory which becomes common ground for all of the schools that spring up around him" (Vlastos, 1991, p. 10).
17. Hesiod focuses on the virtue of (legal) justice, and the Delphic inscriptions advise the virtue of temperance.

18. Solon, in his conversation with Croesus, already held the view that happiness cannot be judged by the wealth and power that one enjoys at some period of life, and it cannot even be determined until the whole life ends. See Herodotus, *History*, 1. 29–32.
19. *Republic*, 354a1; *Nicomachean Ethics*, 1095a18–20; *Eudemian Ethics*, 1219b1.
20. Cf. *Crito*, 47d–48b; *Charmides*, 161a; *Gorgias*, 507a.
21. Angus Graham felicitously calls ancient Chinese philosophers "the Disputers of Dao." We can appropriate his term to call Greek ethical philosophers "disputers of eudaimonia." Graham himself, however, sets the contrast between Western philosophy and Chinese philosophy as one between "What is truth?" and "Where is the Way?" (Graham, 1989, p. 3). He ignores the issue of eudaimonia.
22. For the distinction between ren as a universal virtue and ren as a particular virtue, see Chan (1955, p. 298).
23. *Doctrines of the Mean*, chap. 20, and *Mencius*, 7a/16. These two books, together with the *Analects* and the *Great Learning*, form the Four Books, the core texts of classical Confucianism.
24. It should be noted that *virtue* is used to render another Confucian term *de*. *De* is related to the verb *to get*. The term has appeared in the ancient oracle bone inscriptions, referring to the psychic power that an individual possesses to influence and attract other people and even the surrounding environment. In particular, *de* means the power that a ruler holds that enables him to command others without appealing to physical force. Since the exertion of such a power is associated with the desirable attributes such as kindness, dutifulness, etc., *de* comes to be used to refer to these attributes or qualities as well, and hence it is usually translated in English as *virtue*. *De* is a formal concept, and needs substantial specification. In my understanding, ren is Confucius's specification of *de* (virtue). In other words, his theory of ren is his version of the theory of *de*.
25. Tradition holds that he edits what are later called the Six Classics: *The Book of Documents* (Shujing, records of the legendary sage-kings Yao, Shun, and Yu, of pre-Zhou rulers, and of early Zhou rulers), *The Book of Odes* (Shijing, an anthology of 314 early Zhou poems), *The Spring and Autumn* (chronicle of the state Lu down to the time of Confucius), *The Book of Change* (Zhou manual of divination with appendices), *The Record of Rituals*, and *The Record of Music*. It has been controversial as to how much work Confucius actually did with regard to each of these texts, but there has been wide agreement that Confucius was the first systematic transmitter of ancient classics, and that his work contributed greatly in preserving and continuing Chinese traditional culture.
26. *Apology*, 21b. Cf. also at 41d, where he says that the affairs of a good man "are not neglected by the gods." At *Euthyphro* 6b–d and 7c–9e, gods are said not to quarrel with one another.
27. It is interesting to compare what they say about death. When asked to discuss death, Confucius replies: "You do not understand even life. How can you understand death?" (*Analects*, 11:12). For Socrates, "I have no adequate knowledge of things in the underworld" (*Apology*, 29b).
28. *Apology*, 21e, 22a, 23b–c, 30a.
29. For a helpful discussion of this remark, see Tu (1989, pp. 41–47).
30. Indeed, although Socrates was accused of impiety, there was no formal definition of the crime of impiety in the Athenian legal system. See Allen (1980, p. 28).

31. It has been commonplace that Plato develops his tripartite theory of the soul to correct Socrates' intellectualism.
32. *Charmides*, 166c–d; *Hippias Major*, 298b–c; *Protagoras*, 348c–d.
33. *Crito*, 48b; *Gorgias*, 471e.
34. *Protagoras*, 352a–358d.
35. About the Great Learning, see note 23. However, if a state is completely corrupted and one can do nothing about it, Confucius approves that one can withdraw (*Analects*, 8:13, 15:7). For a helpful discussion of the relation between self-cultivation and political activism in early Confucianism, see Eno (1990, pp. 42–52).
36. For the commentators who are in favor of ascribing a constructive role to elenchus, the best explanation is that the positive result is inductively generated from Socrates' extensive examinations. The interlocutors fail always because they hold beliefs that are inconsistent with the beliefs that Socrates holds to be true. For various versions of this position, see Vlastos (1991, chap. 4), Reeve (1989, pp. 52–53), and Brickhouse and Smith (1994, 1.3.3). This explanation is helpful in understanding how Socrates can claim, despite his disavowal of knowledge, that he knows certain moral truths. But the inductive inference cannot reach the universal definition that Socrates aims for and cannot yield certainty. It should also be noted that, in Plato's view, the elenctic examination only plays the role of "cleansing," and he does not seem to have confidence that elenctic examination can reach positive moral truth (*Sophists*, 230a–231b). He does not seem to have much confidence in Socrates' elenchus as a method of theory building. It is certainly the case that in the dialogues, where positive metaphysical and epistemological doctrines are established, Plato replaces the elenchus with the method of hypothesis (*Meno*, 86e–87b; *Phaedo*, 99d; *Republic*, 596a). Aristotle also seems to believe that Socrates' elenchus is a negative testing and is unable to reach a positive end. As he says towards the end of the SE, "This was why Socrates used to ask questions and not to answer them—for he used to confess that he did not know" (183b6–7).

THE NOBEL LAUREATES DECLARED

In 1988 in Paris, France, the Nobel laureates declared: If we (mankind) are to survive in the twenty-first century we must go back 2,500 years to find the source of Wisdom in the sayings of Confucius, the master teacher, there we will find wisdom.

CONFUCIUS IN HIS OWN WORDS

Studying and learning (*Xue-Er*)

- Clever words and an outwardly gracious appearance is seldom associated with the true virtue.
- A person worthy of respect may be poor yet he is cheerful or he may be rich yet he loves to act according to the accepted standards of behavior.
- Do not worry about not being acknowledged by others; worry about not acknowledging them.

Principles of governing (*Wei-Zheng*)

- The practice of government by means of virtue may be compared with the North Star to which the multitude of stars pay tribute.
- If you lead the people by means of harsh laws, they will try to avoid the punishment and have no sense of shame. If you lead the people by means of being an example of virtue and by means of following the rules of propriety, then they will have a sense of shame and they will follow the rules on their own.
- If one reviews the old as a means of realizing the new—such a person may be considered a teacher.
- The superior man acts before he speaks, and then speaks according to his actions.
- Learning without reflection is lost labor. Thought without learning is dangerous.
- Failing to act on what is seen as right is a lack of courage.

Source: Confucius Said, by Ed Ho & Nick Lambros, Blue River Press, Chicago, IL. Reprinted with permission.

References

Allen, E. E. (1980). *Socrates and legal obligations*. Minneapolis, MN: University of Minnesota Press.

Annas, J. (Ed.). (2000). *Voices of ancient philosophy*. Oxford, UK: Oxford University Press.

Brickhouse, T. C., & Smith, N. D. (1988). *Socrates on trial*. Princeton, NJ: Princeton University Press.

Brickhouse, T. C., & Smith, N. D. (1994). *Plato's Socrates*. Oxford, UK: Oxford University Press.

Brooks, E. B., & Brooks, A. T. (1998). *The original analects: Sayings of Confucius and his successors*. New York, NY: Columbia University Press.

Chan, W.-T. (1955). The evolution of the Confucian concept Jen. *Philosophy East and West, 4*, 298.

Cicero, M. T. (1971). *Disputations* (Vol. 4, pp. 10–11) (J. E. King, Trans. from Loeb ed.). Cambridge, MA: Harvard University Press.

Cooper, J. M. (Ed.). (1997). *Plato: Complete works*. Indianapolis, IN: Hackett.

Eno, R. (1990). *The Confucian creation of heaven*. Albany, NY: SUNY Press.

Graham, A. C. (1989). *Disputers of the Tao*. La Salle, IL: Open Court.

Ivanhoe, P. J. (2000). *Confucian moral self-cultivation* (2nd ed.). Indianapolis, IN: Hackett.

Kahn, C. (1996). *Plato and the Socratic dialogue*. Cambridge, UK: Cambridge University Press.

Kahn, C. H. (1998). Pre-Platonic ethics. In S. Everson (Ed.), *Companions to ancient thought: Vol. 4. Ethics*. Cambridge, UK: Cambridge University Press.

Lau, D. C. (1979). *Confucius's analects*. Hong Kong, China: Chinese University Press.

McPherran, M. (1996). *The religion of Socrates*. University Park, PA: Pennsylvania State University Press.

Nehamas, A. (1999). *Virtues of authenticity*. Princeton, NJ: Princeton University Press.

Reeve, C. D. C. (1989). *Socrates in the apology*. Indianapolis, IN: Hackett.

Tu, W.-M. (1989). *Centrality and commonality*. Albany, NY: SUNY Press.

Vlastos, G. (Ed.). (1971). Introduction. In *The philosophy of Socrates*. New York, NY: Doubleday.

Vlastos, G. (1991). *Socrates: Ironist and moral philosopher*. Cambridge, UK: Cambridge University Press.

4

Ethics: A Conservative Christian Perspective

Gene Anderson

CONTENTS

KEY WORDS: good, image of God, special revelation, general revelation, conscience, moral absolutism, relativism, pantheism, theism, beauty

Ethics: A Conservative Christian Perspective

In John F. Kennedy's 1960 presidential campaign in the United States, many voters feared that his Roman Catholicism would impact his decision making (Guinness, 1993). Today, Christian politicians of all denominations strike fear in the public because people are unsure of what changes in policy could result if devout Christians were to allow their faith to guide their decisions regarding ethical issues while in office. This is also true in policy decisions regarding technology when moral issues are involved. A better understanding of the roots of Christian ethics and what makes Christian moral teaching distinct from that of other faiths may help relieve some of the apprehension the public sometimes feels when it hears about a candidate being a "born-again Christian" or an "ardent Roman Catholic."

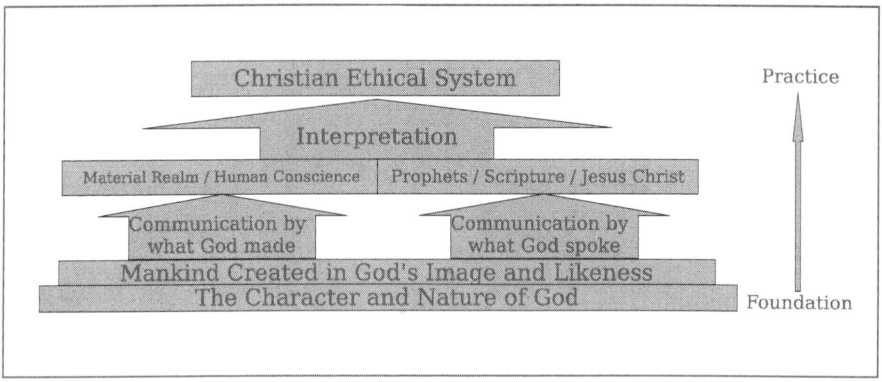

FIGURE 4.1
Diagram of development of Christian ethics from foundation to practice.

Every ethical system must be based on something. There needs to be a basis on which to make moral judgments—a way to distinguish right from wrong and good from evil. In Christian theology, this foundation for the moral law is the nature of God Himself (Newman, 1997).[1] See Figure 4.1 for an illustration of how God's nature serves as the foundation of the Christian system of ethics. Before the human race came into being, before the material universe was formed, even before time itself, there was one Being, and His character would become the standard by which all events would be evaluated. This God is acknowledged by Jews, Muslims, and Christians alike. That which is in conformity with God's character and values is by definition good, and to whatever extent something departs from that likeness, it is to that degree evil. The Bible repeatedly states that God is good: "For the LORD is good" (Psalm 100:5);[2] "God is light, and in Him there is no darkness at all" (I John 1:5); "The Lord is righteous in all His ways, and holy in all His works" (Psalm 145:17—King James version). That His goodness makes Him the rightful standard by which to judge what is good and what is not is clear in the creation account (Genesis 1), in which we repeatedly read, "And God saw that it was good."

In understanding Christian ethics, not only is it necessary to understand who God is, but it is also imperative to recognize who "man" is. Humankind is created in God's own image and likeness and was to bear a resemblance to God in the way in which a child resembles its parents. This likeness is not physical, but involves some internal quality (Baker, 1991) that makes humankind, like God, have the capacity for rational thinking, personality, and a will. This is not true of the rest of the created order. Humankind was not created to be autonomous, a standard unto itself, its own god determining truth for itself. Humankind was made to bring glory to its Creator by freely choosing that which was good. This also makes humans stand apart from the rest of creation as beings who bear moral responsibility. Lastly, humankind is given a position of dominion over all the other living organisms, but this is not to say humans are allowed to mistreat or abuse them. On the contrary, of the Garden of Eden, humankind was told to "cultivate it and keep it" (Genesis 2:15), with the same Hebrew verbs (*abad* and *shamar*) used to describe later how conscientiously the Levitical priests were to care for the tabernacle where they were in charge of the worship of God (Beale, 2011).

If one agrees that it is reasonable to assume that a holy God is our creator and is the standard of goodness, one still has to determine how it is that God's nature can be discovered. Christians assert that God has revealed His nature to us in two ways, special revelation and general revelation (Thiessen, 1986). In special revelation, God reveals Himself to prophets and other messengers and sends them to communicate these truths to the world. The Holy Scriptures are, for the Christian, a compilation of these messages from God and faithfully represent His words to humankind without error. As the Apostle Paul wrote to Timothy (II Timothy 3:16), "All Scripture is inspired by God and profitable for teaching, for reproof, for correction, for training in righteousness."

The second means of ascertaining what God is like is by the observation of our senses combined with our reason. This is what is meant by general revelation. For instance, Paul writes that "that which is known about God is evident within them; for God made it evident to them. For since the creation of the world His invisible attributes, His eternal power and divine nature, have been clearly seen, being understood through what has been made" (Romans 1:19–20). Part of what humankind sees that reveals God is the external world, but part is what is within: the conscience. The cosmos reveals truth about God without giving exhaustive truth, and at the same time the truths therein do risk being misinterpreted or misunderstood. For instance, natural disasters could be misread to imply that God is unjust or capricious. In the same way, our consciences also bear witness to God's nature, though we cannot equate the voice of our conscience with the voice of God or expect it to fully reveal God's nature to us.

Most civilizations come to agreement about basic moral truths, and these conclusions have been arrived at in part because our consciences speak, not quite in unison, but at least with a great measure of unity, about what is acceptable and what should be proscribed. And where it speaks with this unity, we find it is in agreement with what God's revelation has also spoken concerning moral truths. Yet there is a difference between a well-formed conscience and a twisted conscience. Some tribes in Irian Jaya in the 1960s saw betrayal and treachery as positive goods, even to the extent that when they first heard the Gospel, they admired Judas Iscariot and felt little affinity to Jesus Christ (Richardson, 2005). The atrocities committed by developed nations of the twentieth century provide ample evidence that ill-formed consciences are not only the possession of Stone Age civilizations.

Some people do not accept the Christian faith and question the truthfulness of its description of our origins and the legitimacy of its writings to represent the nature of God. It can be expected that they will hesitate to submit themselves to what the Christian religion proclaims regarding ethics if it is

TABLE 4.1
Traits of Christian Ethics and Societal Responses

Traits of Christian ethics	What the public fears about the traits	What is overlooked about the traits
Moral absolutism	Rigidity, lack of flexibility for a changing culture	Stability in a changing culture
Humans are distinguished as special compared to other organisms	Damage to the environment, injustice toward other species	Humankind is elevated; humankind finds a purpose and meaning; vulnerable human lives are protected
Sex is restricted	Too puritanical, too confining, unrealistic demands	Sex is elevated to something sacred, not just something recreational

in conflict with their own wishes. But is it reasonable to regard Christian moral teaching as something dangerous to society? What aspects of a distinctly Christian ethical system could cause alarm to those whose beliefs are outside that system? Table 4.1 lists some distinctive characteristics of the Christian system of morals and possible ways society could respond to them.

One major conflict between Christian ethics and modern civilization is in regarding the nature of moral truth. Since Christians base their ethical beliefs on who God is and what He has revealed, and since this God is changeless and will not alter His nature, then Christian ethics in principle should be without change or variation, even as the societies in which the Church exists undergo changing views on a great number of moral issues. The moral absolutism of the Christian's ethical beliefs is a major cause of discomfort for those who seek a relaxation or reconstitution of a society's ethical standards. This is especially the case for those who are accustomed to the ethical system of moral relativism, in which what is considered morally good can vary from society to society and with changing circumstances. For example, the Bible strongly denounces homosexual behavior, and only the most convoluted misinterpretations of what should be clear passages in the Bible, along with a total disregard for traditional natural law theology, can bring about any advocacy for homosexual acts from a Christian perspective.[3] Moral relativists, on the other hand, would have less problem allowing homosexual acts to be considered good if society brings forward reasons for this change in standards. The tension between society and the fixed standards of moral absolutism will remain unless society brings itself once again into agreement with the Church's stance, because Christian moral instruction on this issue cannot be changed without bringing the entire Christian ethical system's legitimacy into question.

This sounds then as if society does have reason to be concerned about friction with a Christian morality that will not budge. But one must also consider it from another perspective to see the value of this voice for moral absolutism in the culture. Since Christian moral truths are unlikely to change, they at the least will pose no surprises for society. One need not fear, for instance, that Christians will suddenly oppose education as reactionary and dangerous, and call for of all educated persons to work on collective farms or forced labor projects the way the Khmer Rouge did in Cambodia. This is because the Christian faith has acknowledged since its inception the value of learning and the proper role of reason. Similarly, one could rest assured that Christians would not call for a wholesale revamping of the structure of society by deciding the traditional family is a flawed entity and needs to be replaced by some other societal structure. Christian teaching about marriage being ordained by God and the family's validity will never be abandoned by the Church without the Church becoming something that it is not by discarding the standards by which it decides what is true. Thus the Church serves as a ballast to stabilize the ship of society, even on the rough seas of changing philosophies and political systems.

The Christian perspective on the natural order differs from that of some other religions and leads to different moral choices in some circumstances. The Scriptures teach that God created the cosmos outside of Himself. God is not a force or essence that is distributed amongst living organisms or that is resident in nature and its processes. Western religions believe that God created the world out of nothing, but Eastern religions either see god as being part of the world or the world being part of god (Kreeft, 1988). Hinduism, Taoism, Native American, and other folk religions all hold to pantheistic views of nature. If god is present in a forest in a way that he cannot be present in a Walmart store, then it may be not only ecologically harmful to clear a forest to make room for a Walmart store, it is also an act of desecration and morally evil.

Christian and pantheistic ethics come into conflict not simply because they have different understandings of God; their moral values also diverge because they do not agree on the nature of humankind. To the Christian, humankind still carries the image of God, though his choice to sin has marred this image. No other living organism has been elevated in this way by God. As pantheistic views of nature become more commonplace in Western society, more people are questioning humanity's role as dominant over the rest of creation and see any favored treatment of the human race compared to that of other species as something akin to racism, though it could also be called *speciesism* (Land, 2005). The impact of the pesticide DDT on birds and aquatic life led to restrictions on its use in fighting mosquito populations that carry malaria. In recent years, questions have been raised regarding the bans on the use of DDT because of the negative impact this ban has had on the health of people in the less developed regions of the world due to malaria. Dr. J. Gordon Edwards has claimed that 190 people each minute are either dying or suffering directly as a result of the bans on DDT, making the numbers of deaths annually number in the millions (FrogBite Productions, 2011). If humans are to be regarded as innately more valuable than the organisms of other species who are threatened by DDT usage, then it would seem that the outcome of these bans has been disastrous, and to continue these bans could even be morally wrong. On the other hand, if each species is to be treated with equal respect, then to use DDT and cause such severe harm to wildlife could be evil.

Perhaps the portion of Christian ethics that those outside of the Christian faith find the most threatening is that part that speaks regarding sex. The Bible speaks quite plainly in this area of moral theology, but the restrictions on sex contained in the Christian faith are not just arbitrary rules established to limit people from freely enjoying sex. They are inherent in the Christian's own understanding of the relation between the natural and the spiritual realms. Christianity is sometimes misunderstood as a faith that sees the spiritual as good and the natural as bad. This caricature of Christianity presents a faith that frowns on sex because sex is too close to nature, an animal instinct shared with the lower creatures. It would picture God applauding a couple choosing to engage in a study of the Scriptures and prayer together in bed rather than engaging in the marital act. For those who understand the Christian view of nature and the purpose of humanity, this caricature rings false. First, the Christian does not despise nature, if for no other reason than it was made by his God. How could God look down on nature if He decided to create so much of it that is beyond our ability to measure? When God made all things, including humans with their desire for sex, He declared it all very good (Genesis 1:31). In fact, God's direction to humankind to "be fruitful and multiply" (Genesis 1:28), when understood in light of being made in the image of God, shows humankind acting in imitation of and cooperation with God. God by nature is a creator, and by procreating, a man and a woman are creating something new that never before existed, just as God did when He made the world. And the creation of this new life is more than just a natural human action, because it involves the creation of a new being, a new soul, in the universe, which is a spiritual action of God in collaboration with the man and woman.

Seeing human sexual activity as both spiritual and natural, then, means that it is something very sacred, something more than the biology or the psychology involved between a man and a woman. The worldview of scientific materialism, on the other hand, has no basis to elevate sex to such lofty heights, and in the end can find no logic to restrict sex for moral reasons, because it has reduced sex to a physiological urge on the level of having an itch and needing to scratch it. Elevating sex means needing to protect it with moral codes, and this brings restrictions with which modern society is in opposition. Sadly, however, our culture often fails to see the positive good that Christian ethical thinking brings to the morality of sex by making it something more beautiful and more dignified than the degraded and cheap view of sex that has become ubiquitous in society today. What society values it will protect with the aid of limits, restrictions, and guidelines. The movie *National Treasure* showed the lengths to which the government goes to protect the original document of the Declaration of Independence. It is not available to be fingered by the public at large—only the front side of the document is on display—and there is an armed guard to protect it. A materialist would have to say that these restrictions are senseless—after all, it's just ink on paper. But the significance of the document goes beyond the physical, even beyond the content, to the values we hold dear and the people who risked their lives to make certain that these values became the foundation of a new nation. In the same way, because Christians see the sex act as more than physical coupling, it is to be expected that guidelines are established to protect its deeper significance.

The sexual revolution impacted the way society views sex itself, but also changed the way society views reproduction. With sexual activity freed from its traditional restrictions, sex was demoted from something deeply sacred to something that could really be described as simply recreational. It is a short step from that point to wanting to ensure that nothing interferes with enjoyment, even something as intricately linked to sex as conception and reproduction. The recognition of abortion-on-demand as a human right was made the solution to the "complication" of human fertility. From the Christian perspective, however, it was bad enough to demean sex, but now the human life that results from sex is also reduced to an impediment to personal fulfillment that must be dealt with, when it should be respected as the possessor of an immortal soul and bearer of God's own image and likeness. The moral voice of Christendom and the moral values of society at large have clashed more vehemently over the issue of the sanctity of human life than any other issue because it is here that the foundational assumptions of both worldviews cause the greatest divergence in the practice of ethics.

For modern humankind, released from the restraints of traditional sexual morality, to feel the liberty threatened by Christian ethics is not surprising. At the same time, however, I feel that if people could be brought to realize the preciousness of what they have lost in the process of being liberated, they might not be so quick to continue discarding the moral standards that made our existence as humans more humane, more in touch with the eternal, and more in keeping with the glory that God intended for us to experience. Just like Adam and Eve in the Garden of Eden, we are no longer clothed in God's glory; we are wrapped in the skins of animals and appear more like the beasts than like the God we were supposed to be like, and all because we have exercised our liberty in ways God never intended.

How can humankind come to recognize how it has degraded itself when it has aimed to liberate itself? I believe one possible way is through the loftiest aspect of Christian ethics: the moral teachings and life example of Jesus Christ. In Jesus' Sermon on the Mount (Matthew 5–7), He makes clear that Christian ethics goes beyond staying within the bounds of certain rules of personal behavior. It is rather about becoming holy, just as God is holy. And in holiness there is an attraction that appeals to that core of

God's image still deep within us that has not yet been discarded for even wider intemperance. Reading Jesus' instructions to His disciples, we see a way of life that is radically countercultural and yet at the same time highly attractive. This is the reason that Mother Teresa is seen as one of the most beautiful people to have ever lived in modern times, even though she was so distant from what we normally consider to be beautiful. Jesus wins the admiration of countless people of all religious beliefs, or even those with no religious beliefs, because in His life one can see such love, such selflessness, and such deep holiness. Even though people may be repulsed by his teachings on sexual morality, on personal greed, and on the only means to eternal salvation, all are drawn to Him as a person because His life shines forth with what the Psalms describe as "the beauty of holiness" (Psalm 29:2, King James version). When society comes to love the teachings and example of this man, the fear of the ethical system that bears His name dissolves away because, as the Apostle John said, "Perfect love casts out all fear" (I John 4:18).

Questions

1. List the two general means God uses to communicate the truths about Himself and humankind that form the foundation upon which Christian ethics are based. Which means is considered superior and why?
2. Describe how an ethical system based on the character of God results in moral absolutism rather than moral relativism.
3. Given what you have learned about Christian ethics, how do you suppose a Christian would argue in favor of genetically modified (GM) crops? Would there be any particular reason a Christian might oppose GM agriculture?
4. Certain Christian churches have gradually stopped believing in the doctrine of Biblical inerrancy; that is to say, they no longer contend that the Bible is without error. Can you explain why these particular churches might tend to more readily allow their moral teachings to be shaped by the societies in which they exist?
5. Some members of the public state that it is wrong for religion to enter into public policy decisions because of the principle of the separation of church and state. Many Christians contend that all ethical systems are based on fundamental beliefs of some kind. They would add that it is wrong to discriminate against ethical systems simply because they happen to be religious. Is the Christian politician who takes his or her religious beliefs into consideration when advocating a certain public policy overstepping the boundary of the separation of church and state? Why or why not?

6. Compare the Christian understanding of the nature of God and humankind to the view of some other religion or to the view of atheism.
7. Which worldview would tend to result in the most environmentally friendly policy: theism (Judaism, Christianity, Islam, etc.), pantheism, or atheism? Why?
8. Judaism, Christianity, and Islam all accept Moses as a prophet of God and accept his writings in the book of Genesis as truly representing God's nature and humankind's origin and nature. Given this common foundation, why do you think there is not more cooperation among these three faiths in ethical matters? Do you foresee a time when they might band together to support a certain public policy? What modern issue might bring about such cooperation?
9. Summarize your learning of new ideas and information presented in this article by completing the following flowchart:

Flow Chart

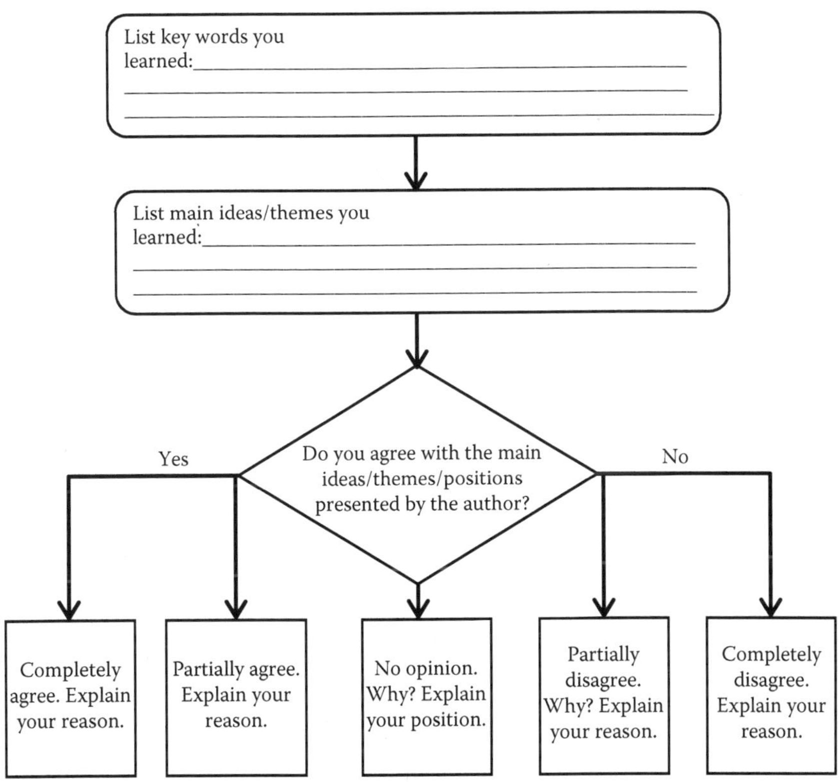

Scenario 1: Brain Control to Suppress Criminal Behavior

California
June 18, 20XX
NanoControl Medi-Tech Inc. has progressed in experiments with neurological behavior control using nanodevices in experiments with primates and is proposing a patent on an electronic control system that can be inserted into the brain in specific locations to limit brain activity related to certain actions and desires. It is specifically aiming to market the system to control violent behavior in repeat offenders of violent crimes. If successful, it would essentially eliminate the desire to act violently by noting the increase of neurological activity in the area of the brain responsible for violent behavior and not allowing it to surpass a certain preset limit, rendering the recipient incapable of a violent response or action. Unlike medications, it would not alter other behavior or retard other brain functions with side effects like lethargy.

Response

1. Discuss the implications of this technology with regard to humans being made in God's image.
2. Should a Christian be in favor of such technology because it might protect innocent people? Should a Christian oppose such a system because it essentially removes free will from the person receiving this device?
3. Would a Christian be more willing to accept the use of such a device if it could be used to control homosexual desires and actions?

Scenario 2: Release of Proprietary Technology for the General Good

Czech Republic
March 15, 203X
A legal battle has erupted in the R&D laboratories of HGIT (Human Genetics Information Technology) in Prague over Dr. Radek Dolezal's insistence that the genetic pattern of the most commonly encountered human cancer cells be released to the public at large. He insists that the information must be made available to biotech companies around the globe in order to enable them to develop systems incorporating this data that will be able to

differentiate genetically between cancer cells and healthy cells. The goal would be the development of nanotechnology devices that would be able to seek out cancer cells and destroy them before they had the opportunity to spread and multiply. HGIT insists that having applied for a patent for this data, it should be legally protected from competition in the development of such equipment. HGIT itself is not a biotech firm, but it has, as a subsidiary, a small biotech company called Health Designs that is in the nanotechnology field and aims to develop just such a device that could be implanted in the human body, programmed with the genetic code to be sought after, and made to destroy the cells with the matching genetic code. Dolezal insists that any delay in designing and manufacturing nanodevices to battle cancer will be the equivalent of murdering those who fall victim to cancer before this technology hits the market. He believes that making this genetic coding "open source" will speed the development of the nanodevices that could be programmed with this code and implemented to prevent or even heal cancer. He threatens to leave HGIT if his demands are not met. HGIT is seeking a court order to prevent him, should he leave, from working for any other biotech or nanotech company.

Response

1. Discuss whether you believe Dr. Dolezal is morally right to insist that this potentially life-saving information be made available to all without restriction in the light of the value placed on human life by the Christian system of ethics.
2. Discuss the potential negative side effects that could arise with the development of such nanodevices. For instance, could such a device, if programmed with other genetic information, pose a health risk or even be misused with genocidal impact on whole ethnic groups? If so, would the health benefits outweigh the risks of accidents and misuse?

Notes

1. In keeping with Christian tradition, when the name of the Christian God is used, or when a pronoun referring to Him is used, the word will be capitalized. The word "god" or "gods" or nouns and pronouns when referring to other deities will not be capitalized.
2. All biblical quotations are taken from the *New American Standard Bible* (Lockman Foundation, La Habra, CA, 1971) unless otherwise noted.
3. See Leviticus 20:13, Romans 1:26–27, I Corinthians 6:9, and I Timothy 1:9–10.

References

Baker, W. H. (1991). *In the image of God* (p. 36). Chicago, IL: Moody Press.

Beale, G. K. (2011). *Garden temple.* Retrieved December 3, 2011, from http://www.kerux.com/documents/keruxV18N2A1.htm

FrogBite Productions. (2011). *Three billion and counting.* Retrieved September 24, 2011, from http://www.3billionandcounting.com

Guinness, O. (1993). *The American hour: A time of reckoning and the once and future role of faith* (pp. 180–181). New York, NY: Free Press.

Kreeft, P. (1988). *Fundamentals of the faith: Essays in Christian apologetics* (p. 90). San Francisco, CA: Ignatius Press.

Land, R. (2005). *Imagine! A God-blessed America: How it could happen and what it would look like.* Nashville, TN: Broadman & Holman Publishers.

Newman, J. H. (1997). *Parochial and plain sermons* (pp. 1050–1051). San Francisco, CA: Ignatius Press.

Richardson, D. (2005). *Peace child* (4th ed.). Ventura, CA: Regal Books.

Thiessen, H. C. (1986). *Lectures in systematic theology* (rev. ed.). Grand Rapids, MI: Wm. B. Eerdmans Publishing.

5

Nanotechnology: Ethical and Social Issues

Fidel M. Salinas, Denise M. Smith, and Shekar Viswanathan

CONTENTS

KEY WORDS: Defense Advanced Research Projects Agency (DARPA), dimensions of technology, ethics, ethnopolitics, nanotechnology, printing in 3-D, quantum computing, stakeholders, Scanning Tunneling Microscope (STM), The Singularity Contents

Introduction

Ethics, by its very nature, invites diverse opinions. When external factors impact decisions that impede the advancement of technology that, on balance, can improve the quality of life, then the balance that sustains the quality of life is potentially compromised.

Any introductory information about nanotechnology must begin with the observation of the late IBM nanotechnology research scientist, Dr. Hans Coufal: "The atoms always knew." While the atoms may have known, it took a while for the human factor to acknowledge and understand the impact of energy at the atomic level. Though technology in general has changed the landscape of many aspects of modern culture, it has also raised questions of ethics and human interactions that have evolved in concert with regulations, and there are yet further guidelines and regulations that remain to be conceived. Corporate culture has become more distant, and the importance of the written word has taken on another dimension, with social media taking center stage in the workplace. The purpose of this chapter is to detail some of the ethical considerations that have resulted in response to the phenomenon of nanotechnology.

Stakeholders of nanotechnology jealously guard their independence of research and development. The two major stakeholders in the United States are the National Science Foundation, which is supported by private and public funds and is also a sponsor of a major portion of nanotechnology research, and the other is the Defense Advanced Research Projects Agency (DARPA), which has had a long history of freedom to conduct research and development in literally any area with military applications. The military-industrial complex would most likely bypass any restrictions it encounters from any guidelines, including ethical guidelines, as a potential conflict with its mission; the military-industrial complex has always been found to prevail in the end by using national security as its chief argument against all objections, including ethical considerations.

On a more global scale, researchers have always guarded their freedoms in research. Hence, they also strive to preserve the same freedom while conducting their research and development in nanotechnology. In this context, it is important to remember that the issue of ethics does vary from country to country and from culture to culture. Potentially, all users of nanotechnology, regardless of the application, would be impacted and, at the most, might slow down or at least ameliorate ongoing nanotechnology development.

Nano Dimensions: A Snapshot

A starting point for putting nanotechnology in perspective must begin with its dimensions, both negative and positive. Figure 5.1 illustrates the scale of natural and human-made things. The term *nano* comes from the Greek word νάνος (*nanos*), meaning *dwarf*. The nano dimension, 10^{-9}, is a nanometer, a value of 10 to the minus 9. To put this into perspective, our fingernails grow roughly at the rate of one nanometer per second. That is about a layer of atoms. As a further comparison, a gene is made up of a few atoms. However, this still might not put the dimension in perspective.

To understand a gene at the nano level, it would be useful to consider a more familiar positive dimension: namely, imagining two humans on the island of Manhattan, the humans being the size of a gene in a human blood cell, with Manhattan roughly being the size of the blood cell. The contrast between 10^{-9} and what we see with technology at this very small dimension certainly demonstrates a dramatic difference. Figure 5.2 illustrates the powers of 10 and shows the negative and positive ranges of these dimensions.

The possibility of "seeing" and "moving" an atom seemed remote for most of human history until IBM fellow Donald M. Eigler, Ph.D., using the scanning tunneling microscope (STM), which he designed and built, demonstrated the ability to manipulate individual atoms with atomic-scale precision. On September 29, 1989, Dr. Eigler became the first individual to successfully manipulate individual atoms on a surface by using the STM. He used it to spell out "I B M" by positioning 35 xenon atoms, and in the process, perhaps created what is perhaps the world's smallest corporate logo (see Figure 5.3). This event was likened by Jack Uldrich in his book, *The Next Big Thing Is Really Small: How Nanotechnology Will Change the Future of Your Business* (2003), to the Wright brothers' first flight at Kitty Hawk.

Using the positive powers of 10 as a comparison, we could estimate that light, at the velocity of 186,320 miles per second, will go around the Earth at the equator about 7.5 times in one full second. If we captured a nanosecond of time, we could say that light travels approximately 11.75 inches in that time span. On the other hand, the speed of sound in dry air at 68°F (20°C), at approximately 768 miles per hour or a fifth of a mile per second, would take roughly 32 hours to go around the Earth at the equator.

The following equation can be used to compute the speed of sound; *c* is computed by

$$c = \sqrt{\frac{C}{\rho}}$$

where

C is a coefficient of stiffness (or the modulus of bulk elasticity for gas mediums)

ρ is the density

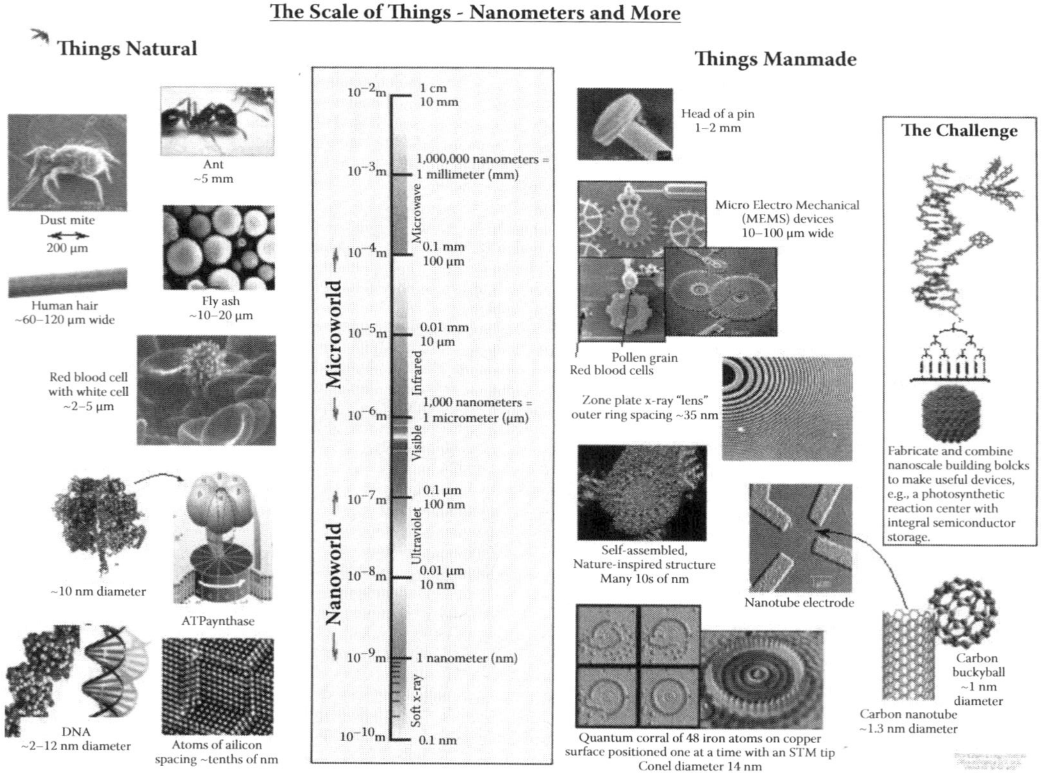

FIGURE 5.1 (See color insert)
Scale of things natural and things made by humans. (Courtesy Office of Basic Energy Sciences, Office of Basic Energy Sciences, U.S. Department of Energy.)

LearningSolutions.3rdPlanet: ™ ***Dimensions of Technology***

Powers of Ten	Negative Powers of Ten
$10^3 \rightarrow$ Kilobyte	$10^{-3} \rightarrow$ mille-technology
$10^6 \rightarrow$ Megabyte	$10^{-6} \rightarrow$ micro-technology
$10^9 \rightarrow$ Gigabyte	$10^{-9} \rightarrow$ nanotechnology
$10^{12} \rightarrow$ Terabyte	$10^{-12} \rightarrow$ picotechnology
$10^{15} \rightarrow$ Petabyte	$10^{-15} \rightarrow$ femtotechnology
$10^{18} \rightarrow$ Exabyte	$10^{-18} \rightarrow$ attotechnology
$10^{21} \rightarrow$ Zettabyte	$10^{-21} \rightarrow$ zeptotechnology
$10^{24} \rightarrow$ Yottabyte	$10^{-24} \rightarrow$ yoctotechnology
$10^{27} \rightarrow$ Xerabyte	$10^{-27} \rightarrow$ xeratechnology
$10^{100} \rightarrow$ Googolbyte	$10^{-100} \rightarrow$ googoltechnology
10^n	10^{-n}

FIGURE 5.2
Dimensions of technology. (Courtesy LearningSolutions.3rdPlanet.)

FIGURE 5.3 (See color insert)
Xenon atoms. (Photo courtesy IBM.)

In continuing with the negative powers of 10, at 10^{-12}, in a picosecond, light travels about the length of a pepper grain and, at 10^{-15}, light travels about the distance across the diameter of a proton or neutron. To continue in the negative direction any further, we would have to follow the progress of the Large Hadron Collider (LHC), the world's largest and highest-energy particle accelerator, where such particles traverse a circular tunnel seventeen miles in circumference that spans the border between Switzerland and France about 100 m underground.

On the positive powers of 10, the difference between a million (10^6) and a billion (10^9) can perhaps be shown with an analogy using money. For example, if one were to win the million-dollar lottery and received the money at the rate of $1,000 per day, it would take just under three years to collect the million dollars. If one were to win the billion-dollar lottery and collected the billion dollars at the same rate of $1,000 per day, it would take approximately 3,000 years to collect the billion dollars. Following the same analogy, adding three more zeros, collecting the trillion (10^{12}) dollar lottery at the same rate would take about 3 million years.

Global Standards on Ethics: An Approach

International standards play a crucial role in supporting global trade and protecting human health and the environment. U.S. government agencies and private sector organizations are becoming more involved in contributing to international efforts in helping to guide standards for nanotechnology. Representatives worldwide must work in concert to unify international standards that are based on sound science.

These standards must be implemented and made available to the international nanotechnology community, and they should be used to support safe and practical research and development in nanotechnology. Certainly, ethical considerations must be incorporated into the global effort by international organizations. Some of these organizations include the Organization for Economic Cooperation and Development (OECD), the United Nations International Labor Organization, the World Health Organization, and the World Trade Organization.

A local-government pioneer emerged early on in the regulation of nanotechnology. The Berkeley City Council in California became the first government entity in the United States to approve a specific ordinance requiring the reporting of nanomaterials used in local laboratories. The resulting legislation is now more well known as the Berkeley regulations on nanotechnology.

The ordinance was initially criticized for the burden it places on researchers and for the inadequacy of the information it collects. However, it was a start and, perhaps, a bit of not-invented-here at play in a community that had no economic incentive to protect the new technology. The Berkeley regulations started by articulating a range of potential human health hazards. A sample of one of the hazards was the quantum dots, which can be absorbed into the skin and dispersed into other tissues. A hazard with more serious problems includes nanoparticles that can be inhaled and could negatively impact the spleen, lymph nodes, bone marrow, and the central nervous system.

Figure 5.4 shows the life-cycle perspective to risk assessment.

Ethics Impact of Nanotechnology: Introduction

The acceleration of technology has brought with it more products and applications in a smaller time frame and, at the same time, additional ethical considerations for society. As sometimes is the case, technological advances bring what at first seems an improvement to our quality of life and, because of a delay of controls and regulations, the perceived benefits can also bring with such improvements the absence of timely protection to society, such as unknown hazards and unintended consequences of technology. With this condition, ethics may suffer as controls and regulations inevitably struggle

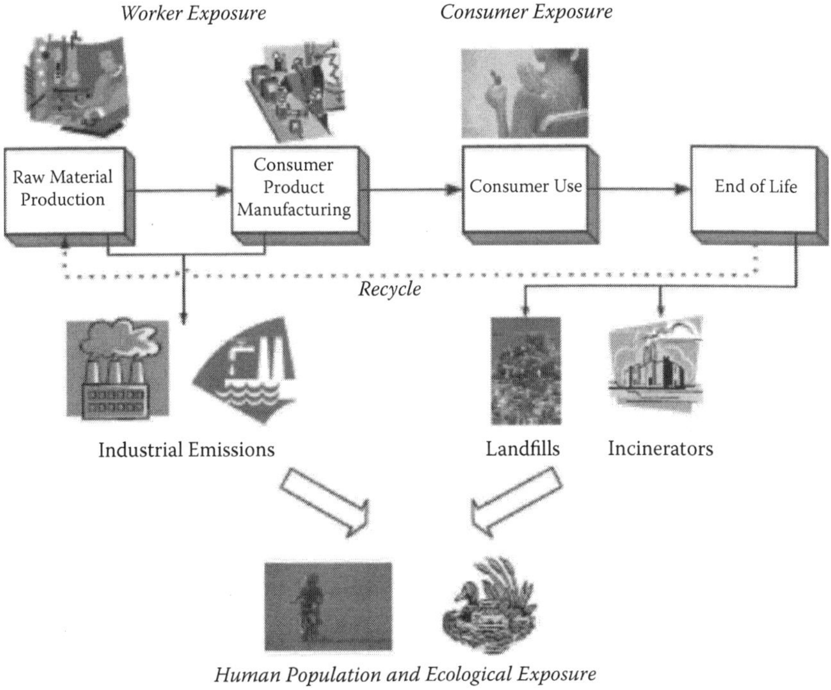

FIGURE 5.4
Life-cycle perspective to risk assessment. (Courtesy EPA, 2007.)

to catch up. This perhaps is what often leads to the notion that these technologies were introduced prematurely into the marketplace.

Dr. Kurzweil cofounded Singularity University with Peter Diamandis, CEO of the X Prize Foundation. Singularity University started its first class at NASA Ames, Mt. View, CA, on Monday, June 29, 2009. Dr. Kurzweil shows that technological change is exponential, contrary to the common-sense "intuitive linear" view. He points out that we will not experience just 100 years of progress in the twenty-first century; we are more likely to experience 20,000 years of progress if we continue to move at the current rate. This is probably because the "returns," such as chip speed and cost effectiveness, have also increased exponentially. Figure 5.5 shows the accelerating returns as we approach the Singularity.

Why Nanotechnology Will Continually Outpace Ethics and Social Issues

As hardware is miniaturized further through nanotechnology advances and takes less physical space, a smaller footprint and more automation will relieve

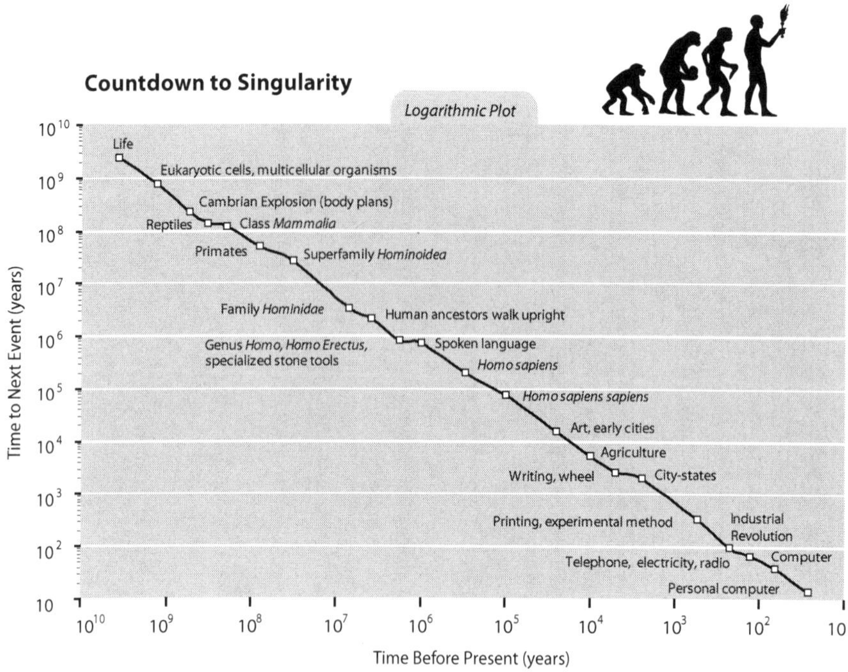

FIGURE 5.5
Exponential countdown to the singularity. (Courtesy Dr. Ray Kurzweil.)

the knowledge worker of repetitive tasks and enhance work processes. See Figure 5.6, which shows the accelerated rate of the implementation of inventions.

Certainly, nanotechnology, with ongoing miniaturization at the atomic level and ever-shrinking, lower-cost creation of computer processors with greater power, is bound to enable telecommunications applications at a higher level of automation and, subsequently, to become available to a wider population.

Technology, which is seemingly changing hourly, is driving this phenomenon, and this plethora of technologies is spawning new applications that, besides offering new advantages and benefits, also contribute to creating hazards and other unintended consequences. To paraphrase the late professor, Dr. Melvin Kranzberg, Callaway professor of History of Technology at Georgia Tech and of WWII intelligence fame, in General Patton's Third Army: "Technology is neither good nor bad; nor is it neutral; it just is."

It is more than likely that the person who discovered fire would have had to face a lot of criticism from his/her peers when he/she first discovered it. This is because they would have found out that though fire could keep one warm and could cook food, it could also burn down our houses, a negative side of this technology. It is this aspect that is discussed in the section entitled "Concerns About Nanotechnology" later in this chapter. Technology has been a part of being human ever since humans discovered fire or picked up a rock or stick to use as a tool.

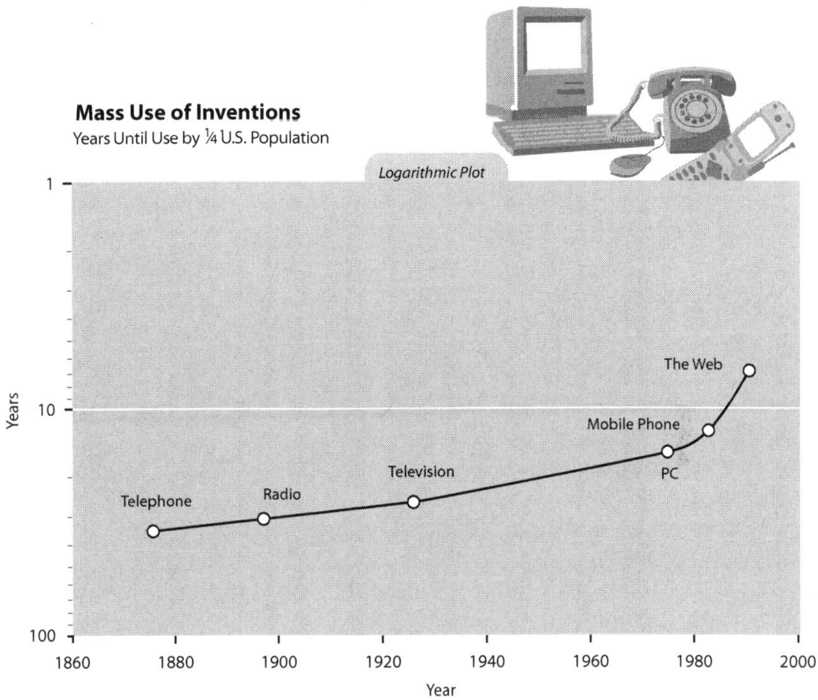

FIGURE 5.6
Accelerated use of inventions. (Courtesy Dr. Ray Kurzweil.)

Certainly, the global Internet has benefited from the developing technology and, in particular, nanotechnology, for it has made it possible to make a smaller "footprint" of hardware installations and create opportunities for new applications so that more of the planet's inhabitants have access to nanotechnology. This is because as hardware becomes smaller and more affordable, it allows for the possibility of making those applications that have been enjoyed by the developed countries for past years to also be accessed by the have-nots of the planet. The cultural dominance of the "haves" will eventually level itself to the rest of the global population, so that all can participate in the information highway in a true global economy. However, this may be at the cost of stepping on ethics.

Nanotechnology will further level the global playing field by localizing and massively reducing the costs of manufacturing anything. With that benefit comes the potential of considerations of an ethical nature in fulfilling the growing needs of society. Communication will become the main conduit of civilization. Eventually, there will only be in-home replication machines that will produce food, clothing, and all the other necessities and wants of humanity. Anything that cannot be replicated at home, such as a car, may be obtained from a local replication center. Consumers will pay royalties for purchased products directly to the designers and innovators. The replication technology will use unprocessed raw materials right out of the ground that are relatively inexpensive at this time.

The cost of the raw materials that make up the average car totals approximately a couple of hundred dollars. This will dramatically reduce the costs of products by cutting out all the waste in production, which will, in turn, eliminate the existing messy processes of traditional manufacturing. Also, the same technology will reverse all the environmental damage traditional manufacturing has caused. In addition, very little energy will be required for replication due to the elimination of all the steps in traditional manufacturing. Traditional manufacturing will go the way of the ugly cocoon that produces the beautiful butterfly. Nanotechnology and the communications infrastructure that will come with it will become the beautiful butterfly, integrating all of humanity in a global village.

However, let us consider a different side of technology and its ethical impact on society. For example, the current technology of three-dimensional (3-D) printing certainly brings ethics to center stage, as discussed in the following scenario.

Consider the still fairly new technology of 3-D printing of just about any three-dimensional object such as tools or a dental mold and actual bridge replication, ready for installation. Such applications with this technology seem unlimited. However, let us suppose an unethical person wants to replicate—print in 3-D—a pistol complete with bullets. Will the pistol fire? Just in case one's position is one of total disbelief and is confident that such a scenario is unlikely, let us inquire if that person will test the theory and stand in front of the pistol to test it. Naturally, ethical considerations play a major role at the forefront of ethical behavior, depending on the person's chosen role in society.

Nanotechnology Regulations, Ethnopolitics, and the Impact on Applications

Agriculture

In 2006, the United States, Argentina, Brazil, Canada, India, China, Paraguay, and South Africa grew 97% of global transgenic foods having genetically modified (GM) material. Ethical considerations come into play when such modified foods are distributed domestically and internationally and consumed by the general public without knowledge of the manufacturing processes, safety factors, and ingredients at the nanotechnology level of DNA technology. Ethical considerations are even more prevalent in GM foods being exported to Third World developing countries. This can also include medicines, vaccines, and tobacco products where nanotechnology modifications have taken place at the molecular or atomic level.

Clearly, ethical considerations play a key part in any perceived exploitation of societies not having the awareness or the communication capabilities of developed countries. Aggressive marketing can create artificial pressures for a sustained revenue performance. This can impede an improving economy for a growing society's awareness of healthy or alternative choices of values and fundamental human rights, including the right to adequate and safe food.

Finance and Insurance

Nanotechnology ethics impact finance and insurance as it does other areas of our lives. As with dealing with insurance for just about every aspect of our lives, insurance with nanotechnology is also based on uncertainty. With the freedoms made possible by nanotechnology due to the capabilities to work at the atomic and subatomic scales, nano-restrictions and nano-risks are created. This condition can have the effect of being in conflict with certainty. The notion of working at the atomic or subatomic levels gives nanotechnology a potential consequence resulting from not being able to calculate the risk of insurance in a predictable business model of profitability.

However, the flip side of the coin of nanotechnology is such that as the technology improves existing and new products; improves materials through advanced manufacturing; enables buildings and other infrastructures to withstand powerful storms, hurricanes, earthquakes, and flooding, then the risks can be spread or diffused throughout new areas as a result of nanotechnology applications. Moreover, just as nanotechnology is an umbrella title that combines diverse technologies, processes, product applications, and even academic disciplines, it also provides the insurance industry more granularity in its potential product offerings and opportunities.

Regulation for the insurance industry is also based on certainty and can provide opportunity to the industry if it is adapted to the risks of nanotechnology. The risks can be ameliorated to multiple classes of clients insured with the same needs and conditions of society, such as third-party liabilities, product quality recalls, human health, the environment, and catastrophes. How we, as a society, consider ethical standards in the context of moral risk in finance and insurance has an equivalent impact on instituting balanced regulations.

A What-If Scenario About Insurance

Let us suppose that one has applied for insurance on a job and a bureaucrat is charged with informing the applicant with the following information: "I'm sorry Mr. or Ms. X, but you have the abc123 gene, and we cannot insure you." This what-if scenario is a look into the potential of regulations out of control or out of balance, and we might consider the potential it would have on society if this should happen.

Health

The breakthrough discovery of DNA along with nanotechnology coincided with the first baby boomers nearing retirement age. While the search for the fountain of youth has always been part of the health landscape—at first, perhaps, more for women—this quest has now become less gender defined. Botox injections as well as face-lifts in a jar have been found to be as effective for men as they have been for women in replacing lengthy surgical procedures. Procedures such as face-lifts and chin implants are becoming less prevalent. Advancements in nanotechnology have perhaps contributed to less-than-ethical marketing of such services and product claims. In many cases, such claims have yet to be realized and documented for the public at large. Evidence and documented proof of the potentially promised benefits remain elusive.

More important than physical appearances is the hope that nanotechnology will help with protecting human life. It is hoped that new, less invasive, and more effective treatments than the traditional chemotherapies or bypass surgery treatments will not only extend human life span, but will also eliminate cancer or heart disease.

Dr. Ray Kurzweil (2006), from the Singularity University, in his work, *The Singularity Is Near*, suggests and more strongly predicts that with technology advancing at exponential rates, humans could reasonably live 100 to 200 years or even indefinitely. In the context of one's belief system, this certainly brings forth a myriad of ethical considerations.

The following two scenarios, one from today and one from the future, suggest the varied potential of nanotechnology.

Scenario for today: Injections of Botox, face-lift in a jar, portable chemotherapy for targeted cells only, laser heart surgery, Lasix eye surgery

Scenario for the future: Injecting one's own stem cells to stimulate the growth of everything from a new heart valve to stimulating collagen production in one's face.

Manufacturing

The late Dr. Richard Feynman, author of *There's Plenty of Room at the Bottom*, foretold in the late 1950s what was to come in nanotechnology development and applications. He envisioned manufacturing at the atomic and subatomic level. His vision predicted molecular nanotechnology (MNT), a technology based on the ability to build structures of complex, atomic specifications by means of nanotechnology manufacturing. Dr. Feynman also envisioned a nanoprocess, with miniature factories using nanomachines to build complex products, including the nanomachines to be used to build other products and machines. MNT would involve combining physical principles demonstrated by chemistry, other nanotechnologies, and the atomic or molecular capability of life with the systems engineering principles found in modern macroscale factories. Dr. Feynman was, indeed, the harbinger of the new industrial revolution of the twenty-first century. Now, we are faced with the ethical issues of the twenty-first century as well. We can use this technology to improve our quality of life or use it to create something that is tremendously dangerous and destructive, which could happen if it fell into the wrong hands or minds.

Organizations such as the Center for Responsible Nanotechnology (CRN) and similar government agencies are crucial to providing guidelines and regulations for responsible research and development of applications. The CRN is charged with ensuring that nanotechnology is being used for productive and beneficial purposes. The promised benefits of nanotechnology cover a myriad of applications.

For example, imagine a world in which cars can be assembled molecule by molecule, garbage can be disassembled and turned into beefsteaks, and people can be operated on and healed by cell-sized robots. For now, it may sound like science fiction. However, with current semiconductor chip manufacturing encroaching upon this scale and the ability to move individual atoms at the IBM Almaden laboratory (see images at http://www.almaden.ibm.com/vis/stm/atomo.html), we are fast approaching the technological ability to fabricate productive machines and devices that can manipulate things at the atomic level.

Universities worldwide are conducting research and development at the nanoscale level. One example is the Stanford Nanofabrication Facility, Stanford University, Stanford, California. The Stanford Nanofabrication Facility (SNF) is a state-of-the-art, shared-equipment, open-use resource. This laboratory serves academic, industrial, and governmental researchers across the country and around the globe. The SNF is more than just a lab; it is a vibrant community of researchers.

A Manufacturing Scenario

Recently, the question has been brought up regarding the validity of having provided Solyndra, a company based in Fremont, California, approximately $500 billion of taxpayer dollars to build cylindrical solar panels when it was widely known by Solyndra employees themselves that China had already developed and was building solar panels of equivalent quality at cheaper prices and was exporting them to the United States.

Were U.S. government officials aware of this? Did they still go ahead and invest taxpayer dollars into this domestic solar-panel manufacturing firm that was doomed to failure from the beginning only because they were under pressure to invest in a U.S. green-technology manufacturing company?

A What-If Scenario

Another uneasy scenario is the possibility that people might use the technology to manufacture money. Nanotechnology offers unethical persons the capability to reproduce every part of a bill, atom by atom, to create counterfeit currency and thus render the latest security printing techniques of color variety, woven-in metallic threads, special inks, paper and cloth aggregates, and other materials all ineffective.

Transportation

Ethical considerations play an important part in transportation in that moral, ethical, and honest professionals are implored to follow honest principles as delineated by the Institute of Transportation Engineers (ITE). Transportation professionals must adhere to strict regard to the safety and health of the public in the advancement of human welfare. However, transportation engineers are also susceptible to succumbing to pressures that may compromise them in following best practices in transportation. Sometimes, pressures to accept certain designs and conditions may be applied by a senior government official in the sense of a quid pro quo. These pressures can compromise the ethical position of the engineer. The following scenario illustrates such a situation.

A What-If Scenario

Electric cars and hybrid cars have become more prevalent in the past five years and have helped to reduce air pollution somewhat. Mass transit is making a comeback in larger cities, especially as work and leisure communities have begun to merge. However, the need for cars and even mass transportation may become obsolete as the office becomes a virtual place of meeting, texting, and social media. We could text a slice of pizza or a batch of margaritas for a quick lunch or office happy hour.

Utilities

Faced with global warming, we look for politically correct solutions to reverse climate change. We faithfully recycle and become more aware of energy consumption, use smart meters, and work to keep our carbon footprints to a minimum. However, what we plan and what we do vary in proportion to our income, vendor availability, and other factors. What really happens is that we go with the cheapest energy source, perhaps at the expense of any proposed ethical considerations, and throw all these good intentions out the window. Given the way costs for both electricity and gas have spiraled, this is certainly understandable. Perhaps, it is possible to be an ethical consumer of energy, too. Utility companies look to become greener and yet remain profitable.

For consumers, there is an implied commitment to using ethical energy, which really means green electricity. However, the current technologies are still in infancy compared to long-established and yet not-so-green energies, such as oil and coal. As a result, green energy is still under development and struggling to be competitive with mature energies. It will take time to develop green energies into competitive offerings. Currently, green energies can even be more costly to produce, deliver, and maintain. It continues to be a challenge to implement green energy on a competitive basis as more and more companies strive to provide such energies for general availability.

Scenario: The San Bruno Gas Pipeline Explosion

San Bruno, CA (KABC)—A federal investigation was completed nearly one year after a deadly gas pipeline explosion rocked a suburban neighborhood near San Francisco. The results were handed over to the National Transportation Safety Board (NTSB) in Washington on Tuesday.

The board chairman says Pacific Gas & Electric is to blame for a litany of failures leading up to the explosion. The five-member NTSB voted that substandard welds and other problems dating to the 1956 installation of a PG&E gas transmission line beneath San Bruno were the direct cause of the accident.

The utility is also being accused of exploiting weak safety oversight by state and federal regulators. "It was not a question of if this pipeline would burst," said NTSB chairman Deborah Hersman. "It was a question of when." PG&E has said it did not know of the welds because its records incorrectly listed the pipe as seamless. The utility also has been unable to produce other key records regarding the pipeline.

The September explosion killed eight people, injured dozens of others, and destroyed or damaged fifty-five homes. It was the worst pipeline accident in a decade. A consideration remains whether potential nanotechnology for leakage detection could have been used for more recent inspections, which could have prevented the failure of the pipeline. If such an opportunity presented itself and its benefits were not recognized in time or bypassed in favor of only cost, then ethical considerations might remain.

Military, Space, NASA

Perhaps more than in any other areas of development, nanotechnology applications have advanced further in military and space applications. The military and space systems developed by industry have realized tremendous benefits from discoveries in nanotechnology, robotics, and artificial intelligence. This is in part from developments in the United States as well as in foreign countries, where some of such work may not be in concert with U.S. goals and objectives.

An example of an emerging military application emanates from current research and development work in quantum technologies in such areas as quantum mechanics, quantum computing, or similar fields. Much of the science behind this fairly new technology is still young; quantum technologies have the promise of wide-ranging applications for the fields of cryptography as well as remote and secure satellite communications. In the near future, such developing applications could be used to send encrypted messages that cannot be cracked or intercepted to securely connected networks, even in remote areas that have no wired infrastructure, including satellites and submarines.

In addition, the fairly new application of cyber warfare can derive from different goals, ranging from the mischievous intentions of antisocial individuals to malevolent goals from loosely affiliated international hacking organizations and nonstate terrorist organizations. This capability may not be limited to any specific nation or group and is a method of attack that would require an ongoing awareness and capacity to respond to changing tactics and strategies with little or no regard for any type of ethical behavior. This real or perceived absence of ethical response has created new roles for our military, for example, the cyber soldier and the nonstate hacker, who must be trained and equipped to deliver attacks against domestic and foreign enemies. To paraphrase the age-old quotation, "All is fair in love and war."

Certainly, in the sense of advancing science and technology, we advocate the development of advanced nanotechnology and subsequently look forward to following the development of current uses for incremental nanoscience and nanotechnology in the expectation that these new technologies will grow and expand our knowledge base, develop new and effective tools, grow profits, and continue the demand for developing advanced nanotechnology applications for humankind.

A Military and Space Scenario

In May 2010, a team of fifteen Chinese researchers from Tsinghua University in Beijing and the Hefei National Laboratory for Physical Sciences, a government-directed research center, published a research paper announcing a successful demonstration of "quantum teleportation" (*liangzi yinxing chuan*) over 16 kilometers of free space. These researchers claimed to have performed the first such successful experiment in the world. The technology on display has the potential to revolutionize secure communications for military and intelligence organizations and may become the watershed of a research race in communication and information technology.

The Promise and Future of Nanotechnology as Ethics Struggles to Catch Up

Figure 5.7 shows a timeline for the field of nanotechnology from the beginning of industrial prototyping and commercialization in 2000 and projected through 2020.

Preparing Students for Nanotechnology and Its Impact on Ethics

Research tells us that college students have chosen to major in math and science long before taking SAT tests and completing college admission essays.

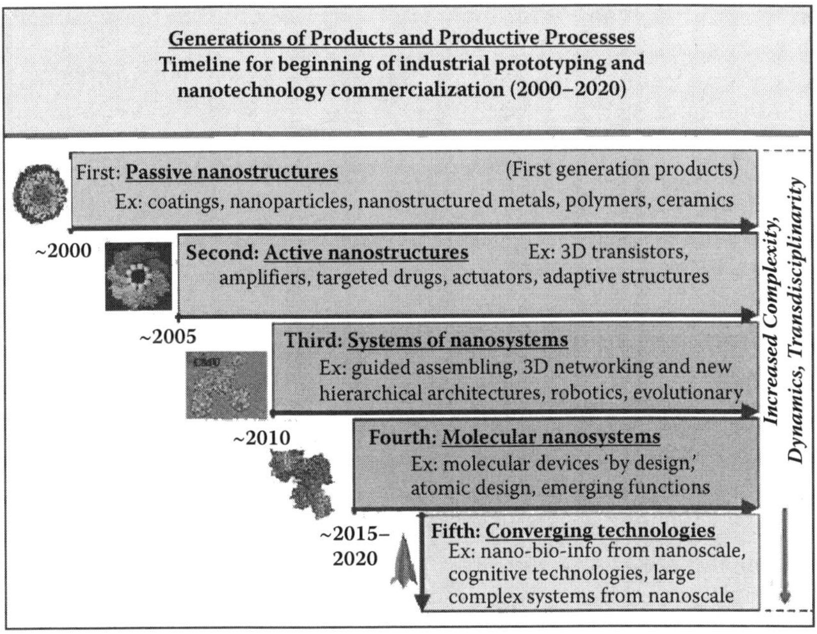

FIGURE 5.7 (See color insert)
Generations of products and processes. (Courtesy of NSF, M. C. Roco, Nanoscale Science and Engineering Education Workshop, "Partnership for Nanotechnology Education" Los Angeles, April 26, 2009.)

There is a strong correlation between math and science instruction during the school years—preschool through high school—and the later career selection. It is true that teacher influence begins in preschool! For too long, math and science instruction emphasized the product, not the process. Details about the process of solving math problems or scientific investigation were absent from the instruction. In more recent years, scores on high-stakes testing and other artificial measures of student learning and school accountability have significantly hampered both teacher creativity and student learning, especially with the instructional content in math and science. Recently it has been found that the first three years of a teacher's tenure determine how effective the teacher is in teaching a particular subject. After all, not everyone is cut out to be a teacher. Increasing pay and incentives to attract people who become these great teachers is imperative to bettering the educational system.

Young children are natural mathematicians; they seek order in their world through sorting objects by size or shape, or by dividing boxes of crayons so everyone has an equal variety of colors. Children start learning about fractions when they cut their first birthday cake; they start learning about the powers of ten even faster when they receive their first money in a birthday card. As children mature, so does their curiosity. Instead of reading about magnets, children should be encouraged to complete experiments in small groups about magnetic fields or to build a battery cell, as these would not only teach them about the magnets or cells, but also promote cooperative work that they could benefit from during their school years as well as in their adult life. This can be an incentive and an early introduction to teamwork.

The foundation for building a "local" workforce strong in math and science starts in the young school years and continues across the life span. While teachers and school administrators bear a great responsibility for school instruction, parents remain their children's first and most important teachers. Parents have a profound impact on every aspect of their children's lives. Parents must take an active role and join teachers and administrators not only to create more interesting and innovative instructional methods in math and science, but also to support their children's math and science learning at home.

This means working in the online environment as well. However, the online world is always changing. The notion of anonymity on the Internet and hiding behind e-mails impacts the way we communicate. Instead of a phone conversation or even an e-mail, a current mode might be "Facebook Me" or to text a message.

The concept of a face-to-face meeting is becoming more foreign, especially to a generation of young people who, most likely, had their own cell phone in elementary school and whose senior pranks included listing their high school on eBay or Craig's List. Their résumé is posted on Facebook or LinkedIn. In this world of faceless interchange, even conference calls become burdensome and too time consuming.

There is a reluctance, perhaps even an aversion, to place phone calls. These are all factors that must be taken into consideration with regard to teaching and learning in the twenty-first century. In such an environment of diminished personal communications, creating almost a free-for-all condition, we should consider how this environment might alter the landscape of ethical behavior and accountable responsibility without the perceived, and perhaps necessary, intrusion of a face-to-face or live voice-to-voice communication.

Conclusion

The benefits, advantages, disadvantages, and outcomes of nanotechnology are all still emerging as we continue research and development of applications. While much is still unknown, undefined, and unregulated, the march toward the promise of nanotechnology remains relentless. Furthermore, we cannot prevent unscrupulous individuals, rogue organizations, or countries or military organizations from developing nanotechnology applications with intentions that result in negative consequences.

Organizations such as the National Nanotechnology Initiative (http://www.nano.gov/); the National Institute for Nanotechnology (http://

www.nsti.org/Nanotech2005/exhibitor.html?id=134); the National Science Foundation (http://www.nsf.gov/); the United Nations Economic and Social Commission for Asia and the Pacific: Asian and Pacific Centre for Transfer of Technology (APCTT) (http://www.apctt.org/); and the European Commission, Community Research: Towards a European Strategy for Nanotechnology (http://ec.europa.eu/nanotechnology/pdf/nano_com_en_new.pdf) help to guide and inform the global community on nanotechnology research and development of applications.

A new technology such as nanotechnology—with undetermined benefits, disadvantages, and consequences—requires a responsible stewardship to develop effective ethical guidelines and policies. This effort is in the beginning stages, and now is the time to promulgate guidelines and policies that articulate the known benefits and disadvantages of nanotechnology. A global goal with respect to ethical standards can help formulate ethical guidelines and policies; to do any less would undercut the goal of ethical behavior.

Questions

1. Compare the benefits with the disadvantages of a medical application using nanotechnology.
2. How would you solve the problem of disposing of waste in nanotechnology at the atomic level?
3. What application in nanotechnology research and development is of the highest importance? Why?
4. What country leads the international effort in nanotechnology research and development?
5. State in one word the number one motivator that drives nanotechnology research and development.
6. Develop an analogy that describes the difference between a nanometer and a physical comparison that a layperson would understand.
7. What is the relationship between local government and international guidelines and regulations with regard to nanotechnology?
8. Fill in the blank:
    - There are ____________________ femtometers in a nanometer.
    - Typical human hair is ____________________ nanometers in diameter.
9. Distinguish the difference between a nanometer and a nanoliter.
10. List an example of nanotechnology use that surprised you.
11. State in your own words why nanotechnology research and development of applications is important.
12. Summarize your learning of new ideas and information presented in this article by completing the following flowchart:

Flow Chart

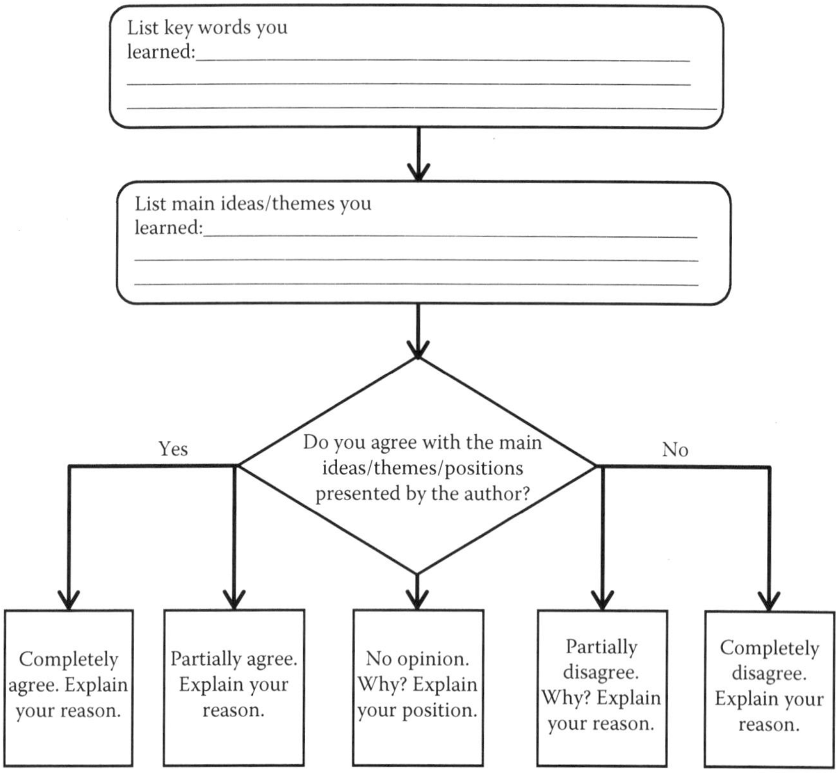

Scenario 1: Where Fraud Took Over

In 1981, a fraud was perpetuated at the Yale School of Medicine when two researchers claimed to have performed research experiments in endocrinology that never took place. A third researcher complained to supervisors and other persons that the two researchers had plagiarized some of the words and ideas of her paper that had not yet been published. They had never actually done the research reported in it.

Scenario 2: Where Fraud Took Over

In 2005, Hwang Woo-suk was South Korea's scientific Superman. He had three institutes, a stamp created in his honor, and, over the years, $60 million at his disposal. His face was plastered on buses in Seoul. *Scientific American* named him its researcher of the year.

Then, late in 2005, Hwang's world began to fall apart. The man lionized for being the first to clone human embryonic stem cells was reported to have paid some women for their eggs (which he denies) and obtained others from staff (which he admits).

Other scientists, in anonymous Web posts, pointed out flaws in research papers that Hwang and his colleagues had published in the prestigious *U.S. Journal of Science*. Hwang confessed to having lied about some of his stem-cell lines and concluded that not a single human embryonic stem cell had been cloned.

As a result, Hwang was fired and indicted for fraud, violating bioethics laws, accepting $2 million based on falsified research, and allegedly embezzling $856,000 in private and government funds. His two chief partners were also indicted, and three others were charged with embezzlement, destroying evidence, and bioethics-law violations.

Case Study: Nanotechnology Waste and Ethics

Appendix B presents a case study from the University of Washington that involves the disposal of nanotechnology waste. We are still struggling with how to dispose of radioactive waste, and here comes nanotechnology with, in some respects, even more dangerous waste consequences. Consider an atom that can literally enter the body at just about any location and come out just about anywhere; it is a dangerous condition, to be sure. It is something like the neutrino particles coming from the sun and entering our body, perhaps at our head, and going right on through our feet as gamma radiation at near the speed of light. The case study is included in Appendix C.

Appendix A: Nanotechnology Pioneers

Francis Crick: (1917–2004) Nobel laureate, codiscovered DNA double-helix structure with James Watson

Eric Drexler: (1955–) Popularized the term *molecular nanotechnology* and wrote *Engines of Creation: The Coming Era of Nanotechnology*

Richard Feynman: (1918–1988) Scientist, teacher, raconteur, and musician on *There's plenty of room at the bottom*: Feynman lectures on physics

Ray Kurzweil: (1948–) Inventor, futurist; speech recognition, synthesis, OCR, the Law of Accelerating Returns

Gordon Moore: (1929–) CEO emeritus, Intel, Moore's Law

Richard Smalley: (1943–2005) Rice University professor, Nobel laureate, codiscoverer of the "buckyballs/nanotubes" with Robert Curl and British chemist Sir Harold Kroto

Norio Taniguchi: (1912–1999) Tokyo University science professor, coined the term *nanotechnology* in 1974

Nikola Tesla: (1856–1943) Wireless technology, Tesla coil, turbine, oscillator electric car, induction motor, rotating magnetic field, particle beam weapon

James Watson: (1928–) Nobel laureate, codiscovered DNA double-helix structure with Francis Crick

Appendix B: Case Study Detail: University of Washington

W University of Washington
Nanoethics Case Study
The Paradigm Shift of Nanotechnology: Consequences of Status Quo Lab Attitudes

Stephanie Vasko and Billy Butler

Alex and Ben are grad students who work in a lab that focuses on nanoparticle development at College X. Recently they have been working on a project with exotically doped quantum dots. Quantum dots have opened a new realm of science, one where the nanoscale properties of a material differ greatly from that of the bulk material. In a bulk semiconductor, most electrons exist in the valence band, while only a few exist in the conduction band. For an electron to move from the valence band to the conduction band, an outside stimulus, such as heat, voltage, or light, must be applied. In bulk conducting material, there are continuous energy levels, but in a semiconductor, these energy levels are not continuous: There is a gap between the valence (filled) electron levels and the conducting electron levels. Only a small fraction of the electrons cross the gap. In insulators, the gap is so large that not even an outside stimulus can provide enough energy for the electron to move from the valence band to the conduction band. Quantum dots get their name from the fact that they are governed by the Heisenberg uncertainty principle, an important relationship from quantum mechanics, which states that you cannot know the location and energy of the electron at the same time. While quantum dots show promise in industries like solar power, in vivo imaging of cancer, and optoelectronic devices, some are made of extremely toxic materials, including cadmium, selenium, tellurium, and lead.

Ben has worked in the lab for five years and is the safety officer. Alex began working in the lab about two months ago, and received his general safety training from a university class and his lab-specific safety training from Ben. Typical waste labeling in labs at College X include chemical composition, percent, and check boxes that are labeled with common associated hazards, including corrosive, oxidizer, and "other." There is no space listed for identification of the material type. In their lab, nanoparticle waste is labeled the same as a bulk waste. When Alex questions Ben about waste processing, Ben tells him that this is standard procedure and that they just need to list the chemical composition. Alex is concerned about the safety conditions of Ben's fume hood, where there is build up from previous reactions and chemicals that are unidentifiable. Alex and Ben share lab space, and Alex is not comfortable with Ben's lab habits. Alex voices his concerns at a group meeting, and his professor tells Ben to clean his space. Ben promises to clean his space, but over the course of a few weeks, nothing is done to remedy the situation.

At this point, Alex feels conflicted. His lab has been creating and characterizing these particles for several years and should have a detailed disposal method for these particles. As the new member of the lab, Alex is torn between discussing the situation with the lab safety director in his department and keeping the problem to himself. He is being financially supported by his adviser and is concerned that if he blows the whistle, his graduate and professional careers might be ruined.

During Alex's first year, the Environmental Protection Agency (EPA) performs spot checks on the labs at College X. The EPA does not examine all labs at the college, but Alex's lab is chosen for the inspection. During the spot check, Alex privately voices his concerns about Ben's hood to the EPA agent. The EPA agent then does a preliminary inspection of the hood and decides that the conditions warrant further investigation. Alex asks the EPA agent what will happen next, and the agent says that the EPA will begin a thorough review of the lab's background. The EPA will gather information about the lab's status as a waste generator, applicable permits the lab may already have, potential violations, and the lab's compliance history (if any). Alex is also told that there may be other inspectors, depending on whether the EPA contacts the Occupational Safety and Health Administration (OSHA) and if results of the preliminary inspection necessitate contacting other departments.

As this is happening, a Resource Conservation and Recovery Act (RCRA) inspector from the EPA is assigned to review the lab's information and prepare for the inspection. Given a concern for hazardous materials storage, the inspector determines that a protective material is necessary. The inspector learns that the lab is using a contractor to dispose of waste and identifies pathways that the waste takes according to the contractor's records. The RCRA inspector determines that one or two more inspectors are necessary because of the size of the facility and contacts OSHA to conduct a joint inspection. This takes a few days as OSHA is informed of the situation and agrees that there is a probable cause for inspection. OSHA and the RCRA inspector decide to make an unannounced inspection of the lab based on Alex's concerns.

Two weeks pass, and Alex continues to ponder whether he made the right decision. Then, that morning as Alex is working in the lab, Ben arrives and asks him if he knows anything about an inspection being done. Alex doesn't say anything about talking to the EPA agent. While Alex and Ben are talking, two inspectors come by with another grad student and the lab safety director. They introduce themselves, and Alex realizes that it is an OSHA and an RCRA inspector. They ask to speak to Alex privately, and the OSHA inspector requests to see Ben's hood. The RCRA inspector asks for details on labeling of the specific chemicals in the hood. Normally the RCRA does not take samples during a compliance inspection, but the inspector decides to after seeing the condition of Ben's hood. Both inspectors leave the hood and continue around the facility, photographing other hoods and containers of material. They continue through the lab, and the OSHA inspector asks about safety and employee working conditions while the RCRA inspector inquires about environmental concerns and human health outside of the lab.

Next, the RCRA inspector documents waste flow procedures and compliance with procedures based on the nature of waste generation in the lab. The OSHA inspector investigates compliance with lab safety practices and verifies that appropriate guidance and regulations are posted. Both inspectors request facility documentation: waste manifests (RCRA) and incident/illness/fatalities reports (OSHA). The RCRA inspector finds all waste management units, documents their condition, checks for appropriate labeling, and examines them for any signs of release. A red flag is raised when the inspector notices that the nanoparticle waste is labeled as bulk. As the generators, students in the lab bear the ultimate responsibility for the life cycle of the nanoparticles after they are discarded. Unfortunately, there are no material safety data sheets that describe the health risk and disposal differences in materials between bulk and nano cadmium and selenium.

The RCRA and OSHA inspectors examine the hazardous and solid waste manifests, inspection logs, training documentation, contingency plans, and the lab's Spill Prevention Control and Countermeasures (SPCC) plan. After the inspection, they meet privately with the lab director and discuss the results. They noted several concerns outside their scope of work that they may need to report to other offices for inspection. For instance, the contractor uses a state-operated landfill to dispose of the waste. The EPA believes that a site inspection of the landfill and contractor's facility is necessary to ensure that the heavy metals involved in the quantum dot process have not leached into groundwater. The lab is informed that, because of their generator status, they will bear the ultimate responsibility for any infractions that can be directly traced back to their products. Alex finds this bittersweet: While the lab will be required to take responsibility for its waste, there are currently no diagnostic tests for quantum dots. He knows that this means there is little chance that the waste products will be traced directly back to his lab. The

EPA and OSHA meet again with the lab director and have a formal discussion about lab violations. Several infractions should be fixed immediately to avoid health issues, such as the cleanup of Ben's hood and relabeling of all containers in the lab that are yellowed, faded, illegible, or otherwise incorrect. EPA and OSHA also mention that their findings are all preliminary and need to be verified before any final conclusions can be made. As they leave, Alex wonders if anything will change.

Questions

1. Is it the university's responsibility to provide nanotech disposal training?
2. What are the implications of labeling potentially hazardous nanosize wastes the same as their "bulk" counterparts? Does the gender/ethnicity role of each character affect the moral responsibility or the views of the respondent?
3. Did the lab have an ethical responsibility to "anticipate" rather than "alert"? (McGinn, 2008, p. 113)
4. What role should the EPA play? OSHA? The federal government in general?
5. What resources does Alex have at your university?
6. Should lab safety also include potential accidents outside of the lab, such as releases from a contractor or other entity that endangers people?
7. What is the order of authority for safety incidents at your laboratory?
8. For the scientists in the audience, how would you develop a nanoparticle test that could be implemented in waste-disposal situations?

Appendix C: Further Reading

Quantum Dots

Alivisatos, A. P. (1996). Perspectives on the physical chemistry of semiconductor nanocrystals. *Journal of Physical Chemistry, 100*, 13226–13239.

Michalet, X., Pinaud, F. F., & Bentolila, L. A. (2005). Quantum dots for live cells, in vivo imaging, and diagnostics. *Science, 307*, 538–544.

Schaller, R. D., & Klimov, V. I. (2004). High-efficiency carrier multiplication in PbSe nanocrystals: Implications for solar energy conversion. *Phys. Rev. Lett., 92*, 186601.

Nanotech Ethics

Bainbridge, W. S. (2004). Sociocultural meanings of nanotechnology: Research methodologies. *Journal of Nanoparticle Research, 6,* 285–299.

McGinn, R. (2008). Ethics and nanotechnology: Views of nanotechnology researchers. *Nanoethics, 2,* 101–131.

Resource Conservation and Recovery Act

What is a hazardous waste? http://waste.custhelp.com/cgi-bin/waste.cfg/php/enduser/std_adp.php?p_faqid=235

RCRA Online, a resource for RCRA questions and answers: http://waste.custhelp.com/cgi-bin/waste.cfg/php/enduser/std_alp.php

Environmental Safety and Policy

Breggin, L. K., & Pendergrass, J. (2007). *Where does the nano go? End-of-life regulation of nanotechnologies.* Project on Emerging Nanotechnologies, Woodrow Wilson International Center for Scholars. Retrieved December 6, 2011, from http://www.nanotechproject.org/process/assets/files/2699/208_nanoend_of_life_pen10.pdf

California Department of Toxic Substances Control. Retrieved December 6, 2011, from http://www.dtsc.ca.gov/TechnologyDevelopment/Nanotechnology/index.cfm

Hoey, M. (2008). *The contribution of nanotechnology, robotics and artificial intelligence to military space systems development, 2015–2025.* Retrieved December 6, 2011, from http://www.spacetransparency.org/Space_Transparency/Home_files/GSWT152025draft-excerpt.pdf

Lippert, I., & Jackson, H. (2006). *The insurance industry and nanotechnology risks discussed taking into account Beck's risk society: The idea of "uninsurability."* Lancaster University, CSEC. Retrieved December 6, 2011, from http://www.erm.tu-cottbus.de/~lippein/horizon/science/contents/working_papers/2006/nanoinsurance4.pdf

Maynard, A. D. (2006). *Nanotechnology: A research strategy for addressing risk.* Project on Emerging Nanotechnologies, Woodrow Wilson International Center for Scholars. Retrieved December 6, 2011, from http://www.nanotechproject.org/file_download/files/PEN3_Risk.pdf

Murashov, V., & Howard, J. *Nanotechnology-enabled quantum computing may fuel a security race.* Retrieved on November 12, 2001, from http://www.foresight.org/nanodot/?p=4331

U.S. EPA. Retrieved December 6, 2011, from http://epa.gov/oppt/nano/stewardship.htm

U.S. Department of Commerce. Retrieved December 6, 2011, from http://www.frtr.gov/nano/pdf/day2/dreher_kevin.pdf

Vasko, S. & Butler, B. University of Washington, Nanoethics Case Study: The Paradigm Shift of Nanothechnology: Consequences of Status Quo Lab Attitudes.

Glossary

Defense Advanced Research Projects Agency (DARPA): The branch of the U.S. government that was founded to maintain a technological superiority and advanced deterrents to prevent technological surprise from harming our national security.

dimensions of technology: The arithmetic or mathematics of working with the powers of 10, both at the positive level and negative level.

ethics: Ethics and nanotechnology (NT) involve a myriad of issues in society. Local and state government initiatives, guidelines, and regulations are leading national and global international initiatives in providing guidelines on research and development, especially in waste disposal of NT products and applications.

ethnopolitics: Ethnopolitics involves the interdisciplinary blending of technologies in the research, development, and deployment of NT applications in the context of emerging global standards of science and technology that follow good ethnoscience and honor global ethnical practices agreed to by convention.

nanotechnology: Technology that works at the atomic or molecular level, this branch of engineering deals with things smaller than 100 nanometers and especially involves the manipulation of individual molecules and atoms.

printing in 3-D: Also called 3-D printing, it is the printing of three-dimensional objects such as prototypes of parts and assemblies composed of several materials by applying successive layers of materials with a variety of mechanical and physical properties.

quantum computing: This is a method of computing that makes direct use of quantum mechanical phenomena, such as superposition and entanglement, to perform operations on data on quantum binary digits (qubits), which will accelerate research and development for nanotechnology applications.

scanning tunneling microscope (STM): The STM was a forerunner of more advanced atomic imagery built and used by Dr. Donald Eigler of

IBM to observe and allow nanomanipulation of the first atom in 1989. The STM led to the development of subsequent atomic imagery such as the atomic force microscope (AFM) and photon-scanning microscopy (PSTM).

stakeholders: This term refers to nanotechnology stakeholders who assume ownership as investors or sponsors of NT research and development and deployment of applications.

the Singularity: A postulated future point when our self-guided evolutionary development accelerates enormously, powered by nanotechnology, neuroscience, artificial intelligence (AI), and, perhaps, uploading what transfers the consciousness and mental structure of a person from a biological matrix to an electronic or informational matrix so that nothing beyond that time can reliably be conceived (Kurzweil, 2006).

Acknowledgment

Grateful acknowledgement is expressed to Professor Usha Viswanathan for her help with proofreading and editing.

References

ASTM. (2011). *Nanotechnology standards*. Retrieved November 11, 2011, from http://www.astm.org/Standards/nanotechnology-standards.html

Barandiaran, J. (2007). *The Berkeley city ordinance on nanotechnology: Shortcomings, improvements, and implications for risk governance*. STEP White Paper Competition. Retrieved December 6, 2011, from http://step.berkeley.edu/White_Paper/Barandiaran.pdf

Center for Responsible Nanotechnology. (n.d.). Retrieved on November 12, 2011, from http://www.crnano.org/

CERN. (2008). *European Organization for Nuclear Research*. Retrieved December 6, 2009, from http://public.web.cern.ch/public/en/LHC/LHC-en.html

Chen, A. (2002). *The ethics of nanotechnology*. Retrieved December 6, 2011, from http://www.scu.edu/ethics/publications/submitted/chen/nanotechnology.html

Eigler, D. (1989, Sept. 29). *Used his Scanning Tunneling Microscope (STM) to position individual atoms*. IBM Corp. Almaden Research Laboratory, San Jose, CA.

EPA & Nanotechnology: National Nanotechnology Initiative & Research Activities. Retrieved November 12, 2011, from http://www.nanotechforest.org/Scopi/Group/nano/scp01nan.nsf/pagesref/KTHN-7A2Q58/$file/02-1%20Savage%202007%20TAPPI%20061107v2.pdf

Food and Agriculture Organization. *Ethical issues in food and agriculture*. Retrieved December 6, 2011, from http://www.fao.org/ethics/ser_en.htm

Foresight Institute. (2011). *Nanotechnology makes possible boat 40% stronger and 75% lighter than metal boats*. Retrieved December 6, 2011, from http://www.foresight.org/nanodot/?p=4481

Fox, C. (2006, December 22). Stem cell scandal shocks South Korea: How one stem-cell researcher and a band of conspirators pulled off the biggest scientific fraud in history. *Fortune Magazine*. Retrieved December 6, 2011, from http://money.cnn.com/magazines/fortune/fortune_archive/2006/12/25/8396782/index.htm

Human Genome Project. (2008). *Genetically modified foods and organisms*. Retrieved December 6, 2011, from http://www.ornl.gov/sci/techresources/Human_Genome/elsi/gmfood.shtml

Jones, C. (2011). *Solyndra employee: "Everyone knew the plant would fail."* Retrieved December 6, 2011, from http://www.thehotjoints.com/2011/09/15/solyndra-employee-everyone-knew-the-plant-would-fail/

KABC News. (2011, August 30). *PG&E faulted for San Bruno gas explosion*. Retrieved December 6, 2011, from http://abclocal.go.com/kabc/story?section=news/state&id=8335664

Kurzweil, R. (2006). *The Singularity is near: When humans transcend biology*. New York, NY: Penguin Group.

Lippert, I., & Jackson, H. (2006). *The insurance industry and nanotechnology risks discussed taking into account Beck's risk society: The idea of "uninsurability."* Lancaster University, CSEC. Retrieved December 6, 2011, from http://www.erm.tu-cottbus.de/~lippein/horizon/science/contents/working_papers/2006/nanoinsurance4.pdf

McGinn, R. (2010). Ethical responsibilities of nanotechnology researchers: A short guide. *NanoEthics, 4*(1). Retrieved November 11, 2011, from http://www.springerlink.com/content/?Author=Robert+McGinn

Neel, W. H. (2004). Striving for ethics in transportation engineering. *ITE Journal, (March)*, 22–24. Retrieved December 6, 2011, from http://www.ite.org/ethics/articles/jb04ca22.pdf

Nickson, C. (2010). *Ethical utility providers*. Retrieved December 6, 2011, from http://www.utilitycharges.co.uk/ethical-utility-providers.html

NNIN. (2004). *Stanford nanofabrication facility*. Retrieved December 6, 2011, from http://www.nnin.org/nnin_stanford.html

Ray, G. H. (2011). *Obama administration exposed to big time fraud in Solyndra collapse*. Retrieved December 6, 2011, from http://mnprager.wordpress.com/2011/09/15/obama-administration-exposed-to-big-time-fraud-in-solyndra-collapse/

Stanford Nanofabrication Facility. (2011). Retrieved December 6, 2011, from http://snf.stanford.edu.

Teague, C., Morgenroth, K., Willis, C., & Haydon, B. (2007 June). Informatics in international nanotechnology standards. Paper presented at Workshop on Nanoinformatics Strategies, Retrieved on November 11, 2011, from http://128.119.56.118/~nnn01/NewFiles/13_Teague_Standards.pdf

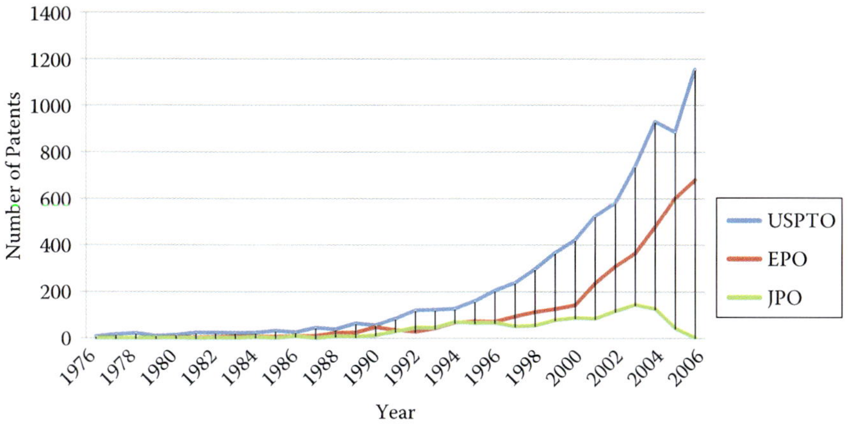

FIGURE 1.1
Trends in nanotechnology patents filed globally, 1976–2006. (From "Trends in Nanotechnology Patents," by H. Chen, M. C. Roco, X. Li, and Y. Ling, 2008, *Nature Nanotechnology, 3*, p. 123. Copyright 2000 by Nature Nanotechnology. Reprinted with permission.)

FIGURE 1.2
The iridescent colors of the Blue Morpho Butterfly's wings are produced by nanostructures that reflect different wavelengths of light. The wing span of a Blue Morpho Butterfly is about 10–15 cm. (Credit: Wikimedia Commons, http://www.nisenet.org/viz_lab/image-collection)

FIGURE 1.4
The feet of the gecko cling to virtually any surface. This image shows the sole of a gecko's foot. The adhesive lamellae on the sole have millions of branching hairs that nestle into nanoscale niches on the contact surface. (Credit: A. Dhinojwala, University of Akron, http://www.nisenet.org/viz_lab/image-collection)

FIGURE 1.6
The lotus effect describes water droplets rolling off leaf surfaces, removing dirt and contaminants. This phenomenon can also be seen in the more common nasturtium. These leaves are covered with wax nanocrystal bundles that trap air and force water to bead and roll off. (Credit: A. Otten and S. Herminghaus, Göttingen, Germany, http://www.nisenet.org/viz_lab/image-collection)

FIGURE 1.8
The corral is an artificial structure created from 48 iron atoms (the sharp peaks) on a copper surface. The wave patterns in this scanning tunneling microscope image are formed by copper electrons confined by the iron atoms. The radius of the corral is about 7 nm. (Credit: D. Eigler, IBM Almaden Research Center, http://www.nisenet.org/viz_lab/image-collection)

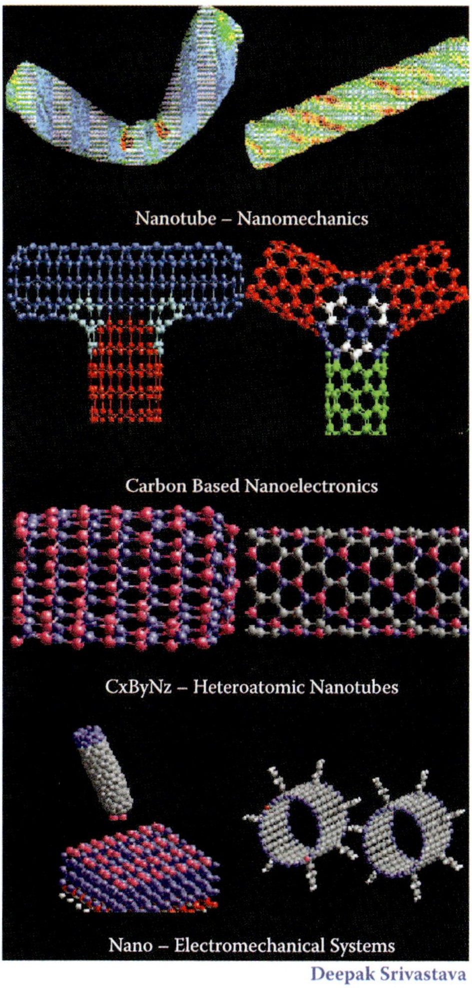

FIGURE 1.16
Nanotube application. (Photo: NASA.)

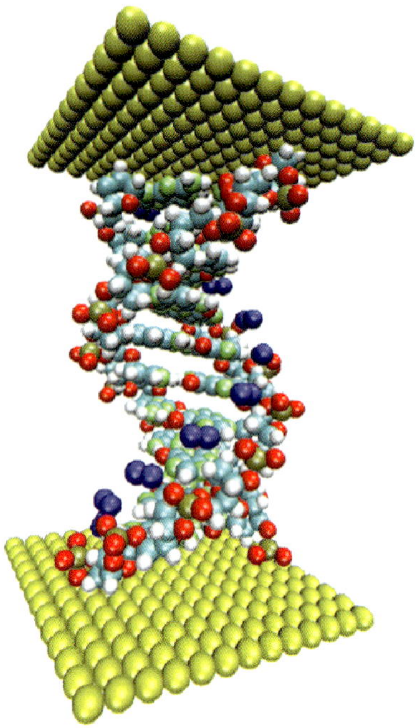

FIGURE 1.17
Engineered DNA strand. (Photo: NASA.)

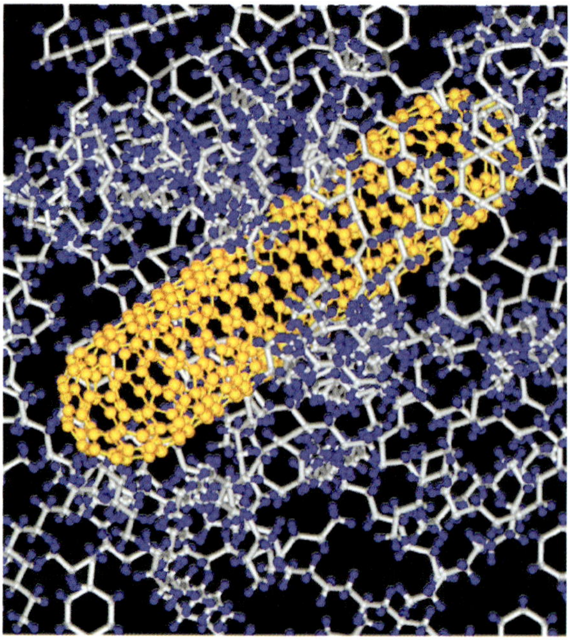

FIGURE 1.18
CNT polymer. (Photo: NASA.)

FIGURE 1.19
Two seals depicting the language of Mohenjodaro (2600–1800 BC). (Photos: Ahmed S. Khan.)

FIGURE 1.20
During the Industrial Revolution, the demand for machine operators was high. The nanotechnology revolution is seeking graduates who have complex thinking skills and possess a multidisciplinary knowledge base.

FIGURE 1.21
CNG-based public-transportation vehicles at Penn State University, State College, PA. (Photo: Ahmed S. Khan.)

FIGURE 1.23
Vacuum coating system: Chemical vapor-deposition system allows fabrication of thin films for various nanotechnology applications.

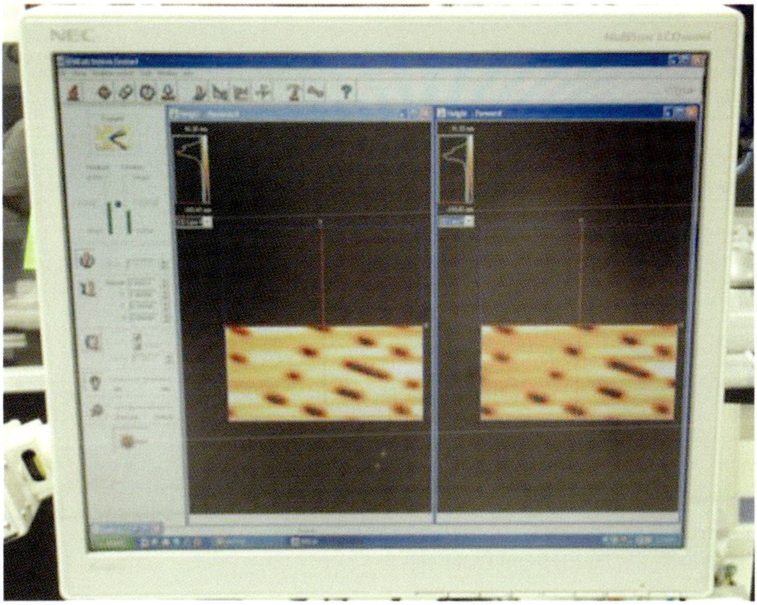

FIGURE 1.26
AFM profile of a sample.

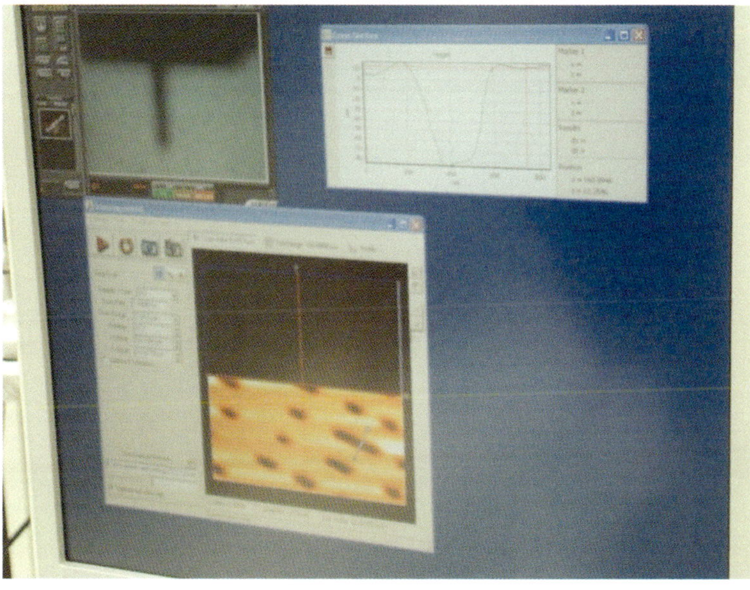

FIGURE 1.27
Another view of AFM profile of a sample.

FIGURE 1.41
Internal view of SEM multiple specimen holder.

FIGURE 1.42
SEM allows loading of multiple specimens for characterization.

FIGURE 1.46
Use of nanoparticles in paint enhances smoothness and appearance.

FIGURE 1.47
Use of nanoparticles in chrome polish enhances shine and covering of surface scratches and defects.

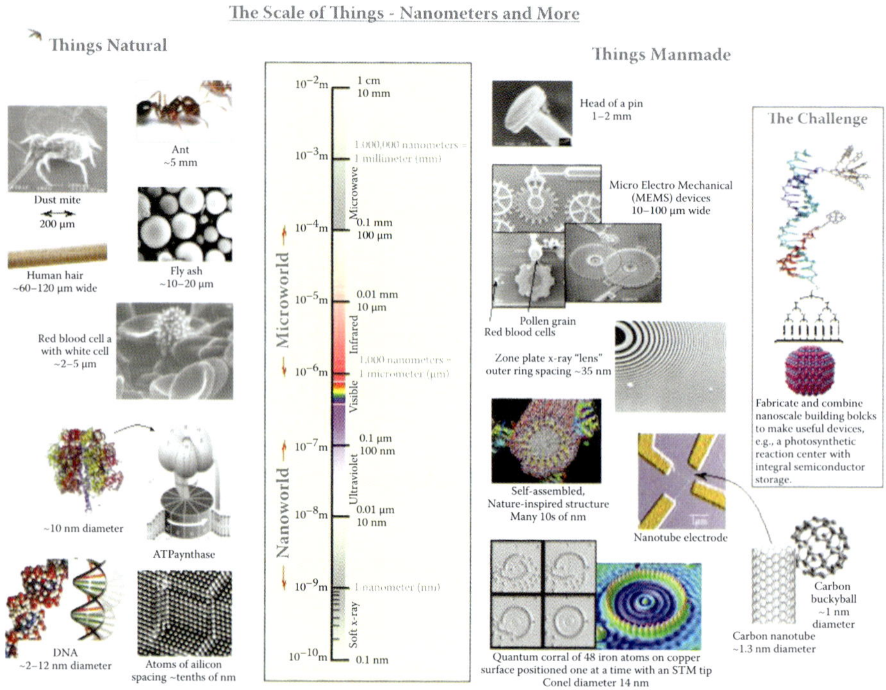

FIGURE 5.1
Scale of things natural and things made by humans. (Courtesy Office of Basic Energy Sciences, Office of Basic Energy Sciences, U.S. Department of Energy.)

FIGURE 5.3
Xenon atoms. (Photo courtesy IBM.)

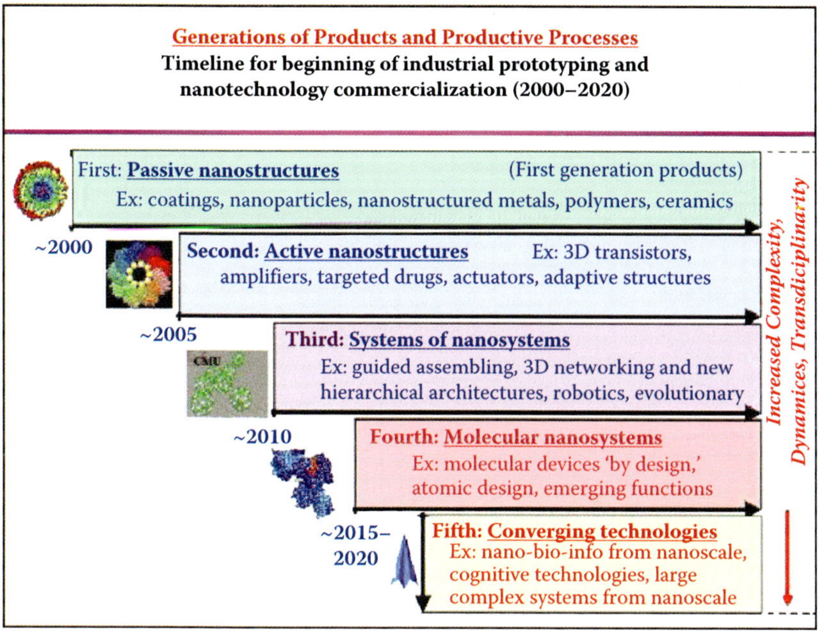

FIGURE 5.7
Generations of products and processes. (Courtesy of NSF, M. C. Roco, Nanoscale Science and Engineering Education Workshop, "Partnership for Nanotechnology Education" Los Angeles, April 26, 2009.)

FIGURE 8.1
Website for International Council on Nanotechnology, EHS page.

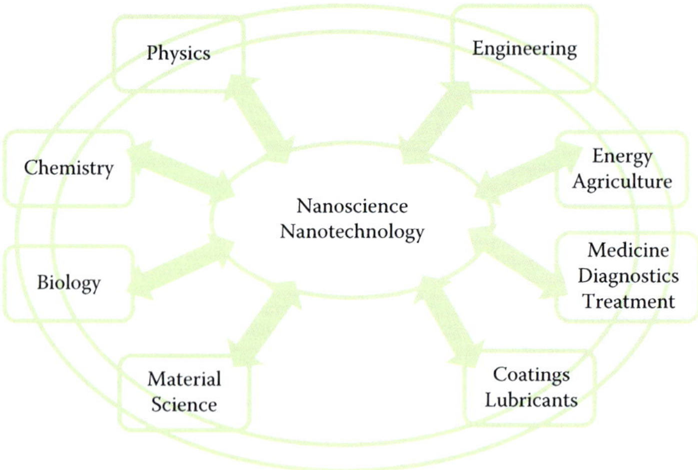

FIGURE 8.2
The inter- and multidisciplinary nature of nanotechnology.

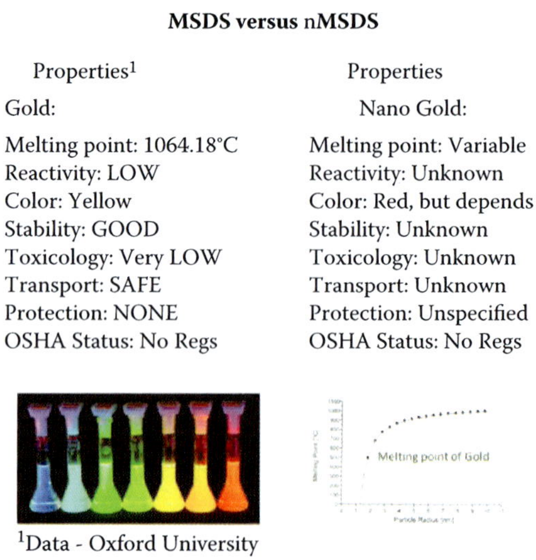

FIGURE 8.3
Macro versus nano-example for some MSDS quantities of gold. (Courtesy of Walt Trybula.)

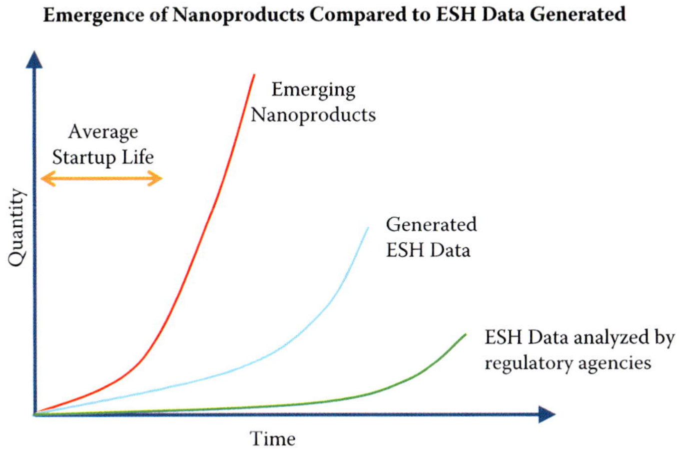

FIGURE 8.4
Representative time scales for development, assessment, and regulatory aspect of nanotechnology. (Courtesy of Walt Trybula.)

FIGURE 14.1
An interior from the ARL facility at Bilkent University, Ankara.

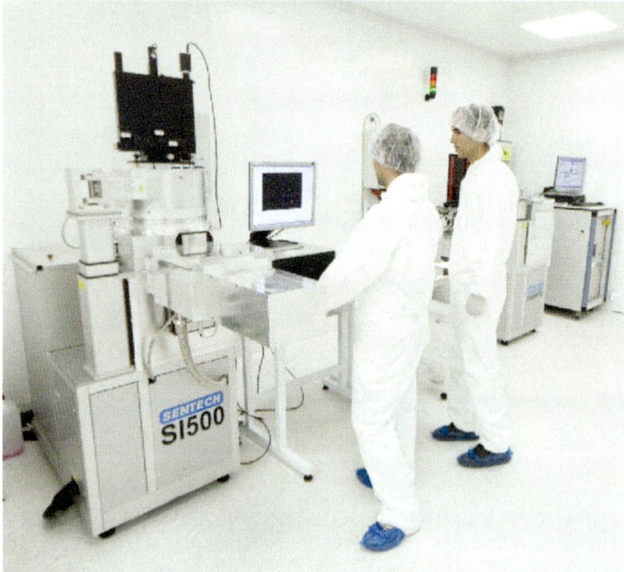

FIGURE 14.2
An interior from the NANOTAM facility at Bilkent University, Ankara.

FIGURE 14.3
An interior from the UNAM facility at Bilkent University, Ankara.

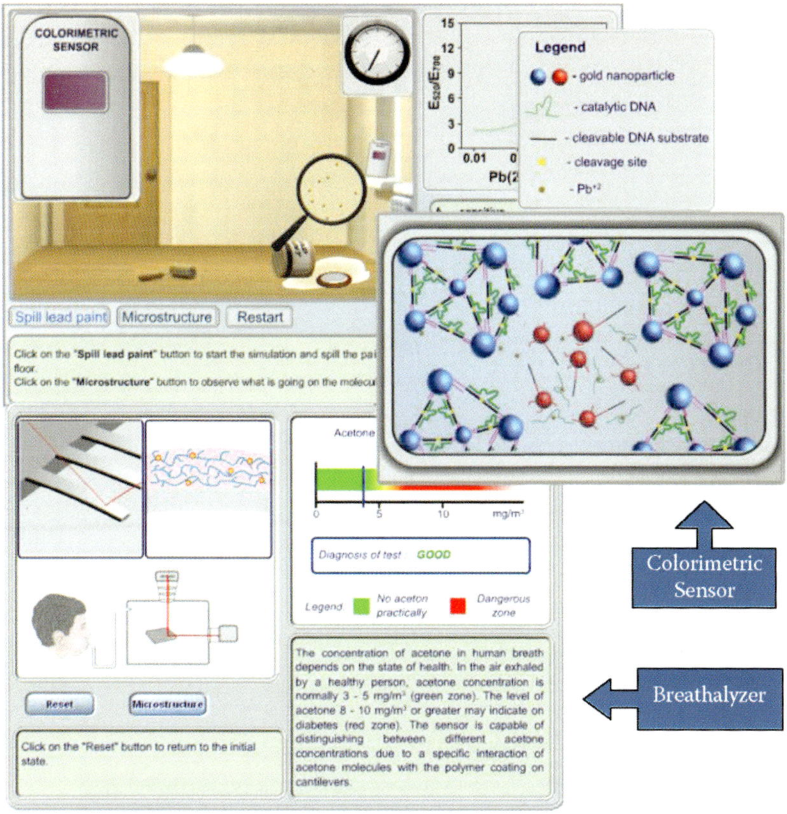

FIGURE 15.1
Application of simulation to help students visualize the interaction between nanoparticles and DNAzymes. (Courtesy Atel LLC.)

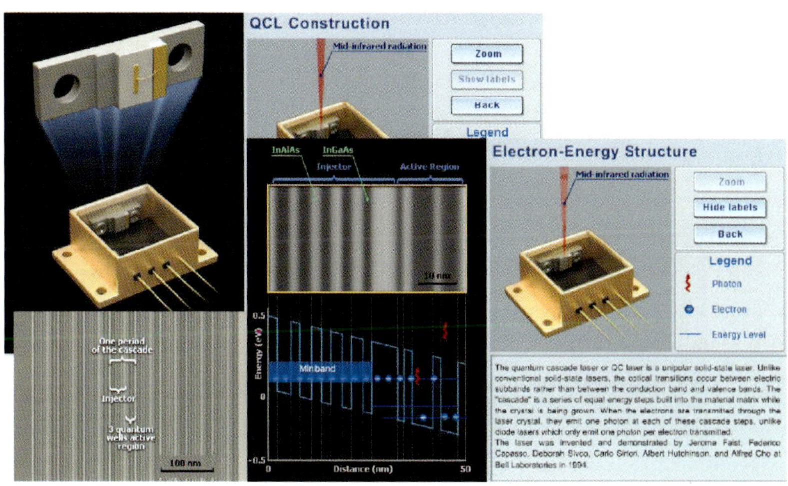

FIGURE 15.2
Application of simulation to help students understand the principles of quantum-well devices. (Courtesy Atel LLC)

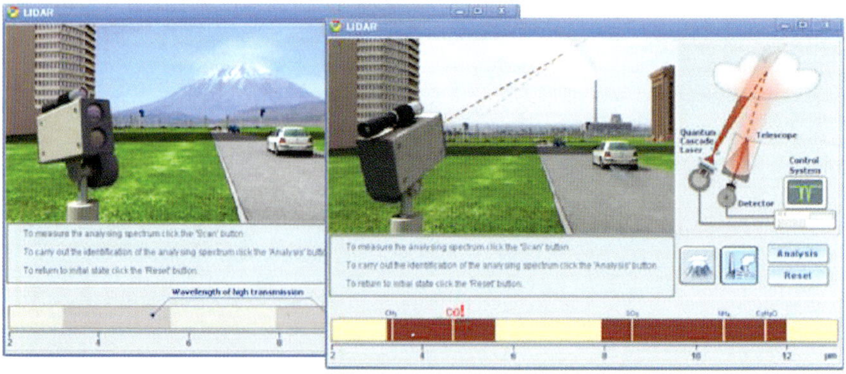

FIGURE 15.3
Application of simulation to help students visualize and measure air pollution. (Courtesy Atel LLC.)

Hubble Ultra Deep Field

A faint glimpse of the cosmos detected by the Hubble telescope. The image shows about 10,000 galaxies. The cosmos contains 12.7 million times more area than the Ultra Deep Field. (Hubble Ultra Deep Field Photo: NASA.)

6

The Unanticipated Consequences of Technology

Tim Healy

CONTENTS

Introduction

When the successful cloning of a lamb called Dolly was announced in February 1997 by Scottish researchers, it set off a spate of anxious questions. Many of them concerned the ethics of cloning, but another set asked about the unanticipated consequences. If we go down the cloning road, where will it lead? The answer is that we don't know. All of our technological roads twist and turn, and we can never see around the bend or through the fog.

The purpose of this chapter is to investigate the ubiquitous phenomenon of unanticipated consequences. We begin with a look at some definitions that shed light on the matter, and then consider the nature of change. This leads to a broadening of the definition of the word *technology*, and a look at what was one of our earliest examples of unanticipated consequences. We then address the crucial question of why we have such consequences. Some additional examples follow, and we then look at what society does in the face of unanticipated consequences. The chapter concludes with a discussion of some of the ethical implications of acting when we know that there can be unanticipated consequences to our actions.

Some Definitions

It is important here to distinguish between unanticipated and undesired consequences. The former are consequences that are not foreseen and dealt with in advance of their appearance. Undesired consequences are those that we know are harmful but that we are willing to accept the risk of occurring. Consequences may be:

Anticipated

- Intended and desired
- Not desired but common or probable
- Not desired and improbable

Unanticipated

- Desirable
- Undesirable

As an example, consider the development of a nuclear power plant at an ocean site. The anticipated and intended goal or consequence is the production of electric power. The undesired but common and expected consequence is the heating of the ocean water near the plant. An undesired and improbable consequence would be a major explosion, and we would associate the term *risk* with this outcome, but not with the heating of the water.

An unanticipated and desirable consequence might be the discovery of new operating procedures that would make nuclear power safer. An unforeseen and undesirable consequence might be the evolution of a new species of predator fish in the warmed ocean water, which then destroys existing desired species.

This chapter concentrates on unanticipated consequences of our technologies. Anticipated negative consequences have been dealt with extensively in the literature on risk. See, for example, Margolis (1996), and Bernstein (1996). The latter emphasizes the role of mathematics in risk assessment.

Two brief points should be made before we proceed. The first is that change is always with us. Even without the intervention of human beings, nature changes constantly. Continents move, weather changes, species evolve, new worlds are born, and old ones die. The second point is that all change seems to involve unanticipated consequences. Hence, the unanticipated is a part of life. There is no absolute security. Unanticipated consequences can be mitigated, largely through the gaining of additional information or knowledge, but not eliminated. That's the nature of our life, natural and human.

A Broader Definition of Technology

Although we focus here on the term *technology* as it is usually taken, it is worth pointing out that human beings do much that has unanticipated consequences, in all areas of life, certainly including, for example: medicine, business, law, politics, religion, education, and many more. Because of the parallels among these fields, it is useful to think of a broader definition of technology, such as "that which can be done, excluding only those capabilities that occur naturally in living systems" (Beniger, 1986). This matter is dealt with in some detail by in *Technopoly: The Surrender of Culture to Technology* (Postman, 1993).

Seen in the light of this broader definition, writing is one of our first technologies. Postman recalls the story of Thamus and the god Theuth, from Plato's *Phaedrus*, as an example of unanticipated consequences. Theuth had invented many things, including numbers, calculation, geometry, astronomy, and writing. Theuth claimed that writing would improve both the memory and the wisdom of humans. Thamus thought otherwise.

> Theuth, my paragon of inventors, the discoverer of an art is not the best judge of the good or harm which will accrue to those who practice it. So it is in this: you, who are the father of writing, have out of fondness for your off-spring attributed to it quite the opposite of its real function. Those who acquire it will cease to exercise their memory and become forgetful; they will rely on writing to bring things to their remembrance by external signs instead of by their internal resources. What you have discovered is a receipt for recollection, not for memory. And as for wisdom, your pupils will have the reputation for it without the reality: they will receive a quantity of information without proper instruction, and in consequence be thought very knowledgeable when they are for the most part ignorant. And because they are filled with the conceit of wisdom instead of real wisdom they will be a burden to society.

It is true in all of our technologies; the discoverer of an art, or the designer of a new system, is not usually the best judge of the good or harm that will accrue to those who practice it. And yet, paradoxically, it is often the designer to whom we must go to ask for the likely outcomes of her work. This is a problem with which society must wrestle, and we shall discuss later how it does so.

Why Do We Have Unintended Consequences?

Dietrich Dorner (1996) has recently analyzed systems in a way that can help us see why they can be so difficult to understand, and hence why consequences are unanticipated. Dorner has identified four features of systems that make a full understanding of any real system impossible. These are:

- Complexity
- Dynamics
- Intransparence
- Ignorance and mistaken hypotheses

Complexity reflects the many different components that a real system has and the interconnections or interrelations among these components. Our system models necessarily neglect many of these components or features, and even more so their interrelations, but there is always a danger in doing so, because it is from such interrelations that the unanticipated may arise. Our economic system is an example of a highly complex system. Not only are there many players, but the players are also interrelated in many ways that are difficult to identify and define. If Player A sets this price, how will Player B respond, and what will Player C think and do when she observes the actions of A and B?

Many devices and systems exhibit *dynamics*, that is, the property of changing their state spontaneously, independent of control by a central agent in charge of the system. One of the most fascinating examples of our time is the Internet, an extraordinarily dynamic system, with no one in charge. There is no way to model the Internet system in a way that will predict its future and the future of the people and things that will be impacted by the Internet. Many of our complex technological systems have this property. Examples might include: a new freeway system, nuclear power, high-definition television, genetic engineering. For example, a freeway system is dynamic because a large number of players initiate actions beyond any central control. Driver A slows down to observe an accident; Driver B responds in an unpredictable way, depending on his skills, state of mind, sobriety perhaps, and other factors. The system, though structured to some degree, is in many ways on its own.

Intransparence means that some of the elements of a system cannot be seen, but can nevertheless affect the operation of the system. More complex systems can have many contributors to intransparence. In the Internet, for example, the list would include almost all of the users at a particular time, equipment failures at user sites, and local phenomena such as weather that affect use of the Internet at other locations. We need to understand that what you can't see might hurt you.

Finally, *ignorance* and *mistaken hypotheses* are always a possibility. Perhaps our model is simply wrong, faulty, misleading. This last problem is particularly interesting and important, because it is the one we can do something about. We can take steps to reduce our ignorance, to increase our understanding, as we shall discuss in the section entitled "Society's Response."

And in the section entitled "Ethical Implications" we argue that we are obliged to do so.

Let us look next at some other perspectives on this problem. Peter Bernstein (1996) has addressed the matter from the viewpoint of probabilities and economics. He points out that economists have sometimes believed that deterministic forces drive our societies and their enterprises. More contemporary economists have seen less order. Bernstein puts it this way:

> The optimism of the Victorians was snuffed out by the senseless destruction of human life on the battlefields (of the First World War), the uneasy peace that followed, and goblins let loose by the Russian Revolution. Never again would people accept Robert Browning's assurance that 'God's in his heaven:/All's right with the world.' Never again would economists insist that fluctuations in the economy were a theoretical impossibility. Never again would science appear so unreservedly benign, nor would religion and family institutions be so unthinkingly accepted in the western world.... Up to this point, the classical economists had defined economics as a riskless system that always produced optimal results.

Such convictions died hard, even in the face of the economic problems that emerged in the wake of the war. But a few voices were raised proclaiming that the world was no longer what once it had seemed. Writing in 1921, the University of Chicago economist Frank Knight uttered strange words for a man in his profession: "There is much question as to how far the world is intelligible at all.... It is only in the very special and crucial cases that anything like a mathematical study can be made."

Edward Tenner (1996) takes still another perspective on the phenomenon of unanticipated and unintended consequences. He sees in some of our technologies a "revenge effect" in which our perverse technologies turn against us with consequences that exceed the good that had been planned.

Security is another window on revenge effects. Power door locks, now standard on most cars, increase the sense of safety. But they have helped triple or quadruple the number of drivers locked out over the last two decades—costing $400 million a year and exposing drivers to the very criminals the locks were supposed to defeat.

We shall return to this issue of perversity in the section entitled "Society's Response," when we see how society attempts to deal with unintended consequences.

For Dorner (1996) on engineering, for Bernstein (1996) on economics, and for Tenner's (1996) perverse technologies the message is the same. The world is not knowable and predictable. Its complexities are too great; its uncertainties beyond our understanding. Some unanticipated consequences are a necessary feature of all of our enterprises. But this does not mean that we should give up the effort to reduce uncertainty. We shall return to this in the section on "Ethical Implications."

In the next section we turn to some examples of such consequences. Then in the section entitled "Society's Response," we consider how society responds to the problem of unanticipated consequences.

Some Examples

In this section we consider some anecdotes and some cases of consequences that were not anticipated. We follow a historical path in this effort. We have already reached back to a time before the dawn of human history for a story of the invention of writing. Now we jump forward to the last 200 years, touching on some of the effects of the Industrial Revolution and moving on to questions that are being asked today about newly proposed technologies.

James Beniger's (1986) *The Control Revolution: Technological and Economic Origins of the Information Society* traces in some detail the evolution of technological development over the past two centuries, particularly in the United States. While Beniger stresses the role and need for control in technology, he does not pay a great deal of explicit attention to consequences. But, the implicit implications of the changes in speed brought on by the Industrial Revolution are clear.

> Speeding up the entire societal processing system...put unprecedented strain on...all of the technological and economic means by which a society controls throughputs to its material economy. Never before in history had it been necessary to control processes and movements at speeds faster than those of wind, water, and animal power—rarely more than a few miles an hour. Almost overnight, with the application of steam, economies confronted growing crises of control throughout the society. The continuing resolution of these crises, which began in America in the 1840s and reached a climax in the 1870s and 1880s, constituted nothing less than a revolution in control technology. Today the Control Revolution continues, engine of the emerging Information Society.

The twentieth century was to bring still another quantum leap in speed with the development of aviation. Perhaps no American is a better metaphor for the growth of technology in that century than Charles A. Lindbergh. Lindbergh's fascination with emerging technologies in the first decades of that century mirrored that of the nation as a whole, though in Lindbergh's case it was tempered by a love of nature.

> I loved the farm, with its wooded river and creek banks, its tillage and crops, and its cattle and horses. I was fascinated by the laboratory's magic: the intangible power found in electrified wires, through which one could see the unseeable. Instinctively I was drawn to the farm,

> intellectually to the laboratory. Here began a conflict between values of instinct and intellect that was carried through my entire life, and that I eventually recognized as inherent in my civilization.

In 1927 Lindbergh symbolized the triumph of technology when he electrified the world by flying alone across the Atlantic Ocean. But the euphoria did not last, as it never does. Two years later the Great Depression hit, and the decade to follow saw the rise of Hitler and the terrible destruction brought on by the Second World War, with its technologies so dependent on aviation. Lindbergh began to question the idea of progress.

> Sometimes the world above seems too beautiful, too wonderful, too distant for human eyes to see, like a vision at the end of life forming a bridge to death. Can that be why so many pilots lose their lives? Is man encroaching on a forbidden realm?... Will men fly through the sky in the future without seeing what I have seen, without feeling what I have felt? Is that true of all things we call human progress—do the gods retire as commerce and science advance?
>
> As a college youth, I thought civilization could never be destroyed again, that in this respect our civilization was different from all others of the past. It had spread completely around the world; it was too powerful, too universal. A quarter-century later, after I had seen the destruction of high-explosive bombs and flown over the atomic-bombed cities of Hiroshima and Nagasaki, I realized how vulnerable my profession—aviation—had made all peoples. The centers of civilization were the centers of targets.

In the end, Lindbergh found a reconciliation between the world of nature and spirit and the world of technology, a balance between what Eliade has called the Sacred and the Profane. He came to see that a balance was essential, and that technology is good when it helps to preserve that balance.

> Decades spent in contact with science and its vehicles have directed my mind and senses to areas beyond their reach. I now see scientific accomplishments as a path, not an end; a path leading to and disappearing in mystery.... Rather than nullifying religion and proving that 'God is dead,' science enhances spiritual values by revealing the magnitudes and the minitudes—from cosmos to atom—through which man extends and of which he is composed.

As the undesired consequences of much of twentieth-century technology became evident to Lindbergh, his response was not a rejection of technology, but rather a turning to the fundamental questions of why we are here. After his death, Susan Gray (1988) put it this way: "Of all the man's accomplishments—and they were impressive—the most significant is that he spent most of his life considering and weighing the values by which he should live."

By the second half of the twentieth century we had become painfully aware that our technologies are not unmixed blessings, that they can have fundamental effects on the way we live. Let us look at some more prosaic examples from the past couple of decades. We'll start with one from Charles Handy's *The Age of Unreason*.

Microwave ovens were a clever idea, but their inventor could hardly have realized that their effect would ultimately be to take the preparation of food out of the home and into the, increasingly automated, factory; to make cooking as it used to be into a matter of choice, not of necessity; to alter the habits of our homes, making the dining table outmoded for many, as each member of the family individually heats up his or her own meal as and when they require it.

Tenner (1996) raises an example that is particularly interesting for two reasons. First, it is not clear which of a number of technologies is causing the unanticipated effects. Second, the issue is intensely political and interpersonal, partially because of the first reason. This is the matter of the effect that various erosion-control technologies have on the condition of coastal beaches. In Tenner's words:

> People concerned about the coasts are likely to dispute when and where environmental revenge effects are happening. Whatever more rigorous research may show, it is clear that the shoreline is a zone of chronic technological difficulty. Just as logging and fire suppression alter the forest's composition and fire ecology, compelling more and more vigilance, so beach protection feeds on itself by establishing a new order that needs constant and ever more costly maintenance.

Now let us turn very briefly to two examples of emerging technologies whose major unanticipated consequences we have yet to experience. We shall look at the Internet and at cloning.

Actually, the Internet has already had a very significant impact on human life, involving the ways in which we meet each other, the ways we transact business, the ways we share information, and many more. Still, all of this is surely only the small tip of a huge iceberg, which seems very likely to change our lives in ways that we cannot today imagine. We cannot begin to anticipate the consequences of this technology.

The other technology that has stirred the public imagination in the waning years of this century is the cloning of animals and the possibility that we may eventually be able to clone human beings. There has been no dearth of questions about the future raised by this subject. Here are just a few.

1. Will cloning of human beings change what it means to be human?
2. What good might come from the development of cloning?
3. Should we halt research on cloning animals because it might lead to human cloning?

4. If the government takes no action to control cloning, could that decision be worse than a decision to take some specific action?
5. Is it possible to control cloning effectively?

Each of these questions speaks to the uncertainty inherent in the actions that we might take in this field.

We have surveyed here just a few examples that make concrete the concerns we may have about the unintended and unanticipated consequences of our actions and our enterprises. In the next section we ask what society as a whole does in the face of uncertainty.

But before we get to society, we really must say something about the individual. It is clear, but nonetheless worth stating, that each of us often has the opportunity and the right to reject the unanticipated consequences of a technology by refusing to use the technology in ways that have undesirable consequences. Whether a technology is for good or for ill must be our choice. We can use it to enrich our lives or to let our lives lose all meaning. Sometimes a technology is so pervasive that we cannot escape it, but often we have the freedom to choose.

The microwave oven is a good example of a technology whose unanticipated consequences can be rejected if we so choose. We don't always have to accept the fast-food approach if we choose not to. The violence of television and the pornography of the Internet are not forced on us. The contribution that the automobile makes to a sedentary life can often be rejected. If we become a slave to our telephone or other like media, it is not the telephone that should accept the blame. Discipline is still a virtue, for ourselves and for our children.

Society's Response

In this section we turn to the question of what individuals and societies do in light of the fact that their actions will have unanticipated consequences. We begin with an expansion of the discussion begun in the section on why we fail to anticipate consequences fully. Then we ask what specific steps can be taken to reduce uncertainty. Finally, we ask how people respond to proposals for new actions and how this helps set the course of our actions.

The first part of this section is based largely on an outstanding study by Robert K. Merton titled "The Unanticipated Consequences of Purposive Social Action." Merton chose his title carefully. His use of the word *unanticipated* rather than *unintended* helped motivate the brief discussion on terminology in the section "Some Definitions." The word *purposive* is meant to stress that the action under study involves the motives of a human actor and, consequently, a choice among various alternatives.

Merton begins by stating that up to the time of the article (1936) there had been no systematic, scientific analysis of the subject. He surmises that this may be because for most of human history we attributed the unexpected to "the gods" or "fate," or divine interference. With the dawning of the Age of Reason, we began to believe that life could be understood. We didn't have to leave it to "the gods." It is curious that in this century that optimism, may we say "faith," in human understanding of the complexities of life began to fade. Later we shall see this view espoused by the economist Frank H. Knight (1921), in the first half of this century, and by another economist, Kenneth Arrow (1992), writing in the second half of the century. In a sense we have come full circle, though today our uncertainty is not generally attributed to the gods, as such.

Merton cautions us that there are two pitfalls to be aware of in considering actions and consequences. The first is the problem of causal imputation, that is, the matter of determining to what extent particular consequences ought to be attributed to particular actions. The problem is exacerbated by the fact that consequences can have a number of causes. Let us consider an example.

Periodically, the Federal Reserve Board changes the short-term interest rate in an attempt to maintain a balance between the health of the economy and the inflation rate. What makes the problem difficult is that inflation rates over a given period of time are dependent on many factors, including, for example: the short-term interest rate, consumer confidence, employment rates, technological productivity, and even the weather. So, if the inflation rate follows a certain path over a given one-year period, to what extent should we attribute the path to the actions taken by the Federal Reserve?

The second pitfall is that of determining the actual purposes of a given action. Suppose that in a given year the unemployment rate drops from 8% to 6%, and the president claims that this drop was due to a series of social programs pushed by the administration. How do we know if this was in fact the cause, or at least a major cause, of the result? This is, of course, an important question because it helps us decide whether this same action in the future may be desirable. It is, of course, a very difficult question to answer. Merton suggests a test: "Does the juxtaposition of the overt action, our general knowledge of the actor(s) and the specific situation and the inferred or avowed purpose 'make sense'?"

It is clear that the major limitation to the correct anticipation of consequences is our state of knowledge. Our lack of adequate knowledge can be expressed in a number of classes of factors. The first class derives from the type of knowledge that is obtained in the sciences of human behavior. The problem is that such knowledge tends to be stochastic or conjectural in the sense that the consequences of a repeated act are not constant, but rather there is a range of consequences, any one of which may arise. Consider again the case of the Federal Reserve increasing the short-term interest rate by, say, 0.25%. This action may be repeated a number of different times over the years, with a number of different consequences to the inflation rate. In this sense, the consequence is stochastic or random, though probably within a fairly small range. The reason that we do not get a dependable result is that many factors influence the inflation rate, as we saw earlier. We do not know exactly how these other factors will influence the rate, nor do we know how the various factors will interrelate with each other, with secondary effects on inflation.

Another class of factors is error. We may, for example, err in our appraisal of the situation. We may err by applying an action, which has succeeded in the past, to a new situation. This is a particularly common mistake. It has been said that we are creatures of habit, and with good cause. Much of our lives are lived repeating the same or very similar actions (eating, driving, walking, etc.), and it is absolutely necessary that we have in place habitual ways of carrying out these actions to accomplish a desired end. And it is natural that we extend our habit of doing things habitually to areas where situations have changed. This is a problem that we must be aware of constantly. We may also err in the selection of a course of action. We may not choose the correct thing to do. We also may not do what we do well. And finally, we may err by paying attention to only one factor affecting the consequences.

Another of Merton's factors that limits our ability to anticipate consequences is what he calls the "imperious immediacy of interest," which refers to situations where the actor's concern with the immediate foreseen consequences excludes consideration of longer-term consequences. One's actions may be rational in terms of immediate results, but irrational in terms of one's long-term interests or goals.

A related phenomenon has to do with one's basic values, and a strange twist of fate that sometimes arises. Suppose that one's immediate values call for frugality and hard work, a "Protestant ethic." Such an individual may well end up accumulating a significant amount of wealth and possessions. On the other side of the coin, one whose values call for spending and conspicuous consumption may well end up with little wealth in the long run. This phenomenon is explored in some detail by Stanley and Danko (1996).

The final point that Merton makes is that the very prediction of a consequence becomes a new factor in determining what will ensue as a result of some action. Prediction is a new variable in the complex of factors that lead to consequences. Consider again, for example, the case of the actions of the Federal Reserve Board. Suppose that the short-term interest rate is increased by 1/2% and that a major financial leader predicts that this will cause the stock market to drop by 10%. This prediction will almost certainly affect, in one way or another, the stock market.

In a beautiful brief biographical essay, the economist Kenneth Arrow (1992) adds a further caution for the anticipator of consequences. Arrow believes that "most individuals underestimate the uncertainty of the world." The result is that we believe too easily in the clarity of our own interpretations. Arrow calls for greater humility in the fact of uncertainty, and finds in the matter a moral obligation as well.

The sense of uncertainty is active; it actively recognizes the possibility of alternative views and seeks them out. I consider it essential to honesty to look for the best arguments against a position that one is holding. Commitments should always have a tentative quality.

A related idea has been expressed recently by Stephen Carter (1996), who believes that *integrity* is more than acting out one's convictions. For Carter, integrity has three parts:

- Discerning what is right and wrong
- Acting on what you have discerned
- Saying openly that you are acting on what you have discerned

The process of discerning requires an active search for the truth. We are not free just to act on our beliefs. We are obliged as well to actively challenge our beliefs, search for more appropriate beliefs, and adopt them (tentatively, of course) as we find them.

Next we turn to an analysis of ways in which we can reduce the uncertainties that are a part of our complex lives. This discussion is based largely on the pioneering work of Frank Knight (1921) in the first decades of the twentieth century. Knight argues that we can decrease uncertainty in four ways:

- Increase knowledge
- Combine uncertainties through large-scale organization
- Increase control of the situation
- Slow the march of progress

Clearly an increase in knowledge can help us reduce uncertainty. We can carry out additional studies, analyses, experiments. The major problem with this approach is the cost in money and time. It also requires, of course, as we have just seen, the recognition that we do not already have complete knowledge of the situation.

The typical way in which we combine uncertainties through large-scale organizations is with one form or another of insurance. A group of people come together to protect each other against serious loss in case of a catastrophic consequence. The cost is in money and perhaps freedom.

It is often possible to reduce uncertainty through control. For example, the government might attempt to reduce inflation as a consequence of Federal Reserve actions through price controls. The use of controls generally has a monetary cost and a reduction of freedom.

Finally, we can reduce the level of uncertainty by slowing the march of progress. This might seem draconian at first, but in fact we do this all the time, when we take time to study a problem, do some more tests, write an environmental report, take the matter to the planning commission. Sometimes it is even more dramatic. Two days after the news was released about the successful cloning of a sheep, the president of the United States announced a moratorium on federal funding for any cloning work until the matter could be studied further.

There is a final point to be made, which has been noted by a number of people. As much as we would like to reduce much of the uncertainty in life, we would not choose to eliminate all uncertainty. A life in which everything was predictable, was known before the fact, would be a boring life indeed. We have been given a world to live in that is inherently unpredictable. That's the bad news and the good news, all at once.

In the final part of this section we consider the question of how we respond or react to proposals for change made by others in the light of the uncertainties and the unanticipated consequences of all of our actions. Such proposals might be projects for massive social change: "Let's eliminate welfare programs"; or they might be very personal individual decisions: "Let's buy that house by the lake." Whether it be in the halls of Congress, or the dining room table, when one person makes a proposal, another reacts.

Albert Hirschman (1991) has written a fascinating analysis of negative reactions to proposals, titled *The Rhetoric of Reaction: Perversity, Futility, Jeopardy.* The latter three terms are types of reaction. Let us look at each in turn.

An argument from perversity says that the opposite will happen from that which you claim. Suppose you suggest that welfare be eliminated in order to save taxpayers' money. An argument from perversity might be that such an action will in fact have just the opposite consequence. If you eliminate welfare, crime will increase, more prisons will have to be built, and the cost to taxpayers will increase, not decrease. The argument that Thamus makes to Theuth in the previous section entitled "A Broader Definition of Technology" is another example of an argument from perversity.

An argument from futility suggests that your action will have no effect on the situation that you are trying to change. You propose a law to regulate pornography on the Internet. An argument from futility would say that your law will not change anything, because it will be ignored.

A jeopardy argument claims that your proposal will place in jeopardy some valuable resource. You propose to put in a new freeway along the creek heading up toward Central City. Your opponent claims that such a freeway will ruin beloved old Riverside Park because of the noise and pollution.

Ethical Implications

Ethics is about what we ought to do. But how do you decide what you ought to do when the outcomes of your actions are uncertain? In this section we consider this problem. There are no simple answers to this very important question, but there are some general principles that we can set forth, and that may or may not be applicable in a particular situation. It is almost certainly not the case that all of them would apply in a given situation, since there will in general be conflicts between them. The purpose of this section is to state these principles and say a little about their application.

One should take advantage of the opportunities to reduce uncertainty that are discussed near the end of the section entitled "Society's Response" to the extent that the costs permit. We are not obliged to exhaust all of our resources—of money or freedom or other—to reduce uncertainty, because this could easily outweigh the good to be gained from the action. But to the extent that reductions can take place with reasonable cost, it seems morally prudent to do so. The judgment of what is an acceptable cost to bear may be very difficult to establish. In some cases it may be practical to use formal mathematical techniques, such as decision theory, to help arrive at an appropriate cost. In other cases, it may be so difficult to quantify goods and costs as to render such an approach useless. Of course, the effort to reduce uncertainty is not always an individual task, but is often a community effort, with the individual recognizing the richer and more diverse views brought by other individuals as well as by the community in a collective sense.

All persons should share equally in the benefits of an action or a project, and they should also share equally in the risks due to unanticipated consequences. This is of course an ideal, since we cannot in general ensure that such equal distribution of benefits and risks is possible. In such cases, other principles must be applied.

Persons who do not share in the benefits of an action should not, as a rule, be subject to its costs and risks. Justice suggests that burdens should not be borne by those who cannot benefit. While this is a good ideal, it is often difficult to apply in a particular case. If taken to the extreme, it could make it difficult or impossible to take many of our actions. For example, it would forbid the building of a coal-burning power plant on the grounds that the emissions from the plant could affect the environment of the entire globe, including that of some individuals who could not expect to benefit from the electric power. There is a corollary to this principle.

Persons who gain some benefit from an action should be able to choose their level of cost and risk. There are many situations in which some stand to gain a great amount from a project, and others to gain relatively little. To the extent possible, each person should be able to choose his or her level of cost. It is in this way that one's rights may be preserved. Of course, joint projects do not always allow this. Communities almost always form themselves in ways such that the majority is able to be a tyrant to the minority. For example, a community may decide to initiate a flood control project. A given individual may be concerned that there are possible undesired consequences to him from the project. A rights approach should support his concerns, but the common good might not.

Projects affecting more than one individual provide the greatest balance of benefits over harms for all involved. This is the utilitarian principle. This of course provides us with a way to approach the flood control project. On the basis of anticipated consequences, we might well decide to go ahead with the project on utilitarian grounds. But it seems particularly important that we try as much as possible to anticipate as many consequences as possible because of the potential threat to the rights of some.

In the assessment of value to others, we ought to recognize that a resource has greater value to one who is poor than to one who is rich. If you give $10 to a poor man, you improve his life much more than if you give $10 to a rich man. The nonlinearity of wealth should be considered in decisions on actions. And, in fact, the principle of justice suggests that our first thoughts should be for those who have the least in our society.

We ought to recognize that consequences of an action may extend over the long term, and that the effects of such consequences on the actor or on others must not be ignored. It is not right to assume that our actions and their consequences are necessarily limited to "here and now." In fact, their effects may extend over great distances, perhaps over the entire Earth, and may extend in time for years, decades, or perhaps even centuries. We are obliged to take these factors into consideration to the extent that we can.

Persons should face the truism that life is extremely complex and that all positions should be tentative. It is inherently dishonest to assert that one knows what is best when this is not the case, and it is almost never the case, again because of the complexity of life. Implicit in this principle is the requirement on all of us to seek to refine and improve our positions, to speak with humility about the consequences of our actions, and to act, when necessary, with a clear understanding that we do not own the truth. Each of us has an inherent right to his or her worldview or mental model of life. I do not have the right to assume that my view of the world is correct and your view is incorrect, to assume that "I am right, and you are wrong."

Conclusions

We have attempted in this chapter to outline some of the salient features of the problem of unanticipated consequences. Because the problem is so common to all of what we do in life, it should not come as a surprise that the study has led to some very general results or positions that we might apply to all of our lives. We close here with a brief list of some of these results.

- Life is very complex, more so than we admit.
- All of our actions have unanticipated consequences.
- We bear a moral obligation to take our positions tentatively, with humility in the light of our ignorance.
- Short-term and long-term values are often different, often contradictory.
- Uncertainty can be reduced, but there is always a cost.
- It is desirable to reduce uncertainty—but not to eliminate it.

In the end we are left with a dilemma, which is hardly surprising. We act with uncertainty about the consequences of our acts, and yet we have to act, for even to do nothing is to act, and there will be consequences. Change is an inherent part of life. Part of that change is natural; part is within our control. We have a right to act, but we also have an obligation to accept some level of responsibility for the unanticipated consequences of our actions. That level of responsibility is as hard to define as the unanticipated consequences themselves, but it is there nonetheless.

Acknowledgment

Reprinted with permission of Markkula Center for Applied Ethics, Santa Clara University.

Questions

1. What are different categories of anticipated and unanticipated consequences?
2. What does the author mean by "a broader definition" of technology?
3. Why do we have unanticipated consequences? List some examples of unanticipated consequences.

4. What is the impact of unanticipated consequences at the personal, societal, and global levels?
5. What can be done to reduce uncertainty in developing new technology? What are some of the ethical implications that should be considered?
6. Summarize your learning of new ideas and information presented in this chapter by completing the following flowchart:

Flow Chart

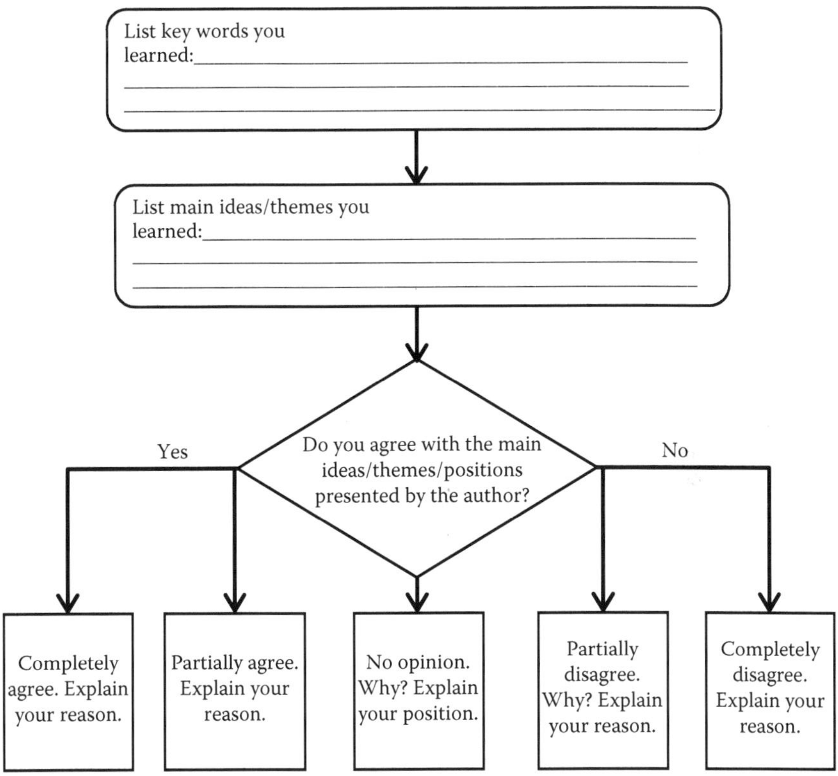

7. Internet Exercise

   Using Internet resources, complete the following table by listing unanticipated consequences of five technologies developed during the past decade and five potential nanotechnology-based products or applications.

Technological products developed during 2000–2010	Anticipated Consequences	Unanticipated Consequences	Nanotechnology products/ applications to be developed during 2012–2022	Anticipated Consequences	Unanticipated Consequences

Other Resources

Lindbergh, Charles A (1953). *The Spirit of St. Louis.* New York: Charles Scribner's Sons.

Eliade, Mircea (1987) *The Sacred and The Profane: The Nature of Religion.* Harcourt Brace Jovanovich.

http://www.guardian.co.uk/technology/2011/dec/01/charlie-brooker-dark-side-gadget-addition-black-mirror

http://ww.psychologytoday.com/blog/the-power-prime/201001/technology-the-law-unintended-consequences

http://plato.stanford.edu/entries/qt-uncertainty/

http://www.whentechfails.com/

http://news.stanford.edu/news/2005/february23/internet-022305.html

References

Arrow, K. (1992). I know a hawk from a handsaw. In *Eminent economists: Their life philosophies*. Cambridge, UK: Cambridge University Press.

Beniger, J. (1986). *The control revolution: Technological and economic origins of the information society*. Cambridge, MA: Harvard University Press.

Bernstein, P. (1996). *Against the gods: The remarkable story of risk.* New York, NY: John Wiley and Sons.

Carter, S. (1996). *Integrity*. New York, NY: Basic Books.

Dorner, D. (1996). *The logic of failure: Why things go wrong and what we can do to make them right*. New York, NY: Metropolitan Books. (Original work published 1989.)

Gray, S. (1988). *Charles A. Lindbergh and the American dilemma: The conflict of technology and human values*. Bowling Green, OH: Bowling Green State University Popular Press.
Handy, C. (1990). *The age of unreason*. Boston, MA: Harvard Business School Press.
Hirschman, A. (1991). *The rhetoric of reaction: Perversity, futility, jeopardy*. Cambridge, MA: Harvard University Press.
Knight, F. (1921). *Risk, uncertainty, and profit*. Boston, MA: Houghton Mifflin.
Margolis, H. (1996). *Dealing with risk: Why the public and the experts disagree on environmental issues*. Chicago, IL: University of Chicago Press.
Merton, R. (1936). The unanticipated consequences of purposive social action. *American Sociological Review, 1*(December), 894–904.
Postman, N. (1993). *Technopoly: The surrender of culture to technology*. New York, NY: Vintage Books.
Stanley, T., & Danko, W. (1996). *The millionaire next door: The surprising secrets of America's wealthy*. Atlanta, GA: Longstreet Press.
Tenner, E. (1996). *Why things bite back: Technology and the revenge of unintended consequences*. New York, NY: Knopf.

7

Portraits in Carbon

Todd Neff

CONTENTS

Forget oxygen—carbon may be the most important element on the planet.

It's hard to avoid the number six in talking about carbon: It's the sixth element in the periodic table, and normally has six protons and six neutrons in its nucleus and six electrons hovering in loose formation around it. This form, called carbon 12, accounts for 99% of all carbon (carbon 13 and 14 account for the rest). As a substance, pure carbon comes in several guises, but among the most important are graphite (which is typically soft, black, slippery, and flaky); diamonds (which can be any of several colors, or colorless, and are as hard as elements come); and amorphous carbon (such as coal and soot). Carbon is abundant in the Earth's crust and biosphere. In the form of coal, carbon costs as little as 1.2 U.S. cents per kilogram. As natural diamonds, however, it can cost thousands or even millions of dollars per gram. The name derives from the Latin *carboneum*, but it is rendered *kohlenstoff* in German, *uhlik* in Czech, and *sekitan* in Japanese. Oh, and carbon is the basis of more than 20 million chemical compounds and thus of life as we know it.

Those are the raw, actuarial facts, perhaps carbon's dullest mugshot face. But carbon is more interesting than these facts suggest, and like a quick-change artist has many faces. Here are five more:

Newborn

Carbon is born in dying stars and scattered when they explode. Our own solar system formed from such remnants, a collector hoarding from interstellar estate sales. Carbon represents less than 1/10th of the Earth's crust by mass, but it is the only element capable of forming robust single bonds with other atoms at ambient temperatures, which is why it is the chemical backbone of the living world. "Organic chemistry" is carbon chemistry. Carbon is proteins, carbohydrates, hydrocarbons, and DNA. Shakespeare was mostly carbon, and so were his quill, ink, and the paper on which he scratched Hamlet. It is perfume and plastic, trash bags and tennis balls and tree stumps. It is also natural gas and oil and coal—the fossil fuels that modern civilization runs on.

Much of this was formed during the aptly named Carboniferous Period, a part of the Paleozoic Era from 354 to 290 million years ago. That period saw untold millions of swamp trees with 2-m-diameter trunks die but not rot—in part, some believe, because herbivores, termites, and certain bacteria had not yet evolved to eat lignin (a new plant innovation at the time), which is the glue differentiating wood from leaves. Unable to decompose, the resulting peat piled up in northern Europe, parts of Asia, and in the eastern and midwestern United States. Seas rose and fell, covering the swamps intermittently and cutting off the oxygen that would otherwise fuel organic decay. Pressure and heat over millions of years converted the material to coal, specifically anthracite, the hardest and most energy-rich form, composed of up to 98% carbon.

Bituminous coal, softer and younger, is only about 60% carbon, but it's lower in sulfur too, and thus emits less acid-rain-forming pollutants when burned. This coal, from places like the Powder River Basin, began as giant ferns and other plants flourishing in the lush effluvial plains of what is now Wyoming about 55 million years ago. Workers listening to satellite radio in enormous dump trucks move the obsidian bulk gouged from 15-m-thick seams to railheads, dispatching 400 million tons every year—40% of the coal burned in the United States—to power plants around the country.

Villain

Which brings us to carbon's dark side. Atmospheric carbon likes to run with pairs of oxygen atoms like some celebrity with arms around identical twins—carbon dioxide (CO_2). It is an association born in fire, through the combustion of trees, dung, and fossil fuels. The fossil fuels are the main culprits. Their carbon, largely buried and sequestered until James Watt came up with the steam engine, is being unlocked and flung into the atmosphere at a rate of 7 billion metric tons of carbon every year. (That's just the carbon; the CO_2 gas weighs more like 27 billion tons. The United States emits nearly 6 billion tons of that total.) The weight of 7 billion tons of carbon is roughly equal to that of a concrete block as tall as the Empire State Building and covering 450 hectares.

The current global average atmospheric concentration of carbon dioxide is about 380 parts per million (ppm). That may not seem like much—the equivalent of less than a nickel in $100—but it's 36% more than the 280 ppm present in the air before the Industrial Revolution. Even such trace concentrations of CO_2 help keep Earth from a state of ice-age frigidity by creating the greenhouse effect, in which carbon is, with water vapor, methane, and a host of other trace gases, a key player.

Carbon dioxide's heat-trapping properties are a matter of chance, the product of the molecule's affinity for certain bands of infrared light. Infrared waves, which radiate invisibly from a good campfire or a sunlight-warmed Earth, make carbon and its escort oxygen atoms dance. As Steve Montzka, an atmospheric chemist with the National Oceanic and Atmospheric Administration's Earth System Research Laboratory in Boulder, Colorado, describes it, oxygen and nitrogen, the dominant atmospheric gases, pay the infrared light no heed. But the bonds in carbon dioxide

> have a certain frequency, like two balls on a spring. If you hit the balls, they'll start vibrating back and forth in a nice, coherent way. That's kind of what happens when infrared light hits carbon dioxide molecules, and that's why it adds energy to the molecule.

The carbon dioxide molecules vibrate, rotate, and stretch for an instant, then spit the infrared light-heat-back out in all directions.

So while 280 ppm was cozy, 380 ppm (and rising) looks like trouble. The U.S. National Center for Atmospheric Research's top climate model has estimated that a doubling of current concentrations—a real possibility by century's end unless we temper our use of coal and other fossil fuels—will raise global average temperatures nearly 6°F by 2100 (3.3°C).

Jailbird

Coal's outsized global-warming footprint has led the U.S. Department of Energy to commit hundreds of millions of dollars to carbon capture and storage research. This is suddenly the Holy Grail of the fossil energy industry. The idea is to sequester CO_2 emissions from coal-fired power plants in underground reservoirs. The carbon came from beneath the Earth, after all: Why not just put it back?

Well, for one thing, it would require the extremely expensive scrapping of conventional coal-burning plants and their replacement with plants using integrated gasification combined-cycle (IGCC) technology. IGCC plants convert coal to a gas of carbon dioxide and hydrogen. The hydrogen is burned to run a turbine, and then the exhaust generates steam to run a second turbine. Such plants are about 10% more efficient than conventional coal-fired plants, which convert less than one-third of their coal energy into electrical energy, and also scrub pollutants such as mercury and sulfur dioxide more effectively. The carbon dioxide would be injected directly into deep saline aquifers or other geological formations, or even pumped down to the sea floor, where the enormous pressures and frigid temperatures would liquefy it and allow it to pool forever on the bottom. In theory.

There are challenges. Only four IGCC power plants exist in the world, including two in the United States. Contrast that with 5,000 large coal plants, with many more proposed or being built (see "Coal Rush," p. 8, and "China and Her Coal," p. 14). Of the roughly 150 new U.S. coal plants currently planned, only a handful would use IGCC technology, and just one (proposed for Colorado by Xcel Energy) is slated to actually sequester the CO_2. It won't become operational before about 2013, Xcel officials say.

Nor is sequestration itself a thoroughly known quantity, although oil companies have been injecting CO_2 underground to force more petroleum out of existing fields for decades. At Chevron's Rangely oil field in western Colorado, for instance, they go through 36 million cubic feet of CO_2 a day, piped 260 km from Wyoming's Green River Basin. It comes with a financial and energy cost. Chevron engineer R. J. Wackowski estimates that the company has spent hundreds of millions of dollars on CO_2 injection. It takes 33,000 kW distributed across eleven howling compressors to squeeze the gas into the 2,100-lb/in.^{2} pressure needed for injection.

That's not cheap, but consumers seem ready to pay almost anything for oil, and it fetches high prices. But coal's historic benefit is its cheapness. The Intergovernmental Panel on Climate Change's (IPCC) Special Report on Carbon Dioxide Capture and Storage, released in September 2005, estimates that a new IGCC plant with carbon storage would be from 20% to 100% more expensive per kilowatt-hour than a new conventional coal-fired plant, assuming no carbon taxes.

And the question remains: Will carbon forced into sequestration actually stay put? Only our descendants will know for sure. There is just one production facility in the world, run by Norway's Stadtoil, that is sequestering CO_2 for sequestration's sake, and it has only been operating for 10 years. The U.S. Department of Energy has spent more than $100 million through 2009 to understand the technical feasibility of storing billions of tons of carbon dioxide under U.S. soil.

Jeff Goodell, author of *Big Coal,* has his doubts about geologic sequestration. He questions the electric power industry's will to implement carbon capture and storage on any significant scale. "They like to pretend it's a really simple thing—pump it down there and cap it off," Goodell says. "But the Earth is like a giant pincushion, and this CO_2 can migrate for miles and come up through fissures." The IPCC report, on the other hand, concludes that 99% of sequestered CO_2 would likely stay put for a millennium. Ken Caldeira of the Carnegie Institution of Washington's Department of Global Ecology, a coauthor of the IPCC report, says that sequestration needs to play a role in future energy portfolios, but that there will be political challenges. "It implies you're going to have some sort of globally coordinated carbon police, part of some incorruptible squad," he said. "My brother works in a food plant that makes kosher food. When the rabbi walks away, they keep putting the label on it even when no one is blessing it."

Displaced Person

Anyone aspiring to carbon neutrality can take action immediately. Drive a mid-sized car 20,000 kilometers a year? A one-time payment of $72 to CoolDriver.org will let you release the resulting six tons of CO_2 skyward with a clean conscience. Flying from New York to Los Angeles? TerraPass.org's Puddle Jumper package can make up for the ton of CO_2 emissions for $9.95.

Both programs deal in carbon offsets, in which fossil fuel burners indirectly pay someone else to make up for their combustion. CoolDriver and TerraPass send the money to renewable energy projects and are among many organizations offering consumers, businesses, and municipal and state governments ways to ease their carbon load by investing in everything from groves of trees to new wind farms.

This is still an unregulated and confusing business. To offset 6 tons of carbon emissions costs $72 at CoolDriver.org, $40 at DriveNeutral.org, and $49.95 at TerraPass.org, where you also get a free luggage tag. There also have been questions about the accountability of outfits promising to offset flights to Singapore by planting trees in Central America. "What was on the land before those trees in Costa Rica and so on were planted? Did they cut other trees down and get paid to replant?" asks Russ Schnell, director of the U.S. National Oceanographic and Atmospheric Administration's global atmospheric gas sampling network. Done properly, Schnell said, such projects do have the benefit of sequestering the weight of the carbon in the tree.

Carbon-offset equivalents of companies like PriceWaterhouseCoopers have emerged to assuage such fears. The San Francisco-based Center for Resource Solutions runs the Green-e verification program, and the Climate Neutral Network has its Climate Cool program to make sure it's all on the up-and-up.

How much impact carbon offsets are having on the market remains a question. Kevin Rackstraw, North American development leader for Clipper Windpower, says it's hard to convince lenders on $200-million wind-farm projects that 100,000 people buying green power today will do the same in fifteen years. Even major buys of renewable energy credits by companies or municipalities wanting to offset their electricity use have a limited effect compared to federal tax subsidies. Rackstraw sums up the current market forces: "If renewable-energy credits go away, we can still do business. If the production tax credit goes away, we're at a standstill, practically."

Superhero?

In the wrong form or place, carbon can be troublesome. But it also has remarkable properties that might be valuable. Consider carbon fiber, for instance.

Spun at high temperature from strands of the polymer polyacrylonitrile, carbon fiber's combination of strength, lightness, and relative imperviousness to heat and cold has landed it a role in the Space Shuttle's nose cone and countless satellites. It is coveted by military aircraft designers and, lately, by those designing warships. Massive wind turbines would be impossible without it. Voices as diverse as former CIA director James Woolsey (now an energy-efficiency champion) and Amory Lovins of the Rocky Mountain Institute want to see carbon-fiber auto bodies, which could weigh one-third less than steel. Given the U.S. automakers' financial situation and the fact that carbon-fiber bodies would cost at least 10 times more than steel, that might be a stretch. But then, the new boss at Ford Motor Co. was formerly CEO of Boeing Commercial Airplanes, and Boeing has certainly seen the carbon-fiber light.

Fully half of the wings and fuselage of the efficient, $150-million Boeing 787 Dreamliner, launched in 2008, is made of carbon-fiber composites. The move lightens the aircraft enough to boost fuel efficiency 3%, but also allows for bigger windows, higher cabin air pressure, and higher cabin humidity because corrosion is not an issue. Boeing archrival Airbus is also using carbon fiber extensively in its gargantuan new A380.

Carbon fiber is so hot that it's hard to come by. Toray Carbon Fibers America, Inc., which is supplying carbon fiber for the Boeing 787, has been running at 100% capacity for three years straight, according to sales manager Buster Tipton. The company opened a new U.S. plant last year and has three more planned in Japan and Europe. "Make no mistake, there's a global shortage," Tipton said, which may explain why high-end carbon-fiber bicycles can cost as much as small cars did ten years ago.

Then There Are Nanotubes

In 1985, scientists at Rice University created a new form of carbon, consisting of 60 atoms vaguely resembling a soccer ball. The buckminsterfullerene, or "buckyball," launched the field of carbon nanotechnology and won its codiscoverers the 1996 Nobel Prize in chemistry. Six years later, Japanese scientist Sumio Iijima reported his creation of carbon nanotubes: cylindrical fullerenes.

Carbon nanotubes are an entirely new and potentially miraculous material: 60 times stronger than steel and capable of conducting electricity 1,000 times better than copper. Carbon nanotube research is among the hottest areas of research in nanotechnology, which refers to engineered structures less than 100 nanometers in size. (A nanometer is a billionth of a meter, or the same part of a meter as the length of your thumbnail is of the Earth's diameter.)

Nanotubes are as thin as 1-nm wide, but can be thousands of nanometers long. Potential applications range from next-generation transistors and solar cells to artificial muscles and blast-resistant clothing. Some firms are using nanotubes in the electrodes of experimental fuel cells, as electron guns in field-emission displays, and in radiation-hardened computer memory for military and space applications. Commercial applications include nanotube-infused plastic that dissipates static electricity, which is an attractive feature in automotive fuel lines and microelectronics manufacturing, where a spark can be costly.

Ken McElrath, vice president at Carbon Nanotubes Inc. in Houston, says that if the first half of the 20th century was the age of steel and the latter half the age of polymers, then the coming decades belong to carbon nanotubes: "We're on the ground floor of a new material."

Can nanotubes possibly live up to the hype? Dean Aslam, director of the Micro and Nano Technology Laboratory at Michigan State University, thinks they will. Aslam is most bullish on nanotubes for use in miniaturized chemical or biological sensors. His group and a University of Michigan team are collaborating in an effort to use carbon nanotubes to make a microchip-based gas chromatograph or an airborne chemical detector. Carbon nanotubes have a huge surface area per unit volume, which is perfect for capturing molecules as they float by. Heat the nanotubes, and they drop their contents into a tiny chamber for analysis. Such a system embedded in a watch could enable the equivalent of food labeling for air. "We know what kind of food we're eating. We don't know what kind of air we're breathing," Aslam says.

What's the downside? So far, nanotubes don't come cheap: Research grades of the tubes cost up to $2,000 per gram, though commercial varieties are far less pricey. Kalman Migler, a scientist at the U.S. National Institute of Standards and Technology who is working on developing nanotube standards, says that nanotubes remain poorly characterized and that the market remains nascent. Another issue, he says, is understanding the possible environmental and human health impacts before nanotubes become the next plastic: "I believe there's a need for a much greater effort to understand the effects of nanotubes and nanomaterials on safety, health, and the environment, and it's clearly an area of concern to great numbers of people thinking about nanotechnology."

Rice University's Center for Biological and Environmental Nanotechnology focuses on health impacts and medical applications of nanotubes. Kristen Kolinowski, the center's executive director for policy, says that nanotubes pried apart in solution have been shown to kill cell cultures, and that nanotubes forced into the lungs of rats and mice caused irritation. A University of Michigan study has shown nanotubes to accumulate in living cells and with that, potentially, food webs. Kolinowski opined:

> We have to think about the impacts of putting these things in the body. There are people who think that carbon nanostructures are not going to be the best materials in the human body unless we can isolate them or they turn out to be pretty nontoxic. I think the jury's still out on that.

Acknowledgment

This chapter was originally published in *World Watch Magazine*, January/February 2007, Vol. 20, No. 1. Reprinted with permission of World watch Institute, Washington, D.C.

Questions

1. Why has the author called carbon the chemical of the living world?
2. What are some of the commercial products based on carbon that we use everyday?
3. How is carbon born? What is meant by the Carboniferous Period and the Paleozoic Era?
4. What is carbon's dark side? How much carbon is released into the atmosphere every year? What is the current global average atmospheric concentration of carbon dioxide compared to the pre–Industrial Revolution period?
5. What does the U.S. National Center for Atmospheric Research's top climate model predict about the increase in global temperature if the present carbon concentration in Earth's atmosphere doubles by century's end?
6. What are the advantages of IGCC plants? How efficient are they? How is CO_2 sequestered? How many such plants exist? How many coal plants are there?
7. What are some of the ways for acquiring carbon neutrality? How effective are such methods?

8. What are the advantages and applications of carbon fiber?
9. Who invented carbon nanotubes? What advantages do they offer? List some applications of carbon nanotubes.
10. Discuss the possible environmental and human health impacts of carbon nanotubes.
11. Summarize your learning of new ideas and information presented in this chapter by completing the following flowchart:

Flow Chart

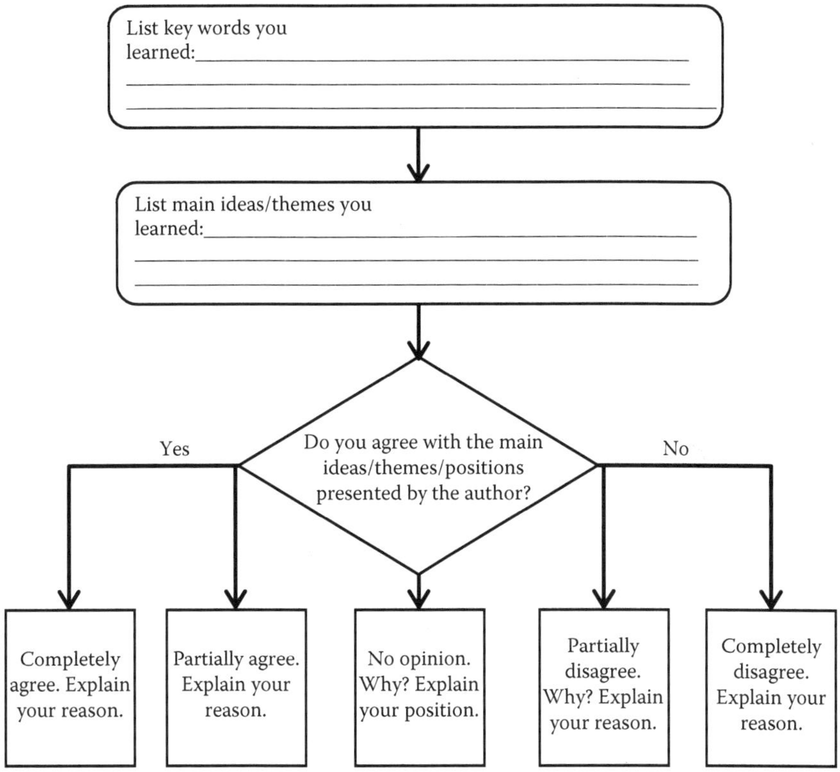

Other Resources

Goodell, Jeff. (2006). *Big Coal: The Dirty Secret Behind America's Energy Future.* New York: Houghton Mifflin Harcourt.

Moran, Susan. (2007). Coal Rush. *World Watch Magazine.* January/February, 2007, Vol. 20. No.1, p.8, Wolrdwatch Institute, Washington, D.C.

Yanli, Hou and Min, Hu. (2007). China and Her Coal. *World Watch Magazine.* January/February, 2007, Vol. 20. No.1, p.8, Worldwatch Institute, Washington, D.C.

http://www.worldwatch.org/

http://www.noaa.gov/

http://www.worldcoal.org

http://www.lasbrisasenergy.com/pdfs/IGCCwhitepaper.pdf

http://www.xcelenergy.com/About_Us/Energy_News/News_Archive/Xcel_Energy_increases_commitment_to_IGCC

http://www.energy.ca.gov/process/pubs/electrotech_opps_tr113836.pdf

http://www.ipcc.ch/pdf/special-reports/srccs/srccs_summaryforpolicymakers.pdf

http://www.nrel.gov/docs/fy08osti/41869.pdf

http://www.esrl.noaa.gov/gmd/co2conference/pdfs/aggi.pdf

http://www.evdriven.com/carbon-fiber/japan/presentations/

http://www.themoneytimes.com/node/1449

http://www.nanocarb.meijo-u.ac.jp/jst/english/lijima/sumioE.html

http://www.phschool.com/science/science_news/articles/nice_threads.html

http://www.egr.msu.edu/~aslam/

http://www.mendeley.com/research/lengthdependent-uptake-of-dnawrapped-singlewalled-carbon-nanotubes/#

http://cnst.rice.edu/uploadedFiles/About_Us/Nano%20DC%20Program-3%20150dpi.pdf

8

Nanotechnology: Environment and Ecology

Deb Newberry

CONTENTS

KEY WORDS: nanotechnology, nanoparticles, nanomaterials, environment, ecology, ecosystem, trade-off, criticality, benefits, risks

Introduction

Nanotechnology will impact ecosystems, the environment, and the living organisms that occupy this world. The nature of these impacts, both positive and negative, is a critical question for developing applications based on nanoscale research. Many chapters in this book discuss the impact of nanotechnology in medicine (diagnostic and treatment), electronics, communication, transportation, and so on. The impetus of this chapter is to raise the awareness of the broader impact of these technology developments on the planet we live on and the organisms that occupy it.

Understanding the impact of nanoscale materials on the researchers, manufacturers, and users of the resulting product is a critical aspect in the development of new products. Also to be considered is the environment in which the product is manufactured, used, and ultimately disposed of. As the potential applications and possibilities are considered for nanotechnology-based products, the topics discussed in this chapter will help the assessment in a broader sense. The technological aspects and advances are to be evaluated, but equally important are the use and lifetime of the product and what may happen to the nanoconstituents after they have performed their duties. These are critical aspects of nanoscale research and nanotechnology that need to be investigated, understood, and perhaps regulated.

As with any technology, there is a seesaw or scale that must constantly weigh the pluses and minuses of any opportunity, product, or technology. For example, in the medical arena, nanotechnology offers ways to create cleaner and more sterile clinics, medical offices, and operating rooms, thereby reducing the possibility of contamination and infections. The ability to detect unhealthy cells in lower concentrations—leading to earlier disease diagnosis and improved prognosis—is another significant health contribution being brought about because of nanoscience. The negative impact is that over a period of time, the nanoparticles used to clean or detect diseases would enter the waste water system and eventually enter the water ecosystem, potentially causing changes to water chemistry or entering fish and wildlife. Nanoparticles used to sterilize hospital surfaces may also sterilize portions of the ecosystem leading to unacceptable imbalances. Developments in nanoscale research are helping to clean air and water, detect toxins at minute levels, and in some cases remove the toxin (Editorial, 2008; Hogan, 2004). Nanoscience is also providing improved methods of harnessing solar and wind energy, thereby reducing our dependence on climate-changing petroleum. And, as with any manufactured product or manipulation of atoms and molecules (either natural or human made), there is the risk that a hazard to the environment or living systems will be introduced.

This chapter provides some examples of the impact of nanotechnology on the environment and ecosystems at a survey level. More importantly, this chapter provides a foundation to discuss and assess the potential pros and cons of current and future nanotechnology applications. A brief discussion of some of the unique aspects of nanotechnology as well as a few definitions serve as a foundation for a discussion of the magnitude of the job ahead in assessing the potential positive and negative impacts of nanotechnology on the environment and ecosystems.

A Foundation for Discussion

Prior to any discussion, it is often wise to ensure that all participants are using the same dictionary, that is, have a common definition of the vocabulary involved. At many a technical conference, confusion was initiated simply because the chemists used one word or phrase to define an interaction, the physicists used a different phrase, and the engineers viewed the interaction via another vocabulary. The words required for the discussion topics within this chapter—*nanotechnology, environment,* and *ecosystem*—are words with multiple definitions in the scientific and popular literature.

Nanotechnology. The term *nanotechnology* has a multitude of definitions—a Google search comes up with over 2 million hits—and most of those definitions involve the length dimension of a nanometer (nm).

The National Science Foundation defines nanotechnology as

> research and technology development at the atomic, molecular or macromolecular levels, in the length scale of approximately 1–100 nanometer range, to provide a fundamental understanding of phenomena and materials at the nanoscale and to create and use structures, devices and systems that have novel properties and functions because of their small and/or intermediate size. The novel and differentiating properties and functions are developed as a critical length scale of matter, typically under 100 nm.

This and other definitions of nanotechnology that use a size scale will become an important aspect of discussion in this chapter. A less specific definition of nanotechnology used by the author is: "Nanotechnology is the observation, study, measurement, interaction, creation and manipulation of materials at the molecular and atomic scale."

Ecology or ecosystem. Over the last few years, the word *ecosystem* has been applied to everything from a pond or river to a corporate organization to global economic growth strategies. An ecosystem, for the discussion in this chapter, will be defined as "a large volume of material in solid, liquid, or gas form that comprises a multitude of interacting biological and nonbiological entities." The major aspect of this definition is that an ecosystem is a very large, integrated system, and ecology is the science that defines the interaction of all of these entities within the system—often a very complex set of interactions and dependencies.

Environment. The term *environment* will be defined as the localized ecosystem around the object of interest. By defining environment in this manner, we provide a limited set of possibilities for interactions and dependencies predominantly by limiting the components in a smaller system.

An example of using this vocabulary: Consider a hypothetical inland factory that produces oatmeal cereal. Periodically, this factory cleans the rollers, steamers, oven grates, and various other pieces of equipment that are used in the making of the oatmeal. Heated water is used to clean the equipment, and the water goes into a treatment system that eventually drains into a small stream. The stream drains into a river that eventually flows into the ocean. As the equipment is cleaned, the cleaning water picks up oat, a natural residue; a combination of vitamins, salts, and preservatives (added to the cereal); probably some oil, dust, and grime that may accumulate on the outside or in the workings of the machines; and perhaps even some very small bits of metal from the machines as they wear. Some of these particulates will be large and can be easily removed with traditional filters or in a water-treatment plant, but some of the particulates will be very small, on the nanoscale. These nanoparticles can be biological and degradable or nonbiological, like a metal.

The job of trying to understand the impact of this oatmeal factory on the global ecosystem—for example, salmon in the north Atlantic—could be done, but it would certainly be a monumental task. However, assessing the impact of the plant and its wastewater on the local ground, water, and air environment is a much more manageable task. Not only is the assessment much easier in a contained, smaller, more manageable environment as defined here, but also any remediation, risk management, or cleanup that has to occur will be much easier at the local environment level. For example, if it is determined at the ecosystem level that the vitamins in the wastewater accumulated from many food-producing plants are impacting fish, it would be very difficult to remove the vitamins from the large, oceanic ecosystem. However, if each plant performed an additional chemical-filtering step before releasing the wastewater, the vitamin level would be reduced—at the local environment—resulting in a total accumulation value that would not affect the fish population.

Assessing the impact at the ecosystem level can certainly be done, and it is necessary. The process of assessing ecosystem impacts can collect data from multiple environments and integrate and assess the combined impact. Note, for example, that what may be thought to be acceptable at the environment level for remediation may prove to be insufficient at the ecosystem level. In this case, changes can still be made at the environment level. For example, each factory within a specific industry may be releasing only a very small amount of a toxic material like arsenic. The amount released by each factory may be well within the limits imposed by local, state, and national regulatory agencies. Further, overall the arsenic may be in such a low concentration within the oceanic ecosystem that a problem does not exist at the large scale. However, it is possible that some mechanism could exist within certain species or within specific regions of the ocean where pockets of higher, hazardous concentrations would accumulate. In this case, it might still be necessary to change the regulations at the factory, individual environment level in order to preclude a localized toxic buildup within the larger ecosystem. Based on the results of this work, informed decisions about trade-offs may be necessary at

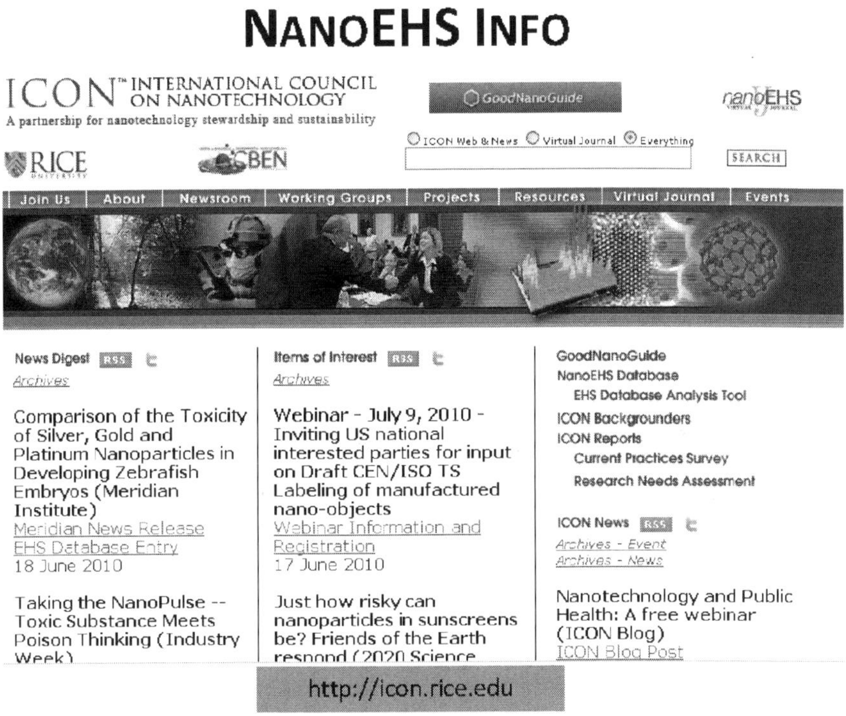

FIGURE 8.1 (See color insert)
Website for International Council on Nanotechnology, EHS page.

either the environment or ecosystem level. For the purposes of this chapter, we can define and categorize the discussion based on the previous definitions.

A final note on a sometimes-confusing acronym: *EHS* is often the acronym used to define the study and assessment of any material on the environment, health, and safety, where *health* and *safety* refer to living entities. People will talk about the "EHS effects of radiation," or the "EHS effects of carbon monoxide." It is a catch-all acronym that not only covers the impact of a material, but often is used to encompass risk-mitigation approaches, specific research topics, and the regulations that may govern that material. EHS may also appear as SHE or HSE, but these terms almost always refer to environment, health, and safety. Figure 8.1 is a representation of one of the many websites devoted to EHS information.

Broad Aspects of Nanotechnology

The multidisciplinarity and the potential global impacts of nanoscience and nanotechnology are key to its huge potential impacts and will require working at levels previously ignored. These two areas are discussed in this section.

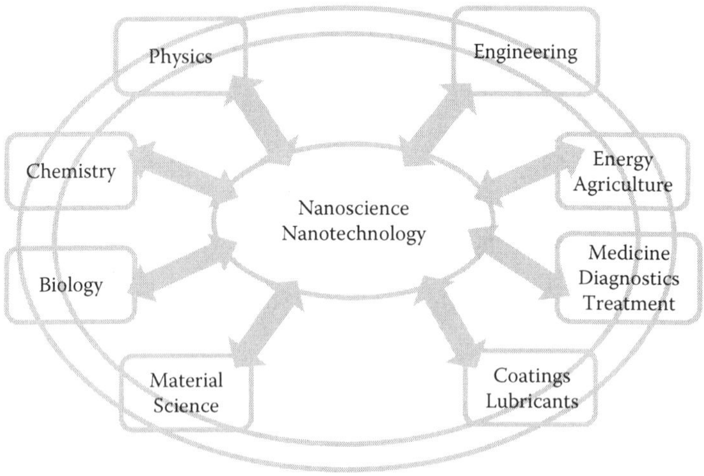

FIGURE 8.2 (See color insert)
The inter- and multidisciplinary nature of nanotechnology.

Multidisciplinary Aspect of Nanotechnology

Nanotechnology by its very nature is multidisciplinary, requiring expertise from many traditional science areas to execute the verbs in the definition of nanotechnology found in the previous section (observe, study, measure, etc.). For example, chemists may be involved in the creation of a specific nanoparticle; physicists can be added for the study and measurement activities; engineers and technicians may be involved in the measurement and characterization, and so on. It not only takes a village to raise a child, but it also takes a university to "raise" a nanoscale material. Figure 8.2 represents the multidisciplinary aspect of nanotechnology as well as the interconnection between all of the technical disciplines required.

This multidisciplinary aspect is one of the most exciting and challenging aspects of nanotechnology. Bringing together researchers from different backgrounds with a common goal can lead to amazing discoveries and products. Over fifteen years ago, corporations such as Motorola and Hewlett Packard led the industry segment with the introduction of multidisciplinary teams to evaluate the impact of nanotechnology on their products and market segments. When this multidisciplinary aspect is brought into an educational institution, the learning experience is enhanced and deepened for the involved students. Note that societal and economic impacts and influences occur within all of the disciplines and relationships implied in Figure 8.2.

The multidisciplinary aspect of nanotechnology also presents challenges. As mentioned in an earlier section, each of these disciplines has a unique vocabulary, and communication can sometimes be a bit bumpy. In the past, this lack of smooth communication has led to engineering or production difficulties, and this communication gap has and will continue to cause problems as nanotechnology moves forward.

This may be especially true with regard to nanotechnology and its impact on the environment or the Earth's ecosystem at large. The issue originates in the traditional educational hierarchy or organizational structure. Chemists and physicists may work together, and often scientists will be engaged with engineers. But often biologists, ecology students, and others involved in the life sciences do not interact with those in the physical sciences. Thankfully, the demarcation and division is slowly fading, but it is a very real societal phenomenon that could potentially impact the assessment of the environmental impact of nanomaterials.

Researchers in the physical sciences are often focused on the process required to produce the material or determining the physical properties and attributes of the material they have created. The impact of this material on flora and fauna may not be a focus of their thinking. Not that they don't care—they are just thinking about other aspects of the material, and such an analysis is beyond their expertise in most cases. The remedy to this situation can range from a quick phone call or e-mail between individuals to meetings between individuals across multiple organizations or disciplines.

Global—Discovery, Regulatory

The global aspect of nanotechnology is creating challenges in the intellectual property arena and also from a regulatory standpoint. Environmental and health-safety philosophies, approaches, and regulations are not uniform throughout the world. This non-uniformity of regulations can result in economic penalties or benefits for companies and/or countries investing in nanotechnology.

One of the most common methods for monitoring, informing, and defining safety requirements for a material is the MSDS or Material Safety Data Sheet. From Wikipedia (http://en.wikipedia.org/wiki/Msds):

> A **Material Safety Data Sheet** (MSDS) (also known as PSDS [product safety data sheet] or a COSHH data sheet in the United Kingdom) is a form with data regarding the properties of a particular substance.

An important component of product stewardship and workplace safety, the MSDS is intended to provide workers and emergency personnel with procedures for handling or working with that substance in a safe manner. It includes information such as physical data (melting point, boiling point, flash point, etc.), toxicity, health effects, first aid, reactivity, storage, disposal, protective equipment, and spill-handling procedures. MSDS formats can vary from source to source within a country, depending on national requirements.

The MSDS originated in the United States but is now a common document for all chemicals used worldwide (although the acronym may vary).

MSDS versus nMSDS

Properties[1]	Properties
Gold:	Nano Gold:
Melting point: 1064.18°C	Melting point: Variable
Reactivity: LOW	Reactivity: Unknown
Color: Yellow	Color: Red, but depends
Stability: GOOD	Stability: Unknown
Toxicology: Very LOW	Toxicology: Unknown
Transport: SAFE	Transport: Unknown
Protection: NONE	Protection: Unspecified
OSHA Status: No Regs	OSHA Status: No Regs

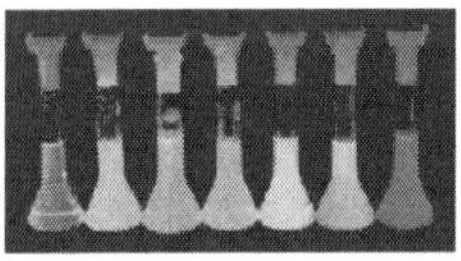

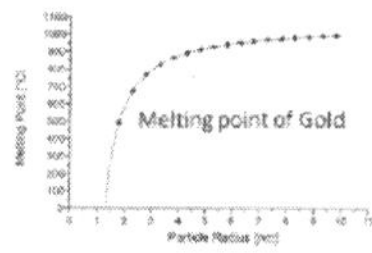

[1]Data - Oxford University

FIGURE 8.3 (See color insert)
Macro versus nano-example for some MSDS quantities of gold. (Courtesy of Walt Trybula.)

Since approximately 2003, there has been a global approach, initiated and supported by the United Nations called the Globally Harmonized System for Classification and Labeling of Chemicals and Mixtures. This program is designed to eventually ensure a uniform approach toward the handling, storage, disposal, and cleanup for chemicals. Although this program is not specifically designed with nanoscale materials in mind, it does provide the framework for a worldwide discussion about the approach to be used for nanomaterials.

It is fairly clear that the existing MSDS documents will need to be modified—at a minimum—to address nanoparticles. Figure 8.3 lists on the left the data from the current MSDS for gold. This data sheet is for quantities of gold in the macro- or microsize scale (volume). At this size scale, we tend to think of gold as being in a ring, on other jewelry, in coins, and so on. When we start creating gold in nanovolumes (nanoparticles), they can result in multiple shapes, have different melting temperatures, and have a higher reactivity and exhibit ungoldlike behavior. Gold particles of different sizes can also create different colors. This was a process that was used in Europe a few centuries ago to create the colors in the stained glass windows found in many of the cathedrals. The list on the right-hand side of Figure 8.3 clearly shows what we don't know about gold nanoparticles. The graph at the lower right shows how the melting temperature of gold changes with particle size, and the figure at the lower left-hand side is solutions of CdSe (cadmium selenide) containing particles of different sizes. This rainbow effect dramatically shows the different interactions with light that different size nanoparticles can have, whether gold, CdSe, or other materials.

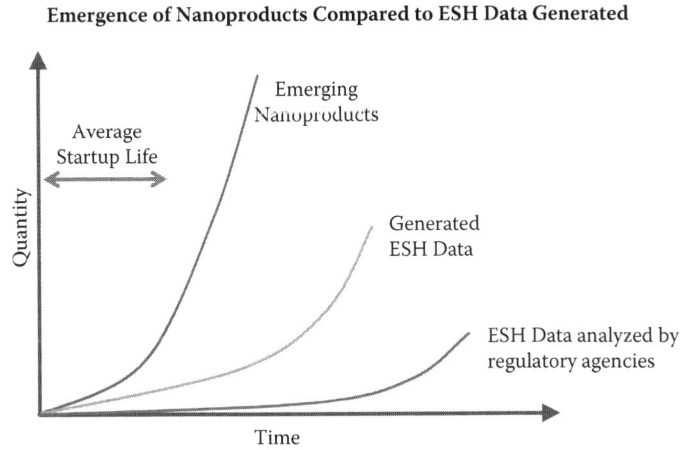

FIGURE 8.4 (See color insert)
Representative time scales for development, assessment, and regulatory aspect of nanotechnology. (Courtesy of Walt Trybula.)

Because of the complexity involved in the evaluation of any newly created entity, there is a time lag between when a material is initially created, when data on its safety can be generated, and when this data is analyzed and acted upon (or not) by regulatory agencies. The relative relation between these time scales is shown in Figure 8.4. Note that this offset of the time lines is true for any newly developed material or use of a material and is not necessarily restricted to nanotechnology materials. The reality that is implicit in this figure—that testing and regulatory aspects lag behind the discovery and often the application of a new material—is a factor that researchers, corporations, and the public need to be aware of (Carpenter, 1976; Editorial, 2007; Nel, Xia, Mädler, & Li, 2006; Salerno, 2004; Service, 2004, 2005). In drug discovery and development, for example, the long process of testing, animal studies, and human clinical trials that are required for drug development is an example of how the United States has addressed this time-lag issue for pharmaceuticals—requiring that questions be answered before the drug can go to market. Sometimes, even the usual seventeen-plus years of analysis and testing of a pharmaceutical is insufficient to validate not only the efficacy of the product, but also to identify the risks and potential side effects.

As nanomaterials continue to be developed and more applications are discovered, companies, regulatory agencies, and the public will need to wrestle with the issue of "How safe is safe?" and "When is there enough data?"

There will always be a point where the benefits of a discovery outweigh the associated risks and vice versa. The difficulty is in determining where that point is located. The exact location of the point could possibly be determined by technical means, or even based on a time scale, but public opinion and acceptance must also be taken into account. It is public opinion that will determine what is good enough and what is bad enough. This opinion is often conflicting. For example, the public desires that our diseases be cured quickly and easily, yet they also do not want any headaches, weight gain, stomachaches, and so on. It truly is a tricky balancing act because what is acceptable for one person, region, or country may be completely unacceptable for others.

The global aspect of nanotechnology and the lack of unified and enforced safety procedures perhaps pose the greatest threat to environments and ecosystems. A manufacturer in any country can potentially damage a large part of the ecosystem by either intentionally or unintentionally depositing nanomaterials into the environment. Concern for the environment has only been in the public awareness in the United States since the 1960s, and many other countries are farther behind in terms of recognizing the impact of any type of pollution.

Both the multidisciplinarity and the global aspects of the nanotechnology phenomenon will influence research, regulatory decisions, and the collective impact, both positive and negative, of nanomaterials.

Impacts of the Research and Products of Nanoscience and Nanotechnology

In 2008, Dr. Barbara Karn of the U.S. Environmental Protection Agency (EPA) described four approaches to the environmental issues of nanotechnology. The four approaches were:

1. Knowledge-gathering approaches carried out in the arena of research and development
2. Governance approaches via government agencies, laws, and regulations
3. Private sector approaches through nongovernmental organizations, academia, and industry
4. Philosophical approaches through public education and cultural evolution

Dr. Karn summarized the positive and negative aspects by discussing the applications and the implications of nanotechnology. In this discussion, positive applications like nanotechnology can be applied in a reactive manner to existing problems. Examples of this reactive approach include smog-reducing paint, water-purification techniques for Third World countries, and electrochemical removal of toxins from factory wastes (Hogan, 2004). Nanotechnology can also propose solutions that are proactive in preventing future problems, for example, by improving the efficiency of solar panels

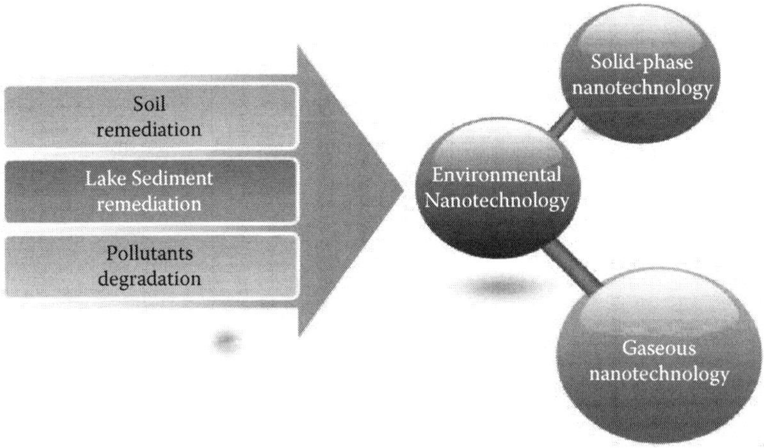

FIGURE 8.5
Environmental nanotechnology process used by the Pan group, Chinese Academy of Sciences.

and creating manufacturing processes that produce less waste. On the other side of the coin, the negative implications of nanotechnology could include unintended consequences of the nanomaterials interacting with the environment and possible risks associated with the nanomaterials themselves (Nel et al., 2006). Acknowledging that there are at least four levels where the impact of nanotechnology from an EHS standpoint can be addressed to assess both the positive applications and the negative implications of nanotechnology provides a good framework for discussion.

There are many examples of research institutions and companies investigating the use of nanotechnology to benefit the environment and the larger ecosystem. One such example is the Pan group, a lake-restoration and molecular environmental science group within the Research Center for Eco-Environmental Sciences, Chinese Academy of Sciences. Figure 8.5 describes their approach and focus areas.

There are many other examples of the potential positive impact of nanotechnology mentioned in other sections of this chapter and other chapters within this text. From an ecosystem standpoint, nanotechnology products can result in cleaner air and water, better use of farm land, earlier detection of toxins, and the removal of toxins.

On the other side of the balance are toxins or pollutants that are a result of nanotechnology. We have recently learned that the human body does not completely metabolize many medications, resulting in an accumulation of medications in rivers and the oceans. This may also be the case for nanoscale-based targeted delivery drugs, diagnostic materials, or preventive or treatment measures. The questions of how much is passed into the water system (or air) and how much is too much will require a significant amount of research and study.

Considerations for Assessment: The Devil Is in the Details

The application, use, acceptance, benefits, and toxicity of any entity will be dependent upon many factors. Historic examples such as plastic, nuclear radiation, and genetically modified plants encompass the range of uses as well as the trade-offs between positive and negative impacts of each of those at the time when they were new technologies. The same is true for nanotechnology-produced entities or products and raw materials that rely on nanoscale materials. This section discusses many of the aspects that must be considered when determining whether a nanoscale object may or may not be harmful to humans, flora, and fauna either in a localized environmental aspect or in a more global ecosystem aspect.

Element type: Knowing or determining the primary compositional elements of nanomaterials is certainly an excellent starting place for the study of these materials and their impact on the environment and ecosystems. Certain elements are known to be toxic to living organisms, and scientists and engineers are not necessarily proposing use of any of these elements for large-scale human use.

However, this is not always the case. CdSe (cadmium selenide) nanoparticles have unique fluorescing properties, where the fluorescent color is uniquely dependent on the diameter of the nanoparticles. Because of the strong color differentiation and the long-lasting fluorescence (when compared to chemical fluorophors), researchers are using CdSe nanoparticles to determine tumor location and size.

Independent of the possible variations, knowing the elemental composition of the nanoparticle is the popular starting place for the assessment of nanoparticles' function, application, and impact.

Object size (volume, mass): Clearly, the mass and/or volume of the nanomaterial that is present can have an impact on both the positive and negative aspects of the interaction. It is important to distinguish between the volume or mass of the individual piece of the nanomaterial and the total volume or mass of all the nanomaterial that may be present in a given location or system.

Total number of objects: This aspect is closely related to the object size, in that the total amount of the material present can have an impact on its effect.

Shape: Whether a nanoparticle is spherical, flat or plated, cylindrical, or any other myriad shapes can have a significant impact on its surroundings. Carbon nanotubes (CNTs), for example, carry the potential to be harmful to some organs within living systems (at high enough amounts) because of their shape (long, cylindrical, and sometimes pointed) rather than their elemental composition, which is biocompatible carbon. It is also interesting to note that it is the shape of carbon nanotubes that make them potential candidates for water purification and blood filtering.

Time in system: The length of time a nanoparticle has been interacting with any system is a determinant in terms of degree of impact, and as with some of the previous factors, the effect can go in either direction. For example, consider a splinter in your finger. Sometimes, after the initial awareness of the splinter, the body will take care of it and remove the splinter without any external influence. On other occasions, the splinter will be removed, but by the application of an external influence. Either one of these approaches can occur within a short or a long period of time. Other times, the splinter may remain in your finger, but as it remains, the immune system tries to respond, but if the response is insufficient and the splinter becomes infected, it is not a pleasant situation. The dwelling time of a nanoentity within a system can also make a difference with regard to its impact on the system. If the nanoparticle dwell time in the system is short, then the potential for impact may be potentially much lower than a nanoparticle that remains for a longer period of time.

Use: The Interaction Quadrants and Surrounding Environment

The composition not only of the nanomaterials (organic or inorganic), but also the environment of interaction can have a significant effect on the impact. A shard of metal on the ground is not an issue, but if that inorganic metal becomes lodged in a finger (organic), then there may be an issue. The temperature and pressure of the surrounding environment may also affect the level of impact. Figure 8.6 represents the four quadrants of interactions.

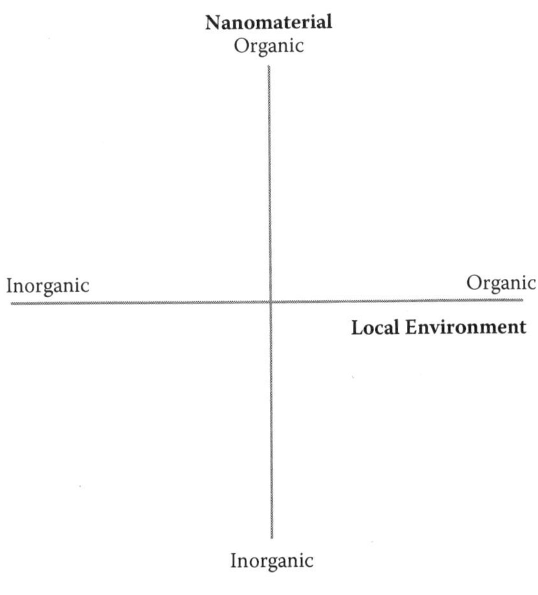

FIGURE 8.6
Representative four quadrants of interactions.

Life Cycle and Length of Exposure

The longevity or life cycle of a nanoparticle can impact its level of risk or concern (or benefit). Often, potential benefits or risks are not only due to that particular aspect, but also based on the probability of the event happening during a certain period of time. A potentially hazardous material that is in existence for a short period of time has a lower overall risk or failure factor than one that has a longer lifetime.

Method of Entry: Air, Fluid, Injected, etc.

How the nanoparticle enters the system is an extremely important factor in determining the magnitude of the impact, either positive or negative. As an example, consider real butter, and let us assume that it is bad for your health based on what we have been told. However, there are many ways that butter can interact with the human body. We can eat it in chunks; we could inhale it when exposed to a butter aerosol; we could smooth it over our skin; and it would probably be possible to inject it into our muscle tissue. The common implied context for the "butter is bad" message is that you are eating it and not using it as a body lotion, for example. When discussing nanoparticles, an awareness of unspoken or assumed conditions must be considered. And if there is one particular method of entry that is deemed hazardous, then regulation and constraints should only be applied to that method.

Purity

Most people wash their fruits and vegetables before eating them to ensure that as much dirt, fertilizer, travel grime, and other undesirable material has been removed from the item before we eat it. We do that because we know that the good or nontoxic item that we want to eat has probably been exposed to toxic or unhealthy substances. The same "cleanliness" is desired for nanoparticles. For example, carbon nanotubes are being considered as scaffold material to help broken bones heal faster. This seems like a good material, since carbon is biologically compatible (Gillette et al., 2008; Dragnea, 2007; Huebsch et al., 2010; Cheng et al., 2009). However, the process to create or grow the carbon nanotubes can involve the use of toxic chemicals, although it should be noted that not all methods of growing CNTs require the use of chemicals. If CNTs are used as a scaffold to aid in bone growth, it is necessary that the CNTs be pure and not carry any residual chemicals used in the manufacturing process.

All of these factors must be considered collectively and are intricately interwoven with each other. For example, carbon is considered biologically compatible; indeed, humans are 16% carbon. Organic molecules are carbon based such as nucleic acids, proteins, and so on. The extracellular membrane is predominantly carbon. Based on this information, one could assume that nanomaterials based on carbon would be harmless to humans. Unfortunately, this is not the case. Carbon nanotubes, which are small cylinders of carbon atoms (similar to rolled up chicken wire), can include shapes that are very small in diameter and pointed at the ends. This shape arrangement of carbon atoms is similar to the shape of the five forms of asbestos that are harmful, and preliminary research shows that these spearlike arrangements of carbon atoms can indeed pose health risks.

> Nanoethics…takes as its subject a science still aborning; many of the ethical and social ills it seeks to address are mere speculations about the hypothetical ramifications of theoretical technologies that may prove technically impossible. It is fair to say that no scientific field or technological innovation has ever faced such intense scrutiny so prematurely. (Adam Keiper, 2007)

Knowing All This

There is no doubt that the materials and products that result from research and engineering efforts at the nanoscale can and will have a positive impact on ecosystems, the environment, and living organisms. By understanding the specific properties of nanoscale materials, we have created smog-reducing building coatings, water purification systems, toxin sensors and detectors and removal methods, and better air filtration and purification methods.

Nanotechnology will also produce processes and products that allow us to better use the resources we have. Energy storage devices based on nanoscale principles will help make fuel cells popular as well as improve battery operation and lifetime. Other approaches will allow solar cells to be cheaper to produce and more efficient at collecting the unlimited energy from the sun and turning it into electricity. Developments in coatings and composite materials will improve wind turbines and wave-energy capture systems.

There is also the possibility of negative consequences of nanotechnology products. Unwanted materials can end up in air and water, where they may interact with humans, animals, and plants. Because of the unique attributes that nanomaterials can have—even dependent on the material size and shape—an entity that may be toxic in one environment can be explosive in another.

For example, aluminum nanoparticles can be made in many different sizes. Aluminum nanoparticles that are 30 nm in diameter are extremely explosive when exposed to air. Aluminum nanoparticles that are 80 nm in diameter, however, are completely inert in an air environment. Similarly, silver nanoparticles have been shown to be antimicrobial and are currently used in bandages, hospital cleaning sprays, and anti-odor socks. In low doses, silver is nontoxic to humans. However, in large doses it can literally turn a person blue. The ramifications of high doses of silver are still under study.

Conclusion

This chapter has discussed the complex issue of determining the impact of nanotechnology on the environment and the larger ecosystem. Aspects of nanotechnology such as its multidisciplinary nature and the global impact of nanoscience will impact how it is used and how it will affect the environment. The myriad factors that must be considered when evaluating this impact have also been discussed.

It is obvious that nanotechnology, like other technologies before it, can have significant impacts, both positive and negative, on the quality of life, the environment, and the global ecosystem. There are many examples for each side of the pro and con scale. The weighing assessment must be made many times with many variations. The best approach is to begin the effort, consider the most probable benefits and risks, and take the first step.

Questions

1. Using the Internet or other resources, research alternative definitions of nanotechnology, environment, and ecology. How could these differences in definitions change the conversation or perception about the interaction of nanotechnology and the environment and ecosystem?
2. Using equations for surface area and volume, create a table or graph comparing a spherical nanoparticle, a red blood cell, a virus, a bacteria, and plankton. Which one has the highest ratio of surface area to volume? Based on your calculations, which one would have the highest reactivity value?
3. Working with a partner, discuss and come to an agreement for the type of measurements that could go into the pros and cons on the two sides of the seesaw.
4. A FMECA (Failure Mode, Effects, and Criticality Analysis) is a method for determining everything that may fail or go wrong in a single item or a whole system. It is also a method to determine the effect and criticality of a given failure. Research a more detailed explanation of an FMECA and then apply it to one of the potential applications discussed in the scenarios in this chapter or select a potential application of your choice.

5. What are some of the factors that must be considered when studying the impact of nanoparticles on a system?
6. Of the factors that must be considered, list the top three most important, in your opinion. State your reasoning.
7. Create your own set of questions to ask your peers or family to determine their familiarity and/or understanding of nanotechnology. Find out about their understanding or perceptions of the benefits and risks of nanotechnology to the environment and humans. Create a list of common questions about nanotechnology and some of the misconceptions. What are some suggestions you have to counteract some of the misconceptions that you find?
8. Provide your own examples for combinations of nanoparticles and local environments that fall into each quadrant of Figure 8.6.
9. The U.S. EPA is tasked with keeping toxins and unwanted materials out of the environment. What do you think its role should be with regard to nanoparticles?
10. Consider the use of nanoparticles for injectable medical purposes. Assume the nanoparticles are intended to be pure gold. You are the researcher trying to develop the product. Create the flowchart of your next several steps. At what point do you think the government regulatory agencies (FDA, EPA, etc.) should be notified of your work?
11. Calculate the ratio of surface area to volume for a sphere, a cube, a torus, and a pyramid. If you wanted to create a highly reactive material, what shape would you use? Why?
12. What do you think is the largest issue facing the Earth's ecosystem today? How would you approach solving it? What would be some steps that you could take at the environment level to help solve this issue?
13. Using the Internet and other sources, find common or not so common natural nanoparticles. How are they used in nature, and are they harmful?
14. Imagine that you are a nanoparticle. Select the material that you are made of, and perhaps your shape and size. What objects or materials would you interact with, and what could be your potential impact on your local environment?
15. Take a peek at your medicine cabinet. How many items can you find that use nanoparticles or are made using nanotechnology?
16. Summarize your learning of new ideas and information presented in this article by completing the following flowchart:

Flow Chart

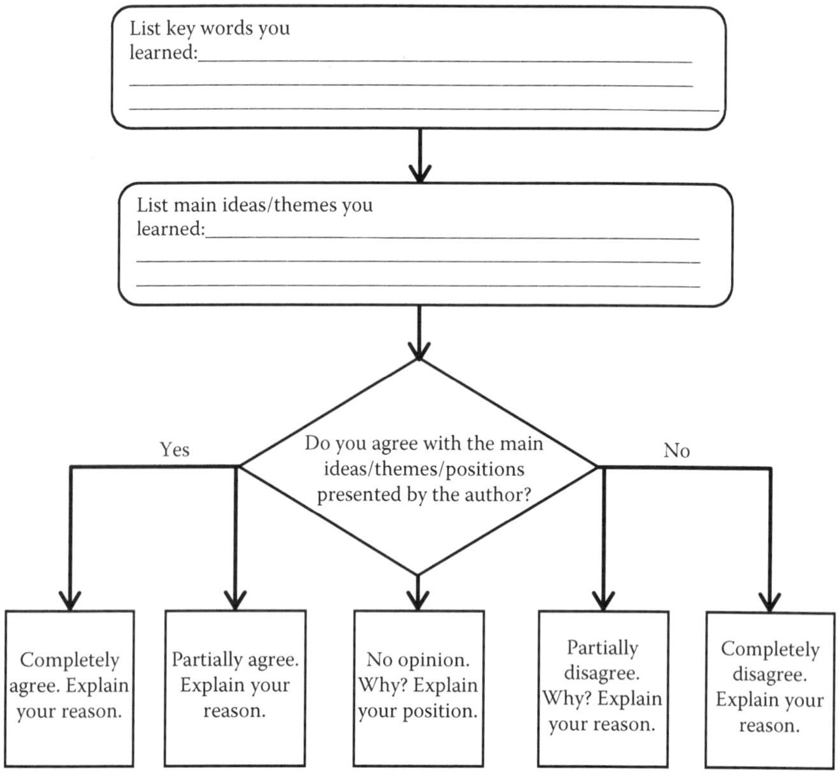

Scenario 1

You own a company that produces fabrics. Your products are predominantly used in automobile and airplane interiors. Your research department has been investigating self-cleaning fabrics and also antimicrobial fabrics, since these are product attributes that would give your company an edge over the competition. During the testing of some research samples, your staff notices that pieces of this fabric not only seem to filter out particulates, but can also kill bacteria and microbes. One of your more socially minded employees proposes that this product would serve a significant humanitarian purpose in Third World countries that are struggling to find clean water. By pouring river water through this material, the water purity could be improved 100-fold. This material could also potentially be used on a larger scale to purify water at the source of the pollution, i.e., factories, or where streams meet rivers, and so on. You determine that it costs about $0.50 to produce a small (8.5 in. × 11 in.) piece of this material. What do you do?

Scenario 2

You are a researcher at a large company. You are tasked with assessing the applications and implications of silver nanoparticles for infection reduction in burn patients. You know that silver nanoparticles that are small (18 nm) are very effective antimicrobial agents. You also know that they can be effective in moderate concentrations. You also know that very small nanoparticles of silver can be harmful to the brain, and high concentrations in the body can cause a patient's skin to turn blue. Devise a set of experiments that would help you determine the optimal volume, shape, and concentration of silver nanoparticles required to help burn victims.

Case Study: Vibrating Gold Nanoparticles to Destroy Tumors

Some nanoparticles will have an observable reaction when exposed to the electromagnetic spectrum. In order for an observable interaction to occur, there needs to be a specific relation between the particle size and the energy, represented by the wavelength of the electromagnetic energy.

For example, at the macroscale, the correct relationship between the length of a piece of string and the frequency at which it is vibrated can result in a standing wave on the string. The same is true at the nanoscale. For some nanoparticles, when there is the proper relationship between the diameter of the particle and the energy (wavelength) of light falling on it, the particle will fluoresce. Each particle with a different diameter will fluoresce with a different color. It has also been shown that when nanoparticles are placed in a radio-frequency environment, they will start to mechanically vibrate. These vibrating nanoparticles can result in heat that could destroy tumors.

If the gold nanoparticles could be located in or very close to the unhealthy tumor cells and a radio-frequency field generated outside of the body, then the nanoparticles could be caused to vibrate, creating heat and thereby killing the unhealthy cells with minimal damage to the surrounding tissue.

Consider that the gold nanoparticles can be directly deposited into the tumor.

- What research is still needed to be done?
- What parameters need to be controlled?
- Create a flowchart with tasks, decision points, and outcomes showing the steps to validate this proposed process.
- What would be some of the risks and benefits that you would weigh?
- Are there one or more points, or pieces of data, that would cause you to abandon this approach? For example, the nanoparticles would need to be injected into the tumor, requiring a syringe to deposit them.

- The radio-frequency activation of the gold nanoparticles would require no incisions or anesthetic for the patient, and the radio waves are not harmful to the human body. Can you think of any other benefits?
- What, if any, would be your initial environmental concerns about using the gold nanoparticles? You need to consider at least two environments.

References

Carpenter, R. A. (1976). Tensions between materials and environmental quality. *Science, 191*, 665–668.

Cheng, W., Campolongo, M. J., Cha, J. J., Tan, S. J., Umbach, C. C., Muller, D. A., et al. (2009). Free standing nanoparticles superlattice sheets controlled by DNA. *Nature Materials, 8*, 519–525.

Dragnea, B. (2007). Bio-inspired materials: Unnatural life. *Nature Materials, 7*, 102–104.

Editorial. (2007). Regulating nanomedicine. *Nature Materials, 6*, 249. Retrieved November 13, 2011, from http://www.nature.com/nmat/journal/v6/n4/full/nmat1875.html

Editorial. (2008). Cleaning up water. *Nature Materials, 7*, 341. Retrieved November 13, 2011, from http://www.nature.com/nmat/journal/v7/n5/full/nmat2178.html

Gillette, B. M., Jensen, J. A., Tang, B., Yang, G. J., Bazargan-Lari, A., Zhong, M., et al. (2008). In situ collagen assembly for integrating microfabricated three-dimensional cell-seeded matrices. *Natural Materials, 7*, 636–640.

Hogan, J. (2004). Smog-busting paint soaks up noxious gases. *New Scientist*. Retrieved November 13, 2011, from http://www.newscientist.com/article/dn4636-smog-busting-paint-soaks-up-noxious-gases.html

Huebsch, N., Arany, P. R., Mao, A. S., Shvartsman, D., Ali, O. A., Bencherif, S. A., et al. (2010). Harnessing traction-mediated manipulation of the cell/matrix interface to control stem cell fate. *Nature Materials, 9*, 518–526.

Nel, A., Xia, T., Mädler, L., & Li, N. (2006). Toxic potential of materials at the nanolevel. *Science, 311*, 622–627.

Salerno, G. J. (2004). Biosecurity and research: Minimizing adverse impacts. *Science, 304*, 687.

Service, R. F. (2004). Nanotechnology grows up. *Science, 304*, 1732–1734.

Service, R. F. (2005). EPA ponders voluntary nanotechnology regulations. *Science, 309*, 36.

Additional Sources

Barnard, A. S. (2006). Nanohazards: Knowledge is our first defense. *Nature Materials, 5*, 245–248.

Berger, M. (2007, August). Water, nanotechnology's promises, and economic reality. *Nanowerk*. Retrieved November 11, 2011, from http://www.nanowerk.com/spotlight/spotid=2372.php

Berger, M. (2008, February). Nanotechnology and water treatment. *Nanowerk*. Retrieved November 14, 2011, from http://www.nanowerk.com/spotlight/spotid=4662.php

Dreher, K. L. (2004). Health and environmental impact of nanotechnology: Toxicological assessment of manufactured nanoparticles. *Toxicology Science, 77*, 3–5. Retrieved November 14, 2011, from http://toxsci.oxfordjournals.org/content/77/1/3.abstract

Gorman, M. E., Groves, J. F., & Catalano, R. K. (2004). Societal dimensions of nanotechnology. *IEEE Technology and Society, 23*(4), 55–62.

Karn, B. (2008, July). Four approaches to the environmental issues of nanotechnology. Slide show presented at Safe Handling of Engineered Nanoscale Materials Workshop, Argonne National Laboratory, Argonne, IL. Retrieved November 14, 2011, from http://www.docstoc.com/docs/40985997/Four-Approaches-to-the-Environmental-Issues-Of-Nanotechnology

Khodakovskaya, M. V., de Silva, K., Nedosekin, D. A., Dervishi, E., Biris, A. S., Shaskkov, E. V., et al. (2011). Complex genetic, photothermal, and photoacoustic analysis of nanoparticle-plant interactions. *PNAS, 108*, 1028–1033. Retrieved November 14, 2011, from http://www.pnas.org/content/108/3/1028

Parks, N. (2002, January). Pollutant-loving bugs munch PCBs. *Science Now*. Retrieved November 14, 2011, from http://news.sciencemag.org/sciencenow/2002/01/16-03.html

Worthington, R. (2009, December). Ecology and nanotechnology. *2020 Science*. Retrieved November 14, 2011, from http://2020science.org/2009/12/17/worthington/

9

Our Biopolitical Future: Four Scenarios

Richard Hayes

CONTENTS

Emerging genetic technologies could radically reshape the world, for good or ill.

Most of us would enjoy being healthier, smarter, more attractive, and longer-lived. But we also know there can be too much of a good thing. If we're overweight, it makes sense to reduce, but anorexia can be lethal. A nice haircut can make us feel good, but repeated, expensive cosmetic surgery can bring more complications than compliments. Most of us understand this and learn to lead full and productive lives within the natural range of diversity that comes with being human.

But what if that natural range of diversity no longer applied? What if it were possible to radically enhance our looks, brains, athletic abilities, and life span with, say, injections of customized genes? What if we could design our children with chromosomes purchased from a catalogue?

Scientists have long speculated that genetic technology would someday allow us to manipulate our own genes. In the past, such musings were dismissed as fanciful, or as so far in the future that they didn't need to be taken seriously.

Now, hardly a day passes without news that researchers have discovered another gene that appears to influence the development of a particular human trait. Genetically modified animals are a staple of laboratory research. Fertility clinics offer an increasing array of procedures to manipulate human embryos. Noted scientists announce, with barely disguised anticipation, that we are about to enter the posthuman epoch, like it or not.

A hallmark of the human species is the ability to intentionally manipulate objects of nature. But not until now have humans been able to intentionally manipulate the biological foundations of human nature. Our common biological nature evolved over many millennia but has been essentially stable over the few thousand years during which modern human values, behaviors, and institutions have developed. The ability to manipulate human nature—in effect, to make the agent of change an object of change—destabilizes both the biological and the social foundations of the human world.

How do we even begin to think about what this might mean? In recent years, developments concerning the new human genetic technologies have been interpreted in many countries largely through the familiar frameworks of abortion politics and the culture wars. Religious conservatives were among the most vocal early opponents of human cloning, stem cell research, and related procedures, and many liberals and progressives reflexively assumed that the enlightened position was therefore to embrace these technologies.

While understandable, this is nonetheless simplistic and misleading. The same genetic technologies that might be used to prevent or cure many widespread diseases and debilitating conditions will allow forms of genetic manipulation that could endanger equality, social justice, human rights, and other core progressive values.

There's no reason we can't draw lines that protect abortion rights and medical research while prohibiting applications of genetic science that open the door to profoundly undesirable outcomes. But to do so we need new interpretive frameworks to help us understand what's at stake and what our options are.

Four Biopolitical Scenarios

The four scenarios of the human biopolitical future presented in this chapter may help us think through these issues. They take place over the fifteen-year period from 2007 through 2021.

A central theme is the tension between libertarian and communitarian values. Humans evolved with tendencies both to compete and to cooperate, and societies have varied in the emphasis they give to one tendency or the other. Environmentalists are familiar with the libertarian/communitarian tension as the tragedy of the commons: An individual may benefit by polluting a river or the atmosphere, but if everyone seeks to benefit in this manner, everyone suffers.

An appreciation of this tension affords us a richer understanding of today's political landscape. The conflict between Left and Right has historically centered on different levels of concern regarding equality. The conflict between libertarians and communitarians centers on different levels of concern regarding solidarity, that is, the willingness to forgo individual desires in the interest of the community as a whole. In the United States, the political landscape thus includes the libertarian right (e.g., Milton Friedman and the Cato Institute), the libertarian left (much of the 1960s counterculture and Hollywood), the communitarian right (religious and social conservatives, and some neoconservatives), and the communitarian left (labor unions, the religious left, social justice advocates, and many environmentalists).

Our four scenarios suggest ways that different combinations of these values might give rise to alternative human biopolitical futures. In "Libertarian Transhumanism Triumphs," both left- and right-libertarian values prevail. In "One Family, One Future" scenario, communitarian values grounded in quasi-religious solidarity and patriarchy prevail. In "A Techno-Eugenic Arms Race," a lethal mix of communitarian nationalism and libertarian techno-capitalism spins out of control. The scenario "For the Common Good" is grounded in communitarian values of the sort historically associated with social democracy and liberal internationalism.

It's unlikely that the future will play out precisely as sketched in any one of these scenarios. But there will be a future. The more we are clear about those futures we wish to avoid and those we would welcome, the easier it will be to figure out what we are called to do now.

Libertarian Transhumanism Triumphs

Individualist social values and free-market economic values combine with powerful new genetic technologies to launch humanity along a new post-human trajectory.

The opening years of the twenty-first century were marked by controversy over cloning, stem cells, and human genetic modification (see sidebar, "The Language of the New Biopolitics," for definitions). Despite concern about fraudulent cloning claims and unethical gene therapy experiments, genetic technology was increasingly seen as part of a progressive vision that rejected outworn, traditionalist values and embraced a bright future of technological innovation and economic growth.

During this same period, libertarian sentiment grew rapidly among many Americans, encouraged by a well-funded network of think tanks, bloggers, and entrepreneurial scientists. By 2009 their ideology of "free markets, free choice, free bodies," was spreading at the expense of both religious conservatism and social democratic liberalism. Democrats and Republicans alike argued in favor of free trade, school vouchers, deregulation, privatization, personal retirement accounts, pharmacological freedom, and repro-genetic autonomy.

With visions of trillion-dollar markets waiting to be served, global biotech conglomerates raced to develop technologies allowing parents to screen embryos for behavioral and cosmetic traits. For the other end of the life cycle, these same firms established high-tech life-extension and cryonics facilities throughout the world, most lucratively in small countries proudly advertising their lack of regulatory oversight.

Among the earliest adopters of genetic modification were athletes, and the public turned out in droves to see gene-doped competitors break one record after another. Despite hand-wringing from an older generation of sports professionals and a short-lived protest movement by concerned parents, by 2011 athletics was fast becoming a contest of competing genetic interventions rather than innate ability, coaching, and practice.

A major threshold was crossed in 2013, when Swedish scientists announced the birth of the first true "designer baby," that is, a child able to pass its modified genes to its own children. Although ostensibly developed to prevent congenital disease, within four years the procedure was being offered commercially for a wide range of aesthetic, cognitive, and performance enhancements. The cost of a designer baby was high (about US$235,000), but affluent couples flocked to the new "better baby" clinics to ensure that their children had the best genes money could buy.

Meanwhile the transhumanist movement, which had started as a fringe group of sci-fi cultists in Los Angeles in the early 1990s, was growing into a major social force. The transhumanists were obsessed with the prospect of reconfiguring the human species and the rest of the natural world through genetic modification, nanotechnology, and synthetic biology. The combination of libertarian politics and transhumanism resonated strongly with ambitious young technophiles throughout the world, and an increasing number of up-and-coming figures in the sciences, commerce, the arts, and politics openly identified themselves as libertarian transhumanists.

In 2015 *Forbes* magazine estimated that flamboyant bioindustrialist and committed transhumanist Dmitri Rastovich had become the world's first person with net assets in excess of US$1 trillion. When asked by reporters to comment on growing fears that biotechnology was giving rise to human genetic castes, Rastovich replied, "There is no alternative. Relax and enjoy it."

One of the earliest casualties of the spread of libertarian transhumanism was the environmental movement. Attempts to channel biotechnology along environmentally friendly paths had succeeded in a handful of instances, such as the time in 2014 when genetically engineered microbes successfully biodegraded a major oil spill off the southern California coast. But at the core of the transhumanist philosophy was a belief that nature, whether in the form of plants, animals, humans, or ecosystems, was an inferior product whose due-date had long since expired. After 2018 the Sierra Club, Greenpeace, and other long-standing environmental groups rapidly began losing membership.

Now, in 2021, it's clear that there's no going back. "Techno" has fully replaced "natural" as a hallmark of excellence. The genetically enhanced elites relax in their gated communities, dine on transgenic squash and cloned beef, dote on their cloned pets, and look forward to receiving the latest GenePak® uploads for their kids. Libertarian transhumanism has become the hegemonic vision of the human future. Few people can any longer imagine a credible alternative.

One Family, One Future

Reaction against the new human genetic technologies is part of a far-reaching neotraditionalist backlash against modernity and postmodernity.

The opening years of the twenty-first century were marked by controversy over cloning, stem cells, and human genetic modification. In 2008 the U.S. biotech industry organized a political action committee to promote an industry-friendly agenda of "Cures for All." Initial success was tarnished, however, when covert human cloning labs were discovered the following year in Thailand. Embryos used for these illicit experiments were traced to fertility clinics associated with the World Stem Cell Consortium, established by scientists in Australia, Belize, and Cyprus, to help themselves and others evade national regulations.

In 2010 a German human rights group documented the deaths of over 300 women worldwide from ovarian hyperstimulation syndrome, the result of aggressive efforts to obtain eggs for cloning research. Meanwhile, wealthy individuals were increasingly outsourcing the entire process of reproduction. Women rated "Grade A" were routinely being offered sums in excess of US$150,000 for their eggs, genetically "superior" sperm could be purchased over the Internet, and young women from Ukraine and Romania were paid little better than minimum wage for the use of their wombs. In 2012 a Scottish gene therapy experiment gone awry left two dozen infants with an incurable form of bone cancer and life expectancies of less than twelve years.

Religious conservatives saw an opening, and began speaking out against the eugenic juggernaut and in support of equality, social justice, human rights, women's and children's health, the sanctity of the natural world, and the precautionary principle. The political tide began to shift. After winning filibuster-proof congressional majorities in the United States in 2014, conservatives quickly succeeded in banning reproductive and research cloning, sex selection, research using human/animal chimeras, physician-assisted suicide, child-accessible Internet pornography, and gas-guzzling SUVs. Protests were heard from the biotech industry, civil libertarians, and the automakers, but the great majority of people in the United States were relieved to find that someone was finally willing to draw some lines.

During these same years, growing repugnance over the dehumanizing impacts of the new genetic technologies, techno-capitalist globalization, and the pervasive tawdriness and superficiality of the postmodern world helped fuel neo-traditionalist movements in Europe, Asia, Africa, and the Americas. The gifted German-Turkish writer Fredericka Musfika, author of the influential book, *Humanity or Transhumanity?* drew on conservative Islamic, Christian, Jewish, Hindu, and Confucian social values to offer a universalist vision of a human future embracing peace, love, and harmony with nature. Her impassioned speaking and writing gave rise to the mass social movement known as "One Family, One Future" (OFOF). It was a secular movement open to people of any (or no) religious faith, but it adopted codes of conduct similar to those found in many traditional religions. In the period after 2016, the practice of wearing a full-length woolen scarf displaying OFOF iconography spread throughout the world as a symbol of the rejection of postmodernity.

Although OFOF endorsed the use of the Supernet, the iWeb, and other new information technologies, it viewed high-tech medical practices with suspicion. By 2018, many countries had abandoned research on genetic modification. The use of naturopathy, aroma therapy, herbal preparations, and a form of massage therapy accompanied by poetry and song had all but replaced conventional medical treatment among significant sectors of the world's population.

As early as 2017, the established religious denominations began losing members to OFOF. In some North American and European cities, as much as 30% of the population would gather for OFOF's Saturday affirmation services. This proportion is certain to increase, because OFOF families shun birth control and now average seven children per couple.

In 2019, OFOF-USA announced the formation of a political party, and in last year's (2020) elections, OFOF candidates—all men, and all wearing the full, luxuriant beards that now designate OFOF clan leaders—won two dozen seats in the House and four in the Senate, taking votes from both Republicans and Democrats. Similar parliamentary gains have been made in about twenty other countries. Earlier this year, OFOF leaders told the tens of thousands gathered at their 2021 annual World Convocation that the human future never looked as promising as it does today.

A Techno-Eugenic Arms Race

Powerful genetic technologies are used by individuals, corporations, and countries in an escalating struggle for superiority and dominance.

The opening years of the twenty-first century were marked by controversy over cloning, stem cells, and human genetic modification. In 2008, biotech enthusiasts in the United States organized a national campaign to "liberate" stem cell research by loosening even the minimal existing state and federal oversight guidelines. Although many scientists worried that this would allow ethically questionable activities to be swept under the carpet, they were reluctant to break ranks and speak out for fear of giving aid and comfort to demands by the religious right that stem cell research be banned entirely.

In 2010, North Korean scientists announced the birth of a child genetically modified to allow an increased respiratory capacity of 18% above the human norm. The scientists involved made no pretense that this was done to address a medical need. Rather, they said, it was the first step toward creating "The New Man" for the twenty-first century.

Just eight months later, China—with an exploding GDP, growing nationalist fervor, and 60,000 freshly trained biotech engineers entering the workforce each year—announced a national initiative to improve the genetic quality of its people. All couples at risk of transmitting genes identified as deleterious were required to take steps to avoid doing so, with the government covering all costs. In addition, couples could volunteer to have their children "enhanced," again with all costs covered. Leading Chinese rock stars and taikonauts were featured in a massive media campaign promoting the program.

Alarms were raised by international human rights and social justice organizations, but to little effect. Other countries knew they had to follow China's lead or risk having their children left behind. A new techno-eugenic arms race rapidly escalated out of control.

In 2014 the CIA reported that Venezuelan scientists had created a virus that turned skin cells containing specified concentrations of melanin carcinogenic. Other countries enacted laws requiring the medical termination of "lives not worth living." Still others approved forms of human experimentation, using prisoners, the disabled, terminally ill patients, orphans, and others, that had been anathema barely a decade earlier.

Some early promoters of human genetic modification argued that its widespread use would result in such a diverse array of genetic types that the concept of "race" would finally be consigned to the dustbin of history. In fact, just the opposite has occurred. With ethnocentrism and nationalism on the rise, right-wing governments have issued genetic profiles of "ideal" racial and ethnic types, and individuals are implicitly or explicitly urged to modify themselves and their children to conform with these profiles.

By 2018, most genetic research was being conducted by secret government and corporate labs. In that year it was reported that scientists in Mumbai had developed a procedure to slow the rate of human cellular aging by as much as 60%. Leading Indian government officials and biotech executives, realizing the havoc this technology could cause if made widely available, moved quickly to limit its use to priority national security resources: themselves.

Today, in 2021, the genetic scientists and their political and military commanders have lost any sense of identification with the larger human community. In their minds, the well-being of any existing human cannot be allowed to stand in the way of the historical transition to a posthuman future. But they differ about who will supply the foundational human stock.

And if it seems that things could not get any worse, just last week a doomsday cult announced that it has perfected and is about to release the "Elysium Virus," a genetically engineered hyper-viroid that inactivates neural calcium ion channels. Its release would rapidly destroy all life on Earth above the level of a sponge. The cult has issued no demands; its members say they are driven by an altruistic desire to relieve "all sentient beings" of the burden of existence. The world is holding its breath, teetering on the verge of panic.

For the Common Good

Liberal democratic and popular institutions, informed by values of equality, social justice, and community, establish cultural norms and legal institutions that support scientific research while precluding its use in ways that endanger human well-being.

The opening years of the twenty-first century were marked by controversy over cloning, stem cells, and human genetic modification. Opinion surveys showed strong support for the development of genetic technology for medical purposes, but controversies involving blackmail attempts using stolen sperm donor records, the deaths of clonal primates at a lab in Oregon, and shady financial practices by leading bioethicists began to raise doubts. Although the new genetic technologies attracted many sincere, socially responsible researchers, by 2009 the field was increasingly dominated by dismissively arrogant scientists, unscrupulous fertility clinic operators, traffickers in clonal embryos, and out-and-out racist eugenicists.

Reaction from the general public and affected constituencies had been building for some time, and by 2010 reached a tipping point. Advocates for women's health, consumer rights, and economic justice raised concerns about risky technologies that put corporate profits above safe, affordable health care. Civil rights leaders warned of a new free-market eugenics that could stoke the fires of racial and ethnic hatred. Disability rights leaders charged that a society obsessed with genetic perfection could come to regard the disabled as mistakes that should have been prevented. Civil libertarians were appalled to learn of plans by global biotech consortia to establish a universal DNA registry. Lesbians and gays were disturbed by reports that prenatal tests for sexual orientation were about to be made commercially available. Environmentalists argued that genetic modification of living organizations, including humans, was a powerfully disruptive technology being deployed before long-range consequences had been considered.

In 2011, liberal and conservative religious denominations put aside their doctrinal differences and convened an international summit that declared the genetic modification of the human species to be a threat to human dignity and the human community. Later that year, the Citizens Health Assembly, representing hundreds of international health, development, and indigenous rights organizations, began a major campaign opposing the global biotechnology industry's drive to have human genomics declared the lead technology for addressing public health problems in poor countries.

The first credible reports of covert attempts to create clonal and genetically modified children appeared in early 2012. The efforts were taking place on a fleet of converted naval hospital ships sailing the South Pacific and guarded by gunboats. The identities of the scientists involved were unclear. Responsible political and scientific leaders realized that a strong response was in order. In late 2012, a group of internationally recognized scientists and health policy experts declared that the new human biotechnologies "carry with them both great promise and great risk," and that scientists must be willing to work within socially determined limits. The declaration received extensive press coverage and commentary.

In 2013 a bipartisan group of U.S. senators began meeting to broker a broadly acceptable, comprehensive package of human biotech regulations. All involved agreed to take the issues of abortion and the moral status of human embryos off the table, and to focus on policies on which it appeared that consensus might be reached. As it turned out, this was easier than had been anticipated. Embryonic stem cell research was allowed, but "designer baby" applications and human cloning were banned, and a new federal commission was established to oversee human biotech research. In 2015 the final bill was signed into law.

The following year, international civil society leaders prevailed upon the United Nations to convene the Extraordinary Summit on Bioscience and the Human Future. Delegates included noted scientists, political leaders, scholars, and representatives of the full spectrum of social and religious constituencies. Negotiations were contentious and frequently threatened to break down. But the delegates realized that this might be the last chance humanity would have to agree upon a common framework for regulating these powerful technologies, and by 2018 success was in sight. In 2019 the UN General Assembly approved the Universal Convention on Biomedicine and Human Rights by a nearly unanimous vote. In 2020 the conventions went into force after having been approved by the parliaments of 110 countries. All involved recognized that they had participated in an undertaking of world-historical import. Just last month, the 2021 Nobel Prizes for Medicine and Peace were jointly awarded to the lead institutions that had made this all possible: the United Nations, the World Assembly of Science, the Global Council of Religions, and the nongovernmental organization (NGO) Network for a Human Future.

Reflection

It is useful to consider ways in which these scenarios might vary. "Libertarian Transhumanism" could prevail without ever being recognized as an explicit ideology. For many people, science, technology, and individual rights are nonideological, self-evident goods, and a de facto libertarian transhumanism could come to pass as the simple extension and deepening of these.

Some transhumanists claim to be motivated by social democratic rather than libertarian values, and suggest that we use genetic modification to bring everyone up to at least the current mean in health, intelligence, and life expectancy, after which all humanity would begin its posthuman journey in unison. But even if such a scenario made sense scientifically, it is imaginable only under the most absurdly authoritarian conditions.

"One Family, One Future" blends elements of the environmental and New Age movements of the 1970s, the multicultural sensibilities of the 1980s, and the social conservatism of the 1990s. It adds a jarring note of patriarchy as well. This scenario is premised on the idea that libertarianism simply doesn't cut it as a mass political philosophy. Much of our experience of purpose and self-worth comes from making and honoring enduring social commitments. If the pendulum swings too far toward libertarianism, people will welcome a communitarian adjustment. The challenge then becomes one of preventing this adjustment from going too far.

"A Techno-Eugenic Arms Race" is one version of a nightmare scenario. The 1995 Aum Shinrikyo nerve gas attacks in Japan and the 2001 anthrax attacks in the United States show the attraction that biological agents hold for fanatics. In 2000, concern about massively lethal applications motivated computer scientist Bill Joy to call for a permanent halt to particular avenues of genetic research. In 2003 the Sunshine Project documented nearly a dozen possible uses of genetic science for biowarfare purposes, including the creation of ethnicity-specific pathogens. In one of his final addresses as UN Secretary General, Kofi Annan urgently called for new international treaties to guard against biological terrorism, and specifically mentioned the dangers posed by new technologies of genetic manipulation.

After the horrific experience of the twentieth century with eugenics and genocide, could any country call for creation of a genetically "superior" population without immediate and massive international censure? One would hope not. But for the past decade, reputable scientists, bioethicists, and others have been actively promoting a revival of eugenic sensibilities and practices, and have received plaudits rather than protests from their peers and the press. In a world that is far from overcoming its propensity for racism, xenophobia, and warfare, this is more than worrisome.

"For the Common Good" incorporates many of my own values and hopes, and is presented as a more-or-less straightforward success story. But the road to any agreements of the sort sketched here will surely be filled with bumps and detours. Conflicts over the new human biotechnologies, like most other conflicts, involve the eternal tension between competition and cooperation among individuals, families, communities, and nations. New technologies developed over the past century have enabled individuals and groups to compete in ways that could endanger humanity as a whole. It is widely acknowledged that humanity needs to develop shared values and institutions that will allow such universal threats to be avoided. Attempts to do this, from the United Nations to bans on nuclear weapons to the Kyoto Accords, have had mixed success.

The noted writer Bill McKibben once said, correctly, that the greatest macroscale environmental challenge is global warming and the greatest microscale environmental challenge is genetic engineering. Technologies that enable humanity to manipulate individual atoms, molecules, genes, and cells are being used to radically transform the fundamental processes of the natural world, including many of those that define what it means to be human.

It is imperative that individuals and organizations committed to a sustainable, just, and truly human future take steps to bring these technologies under effective national and international oversight and control. To do this we need to think of the issues before us and the frameworks through which we interpret them in new ways. There is no greater challenge, and time is short.

THE LANGUAGE OF THE NEW BIOPOLITICS

Twenty years ago, those committed to a just, sustainable world needed to learn a host of new scientific terms, such as *chlorofluorocarbon* and *ozone layer.* Now we need to learn some more new terms.

Cloning: The process of creating a living organism, embryo, or cell that has the same genetic composition as an existing or previously existing individual.

Cryonics: The practice of freezing a body, or just the head, in the hope that biotechnology will allow a dead person to be reanimated sometime in the future.

Eugenics: The attempt to improve the human species by controlled selective breeding, historically, by encouraging the "fittest" to have more children and sterilizing or killing those considered genetically unfit. New genetic technologies could have eugenic applications.

Gene doping: The proposed use of genetic techniques to improve athletic performance.

Gene therapy: In theory, the treatment of a disease by introducing a corrective gene.

Germline modification: An alteration to genes in eggs, sperm, or early embryos that is passed on to succeeding generations; also called inheritable genetic modification.

Nanotechnology: The manipulation of extremely small objects, usually at the atomic or molecular level.

Ovarian hyperstimulation syndrome: An occasionally fatal side effect of hormones women take as part of fertility treatments, or when supplying eggs for research, so that they produce multiple eggs simultaneously.

Sex selection: The deliberate choice of male or female offspring, sometimes performed by the genetic screening of embryos or sperm.

Stem cells: Cells capable of developing into a variety of specialized cells and tissues. Stem cells may have therapeutic value, but could also be used for nonmedical "enhancement" procedures.

Synthetic biology: The construction of novel biological parts or organisms: at its extreme, building from scratch rather than by modifying existing organisms.

Questions

1. How do you feel about people having the choice of "designing children?" What will be the ethical and social implications for society?
2. Expound on the author's statement: "An individual may benefit by polluting a river or the atmosphere, but if everyone seeks to benefit in this manner, everyone suffers," and provide some relevant examples.
3. What does the author mean by "the conflict between libertarians and communitarians"? Explain the political ideologies of these groups.
4. Summarize the author's perspective in the first scenario, "Libertarian Transhumanism Triumphs."
5. What would be the impact of "a designer baby" on society? Would this help narrow or widen the gap between rich and poor?
6. Do you agree that advances in biotechnology and nanotechnology will give rise to human genetic castes? What are the ethical and social implications?
7. Summarize the author's perspective in the second scenario "One Family, One Future." Do you envision that a time will come when there will be a backlash against new human genetic technologies, modernity, and postmodernity?
8. Summarize the author's perspective in the third scenario, "A Techno-Eugenic Arms Race." Do you agree with author's assumption that the use of powerful genetic technologies would encourage individuals, corporations, and countries to strive for superiority and dominance? Discuss ethical and social implications at personal, community, national, and international levels.
9. Summarize the author's perspective in the fourth scenario, "For the Common Good." Expound on the author's statement: "Common conflicts over the new human biotechnologies, like most other conflicts, involve the eternal tension between competition and cooperation among individuals, families, communities, and nations." What do you envision? Will there be more cooperation or competition between all stakeholders?
10. Summarize your learning of new ideas and information presented in this article by completing the following flowchart:

Flow Chart

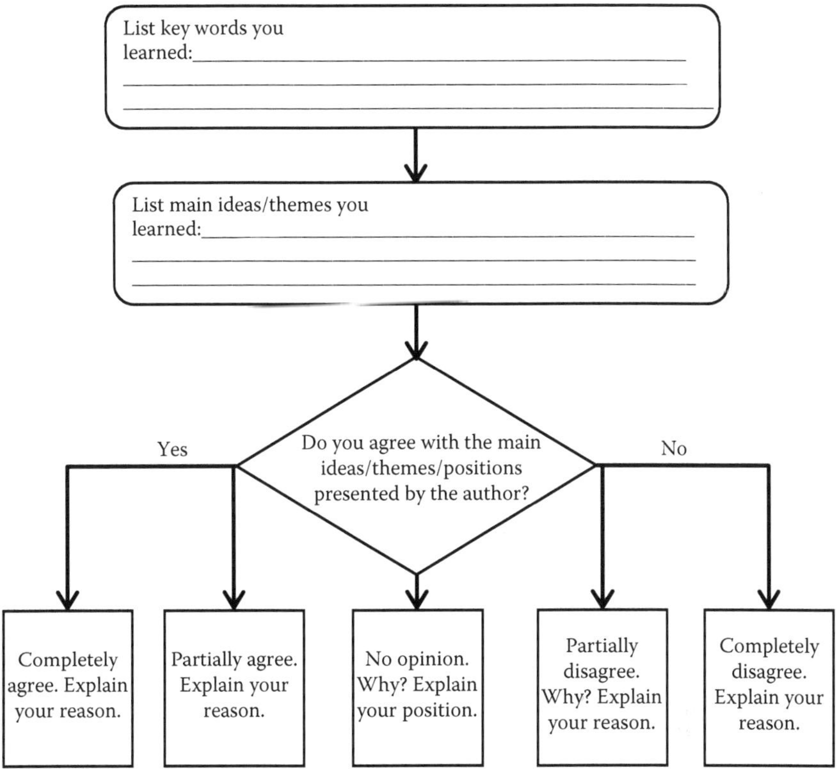
List key words you learned:
List main ideas/themes you learned:
Do you agree with the main ideas/themes/positions presented by the author?
Yes
No
Completely agree. Explain your reason.
Partially agree. Explain your reason.
No opinion. Why? Explain your position.
Partially disagree. Why? Explain your reason.
Completely disagree. Explain your reason.

10

Nanotechnology in Food and Agriculture

Ahmed S. Khan

CONTENTS

KEY WORDS: population, nanoparticles, carbon nanotubes (CNT), engineered nanomaterial (ENM), nanocarrier, nanosensor, packaging, encapsulation, risks, bioavailability, antimicrobial, toxicity

Introduction

As the second decade of the twenty-first century begins, the global population has reached 7 billion. By 2050 it is projected to reach 10 billion. (See Figures 10.1, 10.2, and 10.3). The scientific community is trying to cope with the question: How to feed the growing human population? To answer this

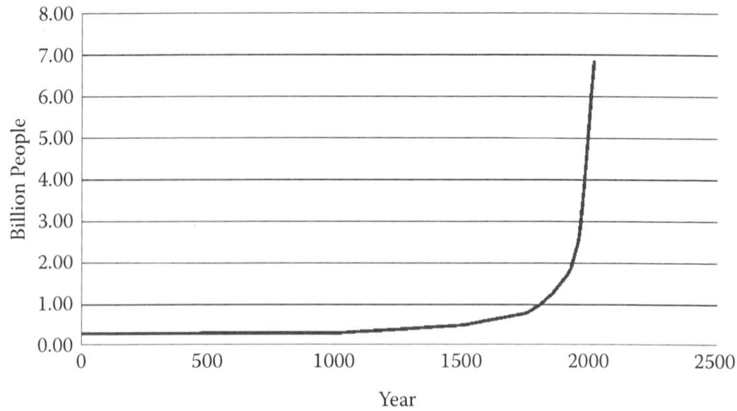

FIGURE 10.1
World population since AD 1.

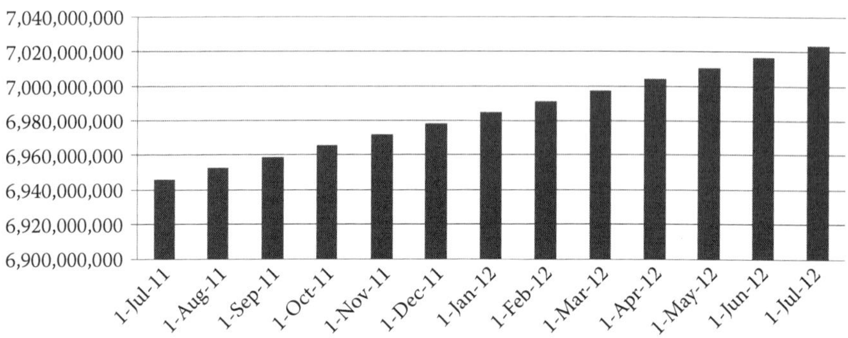

FIGURE 10.2
World population projection.

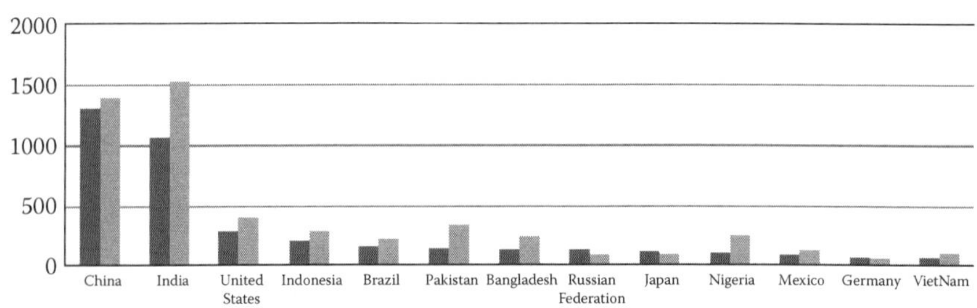

FIGURE 10.3
World's most populous countries, 2003 vs. 2050 projections.

question, the experts are divided into two camps: environmentalists (pessimists) and economists (optimists). The environmentalists believe that the Earth has limited resources, and so our national resources and environment will collapse from the increase in the demand for food. Environmentalists Paul Ehrlich and Anne Ehrlich (1990) presented this thesis in their book, *The Population Explosion*: "Human numbers are on a collision course with massive famines.... If humanity fails to act, nature will end the population explosion for us—in very unpleasant ways—well before 10 billion is reached." The optimists, on the other hand, believe that innovative use of technology coupled with continuous investment in human capital can achieve high levels of food production for supporting the projected population of 10 billion (Bongaarts, 1994).

Today, grain supplies approximately 70% of human calories. In 2009, global grain (wheat, rice, and maize) production reached 2.25 billion tons. This was a decline from the all-time high reached in 2008 and a 160% increase from when record keeping began in 1961 (Jasperson, 2011). Despite advances in the use of technology in agriculture, many countries in Africa, such as Ethiopia and Somalia, continue to experience cycles of famine. The global chronic hunger level has risen above 1 billion. In 2009, an estimated 1.02 billion people were classified as undernourished, 12% more than in 2008. This means that one in six persons on Earth suffers from undernourishment or chronic hunger (an intake of less than 1,800 kilocalories [kcal] per day). In contrast, people in North America and Europe consume on average more than 3,400 kcal per day (Pappas, 2011).

Today, in addition to the population challenge, the world food supply faces a number of other challenges from the use of biofuels, wastage and spoilage, corporate farming, vanishing small farmers, and monopolization of global food chains by large multinational corporations. The five major challenges the world faces today are (Boom, 2011, pp. 40–41):

1. *Food scarcity*: Food is becoming scarcer because the global population is increasing.
2. *Asian economic boom*: Asian countries are experiencing economic growth and gaining affluence; a more affluent population consumes more food.
3. *Scarcity of fossil fuels*: More agricultural land will be used to grow crops for the production of biofuels, and the same land will no longer be available for food production.
4. *Erosion and degradation of agricultural land*: Erosion and degradation of agricultural land due to intensive usage will in the long run decrease the extent of agricultural lands.
5. *Food as a political factor*: Food has emerged as a political factor. The global free market lures many countries to switch from traditional crops to high-value crops.

Ehrlich and Ehrlich (1990), in their book *The Population Explosion*, observed:

> Under extremely optimistic assumptions, a city of a million people would require almost 400 square miles of land in crops to "capture" enough sunlight to feed its citizens a basic, no-frills vegetarian diet. With realistic accounting for...pest losses, wastage between harvest and eating, the feeding of some crops to animals...that number could easily be multiplied five- or tenfold. The city therefore depends on thousands of square miles of farmland just to provide its food. (pp. 88–89)

The production of meat is extremely costly in terms of the usage of land, crops, water, and production of greenhouse gases by cattle. In the United States, growing feed for cattle consumes about 50% of all water and 80% of agricultural land; and cattle raised for food consume 90% of the soy, 80% of the corn, and 70% of the cereals. If all grain currently fed to livestock were fed directly to people, it would feed 800 million people (Boom, 2011, p. 40). However, experts think that in order for developed and developing countries to fulfill the aim of self-sustainability and to enhance food production, the use of nanotechnology at many levels, ranging from nanostructured fertilizers to autonomous nonsensor networks, will be vital over the next twenty years (O'Brien & Cummins, 2011, p. 103).

Applications of Nanotechnology

All stakeholders—academia, industry, and governments—across the globe have identified the potential applications of nanotechnology in food and agriculture, and they are presently in the process of making significant investments to boost food production (FAO/WHO, 2009). In some countries, nanotechnology-based food and health-food products have been marketed, and nanotechnology applications in food packaging are already available to consumers, whereas in other countries potential products and applications are in the research-and-development phase. In the near future, nanotechnology-derived food products will be readily available to consumers worldwide.

Nanotechnology offers unique prospects for the development of innovative products in the areas of agriculture; water treatment; food production, processing, preservation, and encapsulation and delivery of ingredients; sensing of pathogens, chemicals, or other substances; improved packaging; and traceability. The market for nanofoods is growing (see Figure 10.4). However, the transition from research lab to marketplace requires collaboration between academia, research centers, industry, and venture capitalists (Bugusu, Lay Ma, & Floros, 2011, p. 166). Table 10.1 lists applications of nanoparticles.

The application of nanotechnology in food production offers various ways to produce high-quality foods in a much more sustainable way. Experts

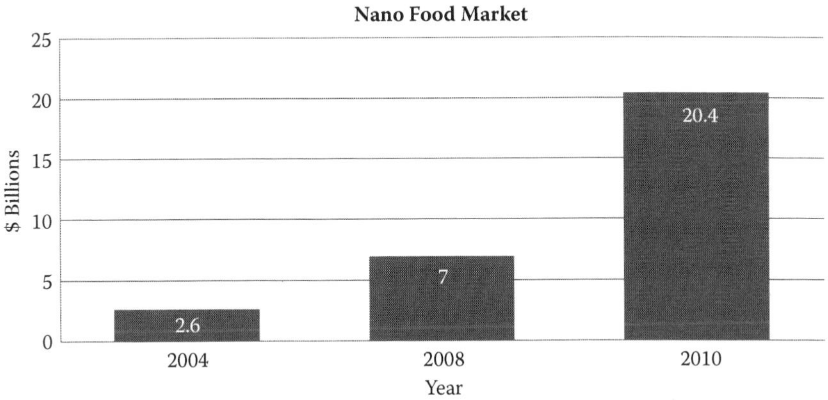

FIGURE 10.4
Nanofood market growth, 2004–2010.

believe the food supply is much safer today compared to 50–100 years ago. Moreover, the food processing industry continues to devise innovative solutions to key issues: How to achieve efficient fractionation of crops? How to do efficient product structuring? How to optimize a product's nutritional value?

The main categories of known and potential applications of nanomaterials for foods and health foods (Chaudhry et al., 2008) can be described as areas where:

1. Food ingredients have been processed or formulated to form nanostructures.
2. Nano-sized or nano-encapsulated additives have been used in food.
3. Engineered Nano Materials (ENMs) have been added into coatings and packaging materials to develop innovative food-contact surfaces and materials, and nano-biosensors for "smart" packaging.
4. Nanomaterials have been used in nanofiltration for the removal of undesirable components from food.
5. ENMs have been suggested for pesticides, veterinary medicines, and other agrochemicals for improved production systems.

Examples of nanoadditives include minerals, antimicrobials, vitamins, and antioxidants. Virtually all such additives and supplements claim improved absorption and bioavailability in the body compared to their larger equivalents. Technological processes involved in nano-sized materials involve fabrication of nano-sized particles or nano-encapsulating them in the form of micelles, liposomes, or biopolymer-based carrier systems (FAO/WHO, 2009).

Nanotechnology can be used to supply high-quality food for an increasing global population through the use of a number of innovative devices and processes.

TABLE 10.1

Applications of Nanoparticles

Nanomaterial	Study/Application Area
Gold (Au)	Virus DNA detection, sample labeling
Silver (Ag)	Virus DNA detection, sample labeling Antimicrobial, deodorant, health supplement in food and health food, food packaging products
Silicon dioxide (SiO_2)	Nanowire transistor, virus DNA detection, sample labeling Porous nanosilica used for removal of undesired components in food and beverages Amorphous nanosilica used for food contact and packaging
Fe_2O_3	Remediation of PCB-contaminated soil (contaminant sequestering and destruction)
Nanotitanium dioxide (TiO_2)	Packaging coating that provides a barrier to moisture and oxygen and improves the shelf life and flavor impact of food
Nanoselenium (Se)	Additive to food products such as green tea to improve uptake of selenium
Microfluids	DNA fingerprinting
$CuCO_3$	Wood preservation
Nanocalcium (Ca)	Salts used in health supplements and chewing gums
Nano-iron (Fe)	Decontamination of water and health supplement
Polyacrylamide	Optochemical nanosensor for detection of chemical/virus-specific analyses
Elastin-like polymers	Drug/gene delivery, lab-on-a-chip
Carbon (C)	Nanotube resonators for RF/wireless sensors
Lipids/natural polymers	Veterinary medicines and pesticides (nano-emulsions and triggered-release nano-encapsulates)
Organic nanomaterials	Encapsulated delivery systems for food and feed products (vitamins, antioxidants, colors, flavors, and preservatives)

Source: From "Nano-Functionalized Techniques in Crop and Livestock Production: Improving Food Productivity, Traceability, and Safety," by N. O'Brien and E. Cummins, in *Nanotechnology in the Agri-Food Sector: Implications for the Future* (L. Frewer, W. Norde, A. Fischer, and F. Kampers, Eds., pp. 94–95), 2011, Weinheim, Germany: Wiley-VCH Verlag.
From *Nanotechnology in Food Production* (p. 39), by R. M. Boom, 2011, Weinheim, Germany: Wiley-VCH Verlag.
From *FAO/WHO Expert Meeting on the Application of Nanotechnologies in the Food and Agriculture Sectors: Potential Food Safety Implications* (p. 26). Retrieved on November 20, 2011, from http://www.who.int/foodsafety/fs_management/meetings/nano_june09/en/

Nanosensor

Nanomaterials have a large surface area and surface reactivity, which allow their application in biosensor labeling, as a sensor substrate, and in electrode enhancement. Nanotechnology could be used to detect specific molecules or bacteria that can lead to food spoilage. Detection and elimination of such spoilage molecules could extend the shelf life of food products. The use of "lab-on-a chip" sensor technology coupled with RFID (radio-frequency identification) tagging and data communications will allow the development of nanosensor-based networks. Such networks will lead to improved data collection and communication for agricultural practices for the efficient use of nano-functionalized fertilizers, pesticides, and herbicides. Nanosensor networks will also improve the traceability of food systems and human-health-hazard warning methods (O'Brien & Cummins, 2011, p. 103).

Packaging

Presently, most applications of nanotechnology in the food and agriculture sectors are in R&D or near-marketing stages. Nanotechnology offers many advantages in packaging food. Contamination and spoilage of food pose a threat to consumer health. Drying or wetting of food can change the properties and hence the quality of food. Efficient food packaging maintains food quality by preventing this contamination and spoilage The physical properties of packaging materials determine the strength, barrier properties, light absorption, interior surfaces, and antimicrobial functionality, and thus safeguard the quality of food. An active package with a built-in nanosensor can be used to gauge the state of freshness and quality. Also, antimicrobial nanoparticles of silver added to the packaging material can increase the shelf life of food products (Kampers, 2011, p. 165). Engineered nanomaterials (ENMs) can be used for packaging of food products. In addition to making the packaging lightweight and strong, ENMs offer a number of other advantages (Chaudhry, Castle, & Watkins, 2011, p. 166):

1. Improved gas barrier properties and temperature and moisture stability
2. Enhanced antimicrobial- or oxygen-scavenging characteristics
3. Augmented monitoring of the quality of food during transportation and storage

Encapsulation

Nanotechnology can be used to deliver active ingredients into a targeted area in the human body without dispersing into other areas. For example, oral vaccines could be developed to reduce the cost of health care (Boom, 2011, p. 47). Nano-encapsulation offers a number of benefits (FAO/WHO, 2009, p. 24), such as:

1. Preserving the ingredients and additives during processing and storage
2. Masking unpleasant tastes and flavors in the food
3. Controlling the release of additives, thereby improving the uptake of nutrients and supplements in the human body

Risks of Nanotechnology

Nanomaterials will be used in food processing as additives to modify food's texture and taste, in food packaging to prevent food spoilage, in nutrients (vitamins) to increase bioavailability, and in agrochemicals to provide novel routes to deliver pesticides to plants. Examples of nanostructured and nano-textured foods include spreads, mayonnaise, cream, yogurt, ice cream, etc. The use of nanostructures will improve texture and give new taste, better consistency, and greater stability in contrast to conventional processed foods. Innovative new low-fat nanostructures are as "creamy" as the full-fat alternative, and thus will offer healthy options to consumers (FAO/WHO, 2009, p. 24).

The application of nanotechnology in food processing offers numerous advantages, but the possibility of a number of potential unintended and unseen risks is present as well. There are differences between the characteristics of a nano-sized material and its bulk counterpart. Due to a combination of their properties, such as size, physicochemical properties, chemical compositions, and surface structure, unexpected toxicological effects might occur (Bernadene, Bouwmeester, & Bouwmeester, 2011, p. 183). The introduction of nanomaterials-based consumer products into the marketplace in various industrial sectors increases the urgency for better understanding of the potential adverse impacts that nanomaterials may have on biological systems.

The impact on human health depends on the mode of exposure and how these nanomaterials behave inside the human body. The impact of nanoparticles will also depend on their sources: Are nanoparticles from nonbiological sources or from biological sources? Researchers have to answer a key question: How does human physiology respond to nanoparticles of nonbiological origin, such as buckyballs or carbon nanotubes? On the other hand, nonstructured components of a biological nature such as fibrillated proteins may be safer because these structures already occur in nature and the human body has perhaps evolved in their presence (Boom, 2011, p. 55). Questions researchers ought to consider: What if lipid-based encapsulates used for delivering drugs through the intestinal wall get contaminated by viruses or other harmful materials? What will be the effect on the human body?

Nanodelivery systems used for targeted drug delivery offer many advantages to users, from increased absorption to the improved bioavailability of nutrients such as vitamins. But they also have the possibility of changing the distribution of substances in the human body. And if a nanocarrier breaks in the gastrointestinal tract (GI), the encapsulated nanomaterial may not be different from its bulk counterpart, and if the nanocarrier delivers the substance to the blood stream, then its ADME (absorption, distribution, metabolism, and excretion) may exhibit different properties than its bulk counterparts (FAO/WHO, 2009, p. 25–30).

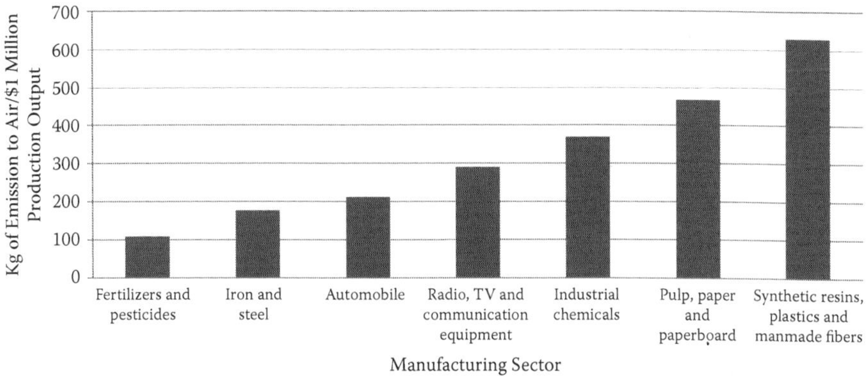

FIGURE 10.5
Toxic intensities of selected U.S. manufacturing sectors, early 1990s.

Regarding the use of nanomaterials in food packaging, there is a concern that the contact between the nanotechnology applied to the packaging (e.g., nanoparticles of silver) and foods may enable nanoparticles to escape into the food, leading to the contamination. And there are environmental concerns, too (Kampers, 2011, p. 71). When the food is consumed, the packaging materials end up in the environment. What will be the impact of nanoparticles on the environment? Will nanoparticles contaminate the environment? What will be the toxic intensity contribution by the nanotechnology sector on the existing toxic intensities generated by the manufacturing sector (see Figure 10.5) in the United States? How much toxic intensity will be contributed by the nanotechnology to that existing today? At present, the answers to these question are not very well defined; they need to be researched further.

Major Concerns

Unlike in the developing countries, many developed countries have a regulatory infrastructure for premarket evaluation of food products. This regulatory framework deals with general food safety vis-à-vis food additives, chemical safety, food contact materials, water quality, amount of chemicals, and protection against biocides, pesticides, and veterinary medicines. However, the application of nanoscience and nanotechnology in the food and agriculture industries is in its infancy, and there is a lack of standardization in the industry. Researchers need to study the effects on humans and the environment to ensure that these benefits do not pose any inherent risks and threats to consumers.

Presently, most of the nanomaterial-based food-processing development is taking place in a regulatory vacuum. There is a lack of global standards and regulations for the use of nanomaterials in food processing and packaging. As multinational corporations work on a global level to cater to the needs of the many developing countries that do not have an effective food-regulation framework, it is essential that global standards and regulations be developed to ensure consumer safety. Stakeholders have recognized the limited size of the knowledge base regarding the effect of nanotechnology on human health. To investigate food safety, experts from FAO (Food and Agriculture Organization) and WHO (World Health Organization) met in June 2009 to address the "application of nanotechnologies in the food and agriculture sectors: potential food safety implications" on a global level.

At this meeting, the experts recognized that present risk assessment methods employed by FAO/WHO and the Codex Alimentarius Commission are suitable for ENMs used in food and agriculture, but the group of experts stressed the need to address the safety concerns and issues that might arise due to unique properties of ENMs. They recommended that FAO/WHO should encourage innovative and interdisciplinary research that can lead to novel risk assessment strategies for the application of nanotechnologies in food and feed while maintaining the current level of protection. The experts also realized that the consumer attitudes toward the application of nanotechnology in food and agriculture are complex; consumers want to understand the potential benefits and risks. Without seeing any obvious advantages, the consumer will not develop favorable impressions of nanotechnology-based food products. So there is a need to promote stakeholder confidence and dialogue. The experts stressed the importance of the need for debate to define the necessity of policy intervention on the use of ENMs and ENPs (engineered nanoparticles) into commercial products as long as potential safety risks cannot be appropriately measured and evaluated. The experts further recommended that the member countries of FAO/WHO should engage the public on the application of nanoscience and nanotechnologies in food and agriculture. Furthermore, FAO/WHO should provide guidance, training, and capacity-building resources for governments to engage all stakeholders (FAO/WHO, 2009, pp. 16–17).

Conclusion

Nanotechnology and nanoscience offer an array of opportunities for the food and agriculture sectors. The use of nanomaterials has a number of benefits for consumers as well for the industries. Many of the potential applications of nanomaterials are in the nascent stages of development, but in the near future, nanotechnology-derived food products should be readily available to consumers all over the globe. The use of nanotechnology in food and agriculture also has a number of potential risks. Given the present lack of knowledge about nanoderived materials and their potential effects on human health and the environment—coupled with the lack of global standards and regulation—it is essential to educate the public in regard to the application of nanotechnology in food and agriculture (FAO/WHO, 2009, p. 28).

Questions

1. Use Internet resources to define the following terms:
   a. ENM
   b. Encapsulation
   c. Nanosensor
   d. RFID
   e. EFSA
   f. FDA
   g. WHO
   h. Nanocarrier
   i. Nanocomposite
   j. Nanomaterial
   k. Nanoparticle
   l. Nanotube
   m. QD
2. List advantages of the application of nanotechnology in food processing by completing the following table:

Application	Advantages (+)	Disadvantages/Risks (–)
Nanoparticles		
Nanosensor		
Encapsulation		
Packaging		

3. Describe the risks of nanotechnology applications in the food and agriculture industries.
4. Formulate a policy to develop a labeling system that informs consumers of the presence of nanomaterials in food products.
5. Examine the effects of nanocarriers getting into the blood.
6. Given I = PAT equation (I = Impact, P = population, A = affluence, T = technology), discuss what will happen to I with advances in nanotechnology.

7. How could nanotechnology be used to promote sustainable economic development around the world?
8. Use Internet search engines to find FDA and FAO/WHO reports, and compare the state and status of food and environmental regulations for the use of nanomaterials in food processing in the United States and the world.

	FDA	FAO/WHO
Examples of nanotechnology standards and regulations for food products		
Examples of nanotechnology standards and regulations for agriculture products		
Measures to educate public about potential risk of nanomaterials in food products		
Labeling requirements for food containing nanomaterials		

9. Evaluate the potential effects of nanocarriers getting into the blood.
10. Use Internet resources to gather data to discuss the status and state of nanotechnology in the following countries:

	State of Nanotechnology R&D (Basic, Moderate or Advanced)	R&D Spending ($)
United States		
Canada		
United Kingdom		
Germany		
France		
China		
India		
Pakistan		
Malaysia		
Turkey		

11. How will the application of nanotechnology in food and agriculture shape the future?
12. What will be the impact of today's nanopollution on future generations?
13. How will the use of nanotechnology and RFID-tag-based sensor networks affect individual privacy?
14. What will be the impact of exposure of human-made nanoparticles and nanotubes on cattle?
15. When using nanotechnology in food and agriculture, what positive and negative consequences must be addressed?
16. What will be the effect of incorporation of nanostructure on religious dietary restrictions of Kosher and Halal foods?
17. Summarize your learning of new ideas and information presented in this article by completing the following flowchart:

Flow Chart

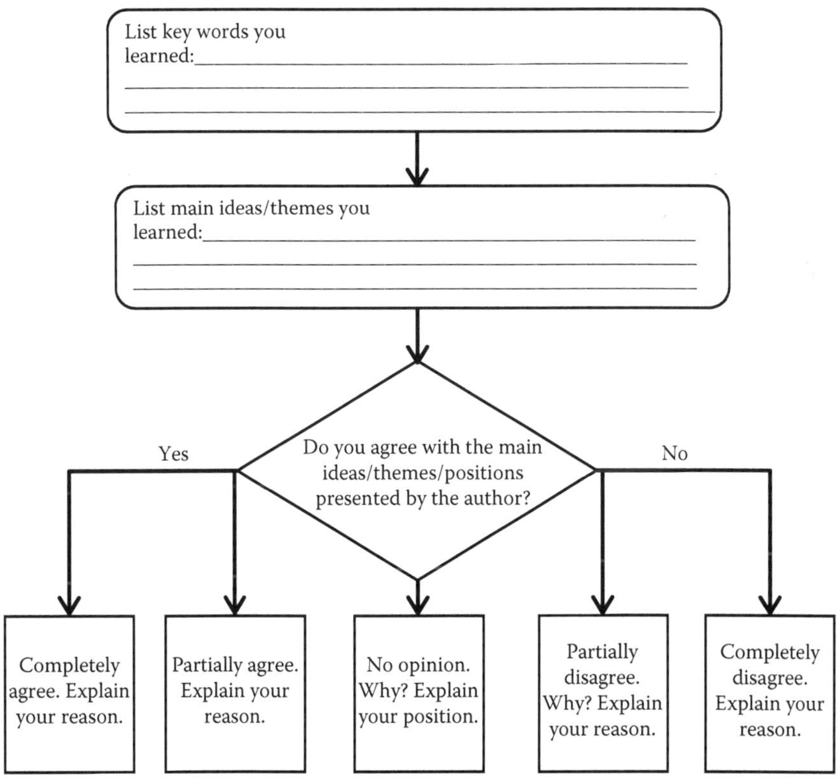

Scenarios

Scenario 1

Julie is a busy executive. She works at an IT (information technology) company 9 to 5 every day, and even when she gets home, she works online solving customer problems in Asia and Europe. She does not have much time to do her weekly errands. She tries to buy her grocery items every three weeks, and she prefers to buy items that have a longer shelf life. Very health conscious, Julie likes low-fat yogurt and ice cream. Recently during one of her trips to the grocery store, she tasted Nanoyugo, a new brand of yogurt and ice cream. She noticed that the Nanoyugo yogurt and ice-cream products were low in fat and that they tasted much better than the yogurt and ice-cream products she currently uses. She also noticed that the Nanoyugo products have a shelf life of three months. The Nanoyugo brand of yogurt, ice cream, and other dairy products offered her a win-win situation: improved taste and longer shelf life. Now she thinks that she can cut down on her trips to the grocery store. She has decided to buy a three-month supply of Nanoyugo yogurt, ice cream, and other food products. Before buying the products, she read the labels and did not notice any change in the typical list of ingredients.

A year after Julie switched to the Nanoyugo brand of products, she discovered that the flakes that peeled off from the lining of the ice cream carton got mixed with the ice cream. This had never happened before with the cartons of other brands of yogurt and ice cream products she used earlier. She became quizzical and did some research and found out that the products she had been using contained nanoparticles to improve taste and texture, and the packaging also contained engineered nanomaterials. For the past three months she has been experiencing unusual body aches. She thinks that nanoparticles in the yogurt and ice cream are the culprits.

Questions

1. What should Julie do regarding her body aches?
2. Is she correct in assuming that her body aches are due to nanoparticles contained in the yogurt and ice cream?
3. Do you think that she should have done some research about the Nanoyugo products *before* switching to them?

Scenario 2

Ehrlich and Ehrlich (1990), in their book *The Population Explosion*, have proposed that a global system should be formed to deal effectively with the growing super problems of overpopulation, environmental degradation, resource allocation, and economic stagnation that loom in humanity's future:

> It may not be possible—or desirable—to create a strong centralized world government patterned after national governments, but a great deal of scope exists to build on the present model of diffuse, semi-independent agencies with different (though often overlapping) responsibilities, which tend to solve problems through consensus and international agreement. A global commons regime, using the law of the sea as a model, might be established to regulate human interactions with the atmosphere. Such a regime would, of course, have to be given some authority over such diverse activities as agriculture, the cleaning of tropical forests, the generation of power, the use of internal-combustion engines, the manufacture of chemical, plastics, and hair sprays, and the disposal of toxic wastes

Questions

1. What is your take on their proposal?
2. Is this idea viable? What are the pros and cons of this idea of global governance?
3. What is the probability that this global common regime could pull humanity through coming crises?
4. Do you think that such a system would promote a balance between population growth and food supply?
5. Do you think a global agency could be formed to regulate the use of nanotechnology in the food and agriculture sector?

Case Studies*

Case Study 1: Zinc Oxide as an Antimicrobial in Food-Contact Material (Hypothetical)

The antimicrobial properties of certain metals and metal oxides (e.g., Ag and ZnO) are well known in the literature. The ability of nanotechnology to produce engineered nanoscale metal and metal oxide particles, thereby increasing surface area and potential activity, has increased interest in the use of metals and metal oxides as antimicrobials in food-processing and packaging applications. Applications for such materials include antimicrobials to protect the integrity of packaging or food-contact materials or to act as antimicrobials on the food-contact surfaces of packaging or food-processing equipment.

* Case Studies 1 and 2 are reprinted with permission of FAO/WHO.

Some such materials are already commonly found in or added to food or food-contact material in their macroscale form. For example, ZnO is used as a nutrient supplement in food and as a colorant and filler in food-contact material. The safety of the use of ZnO on a macroscale version is well established.

Exposure to components of food-contact materials is typically estimated using experimental migration data or migration modeling or by assuming complete migration to an estimated quantity of food. For applications in materials at the surface interface with food, migration modeling will not typically work well. Moreover, existing migration models will typically incorporate few data on nanoparticles, necessitating that care be applied in their application to the evaluation of nanoscale migrants.

Because of the challenges in analyzing migrants in food matrices, migration experiments are typically performed using food simulants. Reactivity of the migrant with the simulant is always a concern, but may be more so with nanoscale materials engineered to have a greater surface area and a higher activity. Therefore, in order to address exposure, it may be necessary to determine the form of the nanomaterial entering food. Alternatively, assumptions may be made that any ZnO detected migrating from the food-contact material is in its most toxic form. The latter assumption may require additional safety-testing data on nanomaterial forms to address a higher estimated exposure.

In cases where a more precise exposure profile may be desired to reduce the need for toxicity testing, the identity and physical characteristics of migrants would need to be further analyzed and established. For example, it would be necessary to demonstrate whether the ZnO particle migrates into food and remains suspended as a nanoscale particle, or whether the nanoscale substances are ionized, aggregated, or agglomerated in the process of migration or in food. Such information will also be used to determine the applicability of any toxicity data on the macroscale version of the material.

Whether the tested substance is the macroscale version or the nanoscale version of ZnO, the relevance of any planned or existing toxicity test must rely on the relationship of the substance tested to the substance to which consumers are exposed. Toxicity testing recommendations for components of food-contact materials are typically tier-based judgments related to migration levels or likely consumer exposure. The expectation of greater bioavailability for a nanoscale ZnO will, at a minimum, suggest more intensive testing at a lower level of mass–mass exposure. Alternatively, more sophisticated analysis of exposure or internal dose may be necessary to relate toxicity data on different versions of the substance or to relate tests using different routes of exposure. In addition to relatively straightforward questions regarding internal dose, assessment of nanomaterials needs to consider alternative mechanisms or sites of toxicity resulting from new properties that alter biological transport. It should be possible to address such questions through bridging studies such as ADME studies or modeling. Finally, depending on the nature of the ZnO migrating into food, it will be necessary to assess the potential for antimicrobial activity of the material in the body.

Questions

1. What is the application of ZnO in food packaging?
2. Is ZnO safe to be used in food packaging? What is your opinion?
3. What do ADME studies or modeling suggest about the toxicity of ZnO?

Case Study 2: β-Cyclodextrin as a Nanocarrier

β-Cyclodextrin is a cyclic heptamer composed of seven glucose units joined head-to-tail by alpha-1,4 links. It is produced by the action of the enzyme cyclodextrin glycosyl transferase (CGT) on hydrolyzed starch syrups. CGT is obtained from *Bacillus macerans, B. circulans*, or related strains of *Bacillus.*

As a result of its cyclic structure, β-cyclodextrin has the ability to form inclusion compounds with a range of molecules, generally of molecular mass less than 250. Consequently, it is used as a carrier and stabilizer of food flavors, food colors, and some vitamins.

Available data demonstrate that β-cyclodextrin is resistant to metabolism in the upper GI tract, but biokinetic studies in rats using radiolabeled β-cyclodextrin demonstrate that the nanomaterial is readily metabolized in the large intestine to open-chain dextrins/glucose through the combined action of endogenous microflora and amylases.

Available data demonstrate that β-cyclodextrin is not absorbed to a significant extent from the stomach or small intestine of rats and that excess β-cyclodextrin is expected to be excreted in feces. The intestinal absorption, digestibility by the colonic microflora, and urinary excretion of ß-cyclodextrin have been studied. Using everted sacs of rat small intestine in vitro and ligated gut loops in vivo, absorption was shown to be slow, concentration dependent, not saturable, and not inhibited by phloretin; this indicates that a passive transport process is involved.

Rat and human cecal microflora were able to utilize β-cyclodextrin under anaerobic conditions in vitro, indicating that the compound is probably hydrolyzed to glucose by bacterial enzymes. Based on these findings, it is believed that β-cyclodextrin may be utilized but only indirectly by the activity of the gut flora. Moreover, additional high-dose (>3 g/kg bw/day) experiments in the rat demonstrate that only negligible amounts of β-cyclodextrin remain in the GI tract, and over 95% of ingested β-cyclodextrin is metabolized to glucose in the intestine. Absorption of β-cyclodextrin was also negligible in a study in beagle dogs, and it was shown to be excreted in the urine.

Expected consumer exposure to β-cyclodextrin was estimated to be between 1.0 and 1.4 mg/kg bw/day from its use as a carrier for flavors, colors, and nutrients based on intended use levels in food and data on food consumption patterns. An acceptable daily intake (ADI) of up to 5 mg/kg bw/day of β-cyclodextrin was estimated from a very large and comprehensive database of toxicity testing. Additional in vitro and in vivo testing demonstrate that β-cyclodextrin has little or no potential to affect the absorption of nutrients or to serve as a carrier for unintended substances (JECFA, 1995).

Questions

1. Do you think that β-cyclodextrin can safely be used as a nanocarrier to stabilize food flavors, food colors, and some vitamins?
2. What does current experimental data suggest about the absorption of β-cyclodextrin in the GI tract of rats?
3. What is your opinion about the use and safety of β-cyclodextrin in food products based on the present in vitro and in vivo testing data?

Case Study 3: China

In China, despite double-digit economic growth and a rising standard of living, the gap between rich and poor is widening in terms of the Gini coefficient (a statistical measure of inequality of income or wealth). According to the World Bank, in terms of per capita incomes, China barely ranks in the top 100 countries list.

Many rich Chinese are migrating to the United States, Canada, and Australia. According to a report by China Merchants Bank and U.S. consultants Bain & Co. published in April 2011, among the 20,000 Chinese with at least $15 million in individual investment assets, 27% of them have already emigrated, and 47% are considering it. The primary reason for their migration is better education for their children; the secondary reason is that many couples want a second child. People are unhappy about China's education system, where students can pass exams but fail to acquire the skills and knowledge required for the global market place. They also feel unfairly treated by the China's one-child policy.

For the last three decades, China has enforced a one-child policy to control population growth. Couples are penalized for having a second child. China's one-child policy has been successful at maintaining population numbers, but it has created a plethora of social and demographic problems. The preference for a male heir to the family name has caused an imbalance in the ratio of males to females in Chinese society. As a result, many Chinese males must look abroad to find a female to marry. Couples who want to have a second child are forced to emigrate to other countries (Watt, 2011). Many able-bodied farmers have moved to cities to work in industries, and that has created a decline in the helping hands on farms in the rural areas.

Questions

1. Discuss the pros and cons of China's one-child policy.
2. What will be the impact on China's population and demographics should it decide to move from a one-child to a two-child policy?
3. What does China need to do to increase its food supply and improve the safety of food products? How can China use nanotechnology to improve the safety of food products?
4. What is the state and status of the use of nanotechnology for food and agriculture sectors in China? (Use Internet resources to answer this question.)

References

Bernadene, A., Bouwmeester, M., & Bouwmeester, H. (2011). Toxicology of nanomaterials in food. In L. Frewer, W. Norde, A. Fischer, & F. Kampers (Eds.), *Nanotechnology in the agri-food sector: Implications for the future*. Weinheim, Germany: Wiley-VCH Verlag.

Bongaarts, J. (1994). Can the growing human population feed itself? *Scientific American, 270*(3), 36–42.

Boom, R. (2011). Nanotechnology in food production. In L. Frewer, W. Norde, A. Fischer, & F. Kampers (Eds.), *Nanotechnology in the agri-food sector: Implications for the future*. Weinheim, Germany: Wiley-VCH Verlag.

Bugusu, B., Lay Ma , U. V., & Floros, J. D. (2011). Products and their commercialization. In L. Frewer, W. Norde, A. Fischer, & F. Kampers (Eds.), *Nanotechnology in the Agri-Food Sector: Implications for the future*. Weinheim, Germany: Wiley-VCH Verlag.

Chaudhry, Q., Castle, L., & Watkins, R. (2011). Nanomaterials and food contact materials: Potential implications for consumer safety and regulator controls. In L. Frewer, W. Norde, A. Fischer, & F. Kampers (Eds.), *Nanotechnology in the agri-food sector: Implications for the future*. Weinheim, Germany: Wiley-VCH Verlag.

Chaudhry, Q., Scotter, M., Blackburn, J., Ross, B., Boxall, A., Castle, L., et al. (2008). Applications and implications of nanotechnologies for the food sector. *Food Addit. Contam., 25*(3), 241–258.

Ehrlich, P., & Ehrlich, A. (1990). *The population explosion*. New York, NY: Simon and Schuster.

FAO/WHO. (2009). *Joint FAO/WHO expert meeting on the application of nanotechnologies in the food and agriculture sectors: Potential food safety implications*. Retrieved November 15, 2011, from http://www.who.int/foodsafety/fs_management/meetings/nano_june09/en/

Jasperson, A. M. (2011). Grain production strong but fails to set record. In *Vital Signs 2011*. Washington, DC: Worldwatch Institute.

JECFA. (1995). Evaluation of certain food additives and contaminants. 44th Report of FAO/WHO Joint Expert Committee on Food Additivies. Geneva.

Kampers, F. (2011). Packaging. In L. Frewer, W. Norde, A. Fischer, & F. Kampers (Eds.), *Nanotechnology in the agri-food sector: Implications for the future*. Weinheim, Germany: Wiley-VCH Verlag.

O'Brien, N., & Cummins, E. (2011). Nano-functionalized techniques in crop and livestock production: Improving food productivity, traceability, and safety. In L. Frewer, W. Norde, A. Fischer, & F. Kampers (Eds.), *Nanotechnology in the agri-food sector: Implications for the future*. Weinheim, Germany: Wiley-VCH Verlag.

Pappas, S. (2011). Global hunger rises above 1 billion. In *Vital Signs 2011*. Washington, DC: Worldwatch Institute.

Watt, L. (2011). Top of Chinese wealthy's wish list? To leave China. Retrieved November 15, 2011, from http://my.news.yahoo.com/top-chinese-wealthys-wish-list-leave-china-065826880.html

11

The Future of Food: An Introduction to the Ethical Issues in Genetically Modified Foods

Margaret R. McLean

CONTENTS

Introduction

Let's begin with a pop quiz—True or False:

1. All plants contain genes.
2. Only genetically modified plants contain genes.
3. Plants can be modified to contain animal genes.
4. A tomato containing a jellyfish gene would taste like squid.
5. Genetically modified foods are available at Safeway.
6. I have never eaten a genetically modified food.

The answers are true, false, true, false, true, and, most likely, false. The truth is that we have been eating genetically modified (GM) foods for a decade. About 75% of processed food that is produced in the United States contains some GM ingredients. This includes crackers, breakfast cereals, and cooking oils. Almost everything that contains soy or corn—including the nearly ubiquitous high-fructose corn syrup—has been genetically modified.

Humans were modifying crops long before the advent of genetics and modern biotechnology. Once humans began to practice settled agriculture some 8,000 years ago, they selected which plants to plant, grow, and harvest—first choosing from the wild and then from cultivated crops. These first farmers chose plants that grew well *and* demonstrated resistance to disease, pests, and shifting weather patterns. Ever since, farmers have bred, crossed, and selected plant varieties that were productive and useful. These age-old techniques can now be complemented, supplemented, and perhaps supplanted by an assortment of molecular tools that allow for the deletion or insertion of a particular gene or genes to produce plants (and animals and microorganisms) with novel traits, such as resistance to briny conditions, longer shelf life, or enhanced nutrient content. A change in a plant's genetic sequence changes the characteristics of the plant. Such manipulation of genes—genetic engineering—results in a genetically modified organism (GMO).

Both traditional and modern biotechnology result in crops with combinations of genes that would not have existed absent human intervention. A drought-resistant crop can be developed through traditional methods involving crosses with resistant varieties, selection, and backcrossing. Modern biotechnology can speed up this process by identifying the particular genes associated with drought resistance and inserting them directly. Whether developed through traditional or modern means, the resultant plants will resist drought conditions, but only the second, the genetically engineered one, is a GMO or, if meant for human consumption, a GMF.

Genetic engineering has both sped up the process of developing crops with enhanced or new characteristics and allowed for the transfer of genes from one organism to another, even from great evolutionary distances, such as the insertion of a gene from an African frog into rhododendrons to confer enhanced resistance to root rot. Moving genes between species creates transgenic plants and crops.

Importantly, genetic engineering is not the whole of agricultural biotechnology, which also includes techniques such as tissue and cell culture. This chapter primarily concerns itself with a small piece of agricultural biotechnology, the genetic engineering of food crops.

The most commonly grown GM food crops are those that have been engineered to withstand herbicide spraying (e.g., Roundup Ready soybeans and corn) or to produce substances toxic to insects (e.g., Bt corn). Crops that can tolerate herbicides have been an economic success story—approximately 80% of the U.S. market in soybeans and cotton is in plants that can withstand the popular herbicide Roundup.

To date, most of the development of GM crops—dubbed "first generation crops"—has been aimed at benefiting the farmers' bottom line—increasing yields, resisting pests and disease, and decreasing the use of herbicides. Over 80% of the soybeans and 40% of the corn grown in the United States is genetically modified. Worldwide, close to a billion acres are planted in GM crops, mostly corn and soy for animal consumption.

The first GM food produced was the Flavr Savr tomato in 1994, touted for its flavor and long shelf life. Interestingly, the Flavr Savr tomato did not contain an alien gene; rather, a gene normally present in the tomato was blocked so that a normal protein involved in ripening was not produced, giving the tomato a longer shelf life and, theoretically, better flavor. It failed to attract consumers.

Despite the tomato's flop, so-called second generation crops will one day line supermarket shelves. These include products such as Monsanto's Roundup Ready soybeans with reduced trans fats and increased heart-healthy mono-unsaturated fats; Syngenta's StayRipe banana, which ripens slowly and has a prolonged shelf life; potatoes and peanuts less liable to trigger life-threatening allergic reactions; and tomatoes that help prevent cancer and osteoporosis (Stokstad, 2004; Associated Press, 2005).

Also in the pipeline are GM crops designed to produce pharmaceuticals, so-called pharma crops. Last year, the California Rice Commission advised the state Food and Agriculture Department to allow Ventria Bioscience of Sacramento to grow 50 hectares of GM rice near San Diego. Ventria planned to grow two types of rice modified with synthetic human genes: one to make human lactoferrin to treat anemia and the second to produce lysozyme to treat diarrhea (Dalton, 2004). Anemia and diarrhea plague children under 5 in developing countries. But the California Food and Agriculture Department denied Ventria's request after rice growers expressed concern that international customers would refuse their rice out of fear of contamination. In April 2005, brewer Anheuser-Busch threatened to boycott rice from Missouri if Ventria were allowed to set up its "biopharming" practices there. Again, the concern is the potential that the GM rice could cross-pollinate other crops and introduce foreign genes and proteins into the human food chain.

INB Biotechnologies (Philadelphia) is developing a nontoxic anthrax vaccine through the transgenic modification of petunias, causing the plant to manufacture new proteins, which when eaten prompt the development of anti-anthrax antibodies. So, instead of "eat your peas," the imperative will be to "eat your petunias"!

The advent of GM crops provides new opportunities for increasing agricultural production and productivity, enhancing nutritional value, developing and delivering pharmaceuticals and vaccines, and feeding the world. But, it is far from easy sailing for GM foods in light of the public concern for associated risks—risks to human and animal health; risks to biodiversity and the environment—and intermittent consumer outrage at not knowing if "the breakfast of champions" has had a genetic boost or not. GM foods are not labeled as such, and the industry game of *I've Got a Secret* has bred distrust among consumers and fuels an inherent skepticism about the safety of GM foods.

A common approach to thinking about the ethics of the genetic engineering of food crops and the appropriate regulatory environment starts by evaluating safety and weighing potential risks and benefits.

Risks

According to the Food and Agriculture Organization of the United Nations, the risk side of the ledger includes:

1. Potential risks to the environment and wildlife:
   a. Genes may escape and find their way into other members of the species or other species. Imagine the trouble if herbicide-resistant genes found their way into weeds.
   b. GM crops could compete or breed with wild species, threatening biodiversity.
   c. Monogenetic crops may not react sufficiently to environmental stresses, posing the danger of a reenactment of Ireland's potato famine.
   d. What are the risks to birds, insects, and other nontarget species that come into contact with or consume GM plants?
2. Potential risks to human health:
   a. There is the potential that allergy-producing genes will be inserted into unrelated foodstuffs. Since GM foods are not labeled, a person could suffer a potentially fatal allergic reaction. For example, an allergenic Brazil nut gene was transferred to a soybean variety, but the resultant modified crop was never released to the public.
   b. GM products may inadvertently enter the human food supply, as evidenced by the settlement earlier this month between Syngenta and the U.S. government over the accidental sale of unapproved GM (Bt10) corn seed to farmers.
3. Potential socioeconomic effects:
   a. Small-scale farmers could be negatively impacted by the market dominance of a few powerful seed companies. Some worry about the potential loss of traditional farming practices such as collecting, storing, and replanting seed.
   b. The proprietary nature of biotechnology may slow basic research, and patent protection may hinder the entry of GM foods into developing countries as has been the case with pharmaceuticals.
4. Potential risk to public trust generated in part by industry refusal to label GM foods as such.

Benefits

The benefit side of the ledger stresses:

1. There are potential benefits to agricultural productivity through the development of crops more resistant to pests, disease, and severe weather, decreasing the risk of devastating crop failure.

2. There are potential benefits to the environment, including:
    a. Improved productivity could result in more food from less land and a decreasing reliance on the cultivation of marginal land.
    b. Genetically engineered pest and disease resistance could reduce the need for pesticides and other chemicals, thereby decreasing the environmental load and farmer exposure to toxins.
    c. The potential longer shelf life of fruits and vegetables could decrease the gross waste associated with transportation and storage.
3. There are potential benefits to human health and well-being.
    a. Genetic engineering could be used to remove genes associated with allergies, for example, the blocking of the gene that produces the allergenic protein in peanuts.
    b. The insertion of genes into crops such as rice and wheat can enhance their nutritional value, for example, Golden Rice.
    c. Genetic modification could be used to produce healthier foods, for example by eliminating trans fats or caffeine.
    d. Genetic engineering could be used to develop pharmaceuticals and vaccines in plants, decreasing the risk of adverse reactions and enabling faster vaccination of large populations.

Conclusion

Although weighing risks and benefits is necessary, it is neither easy nor the sole concern in considering the ethics of agricultural biotechnology. Certainly, both human well-being and environmental safety are of primary concern, but our ethical obligations are not discharged solely by a guarantee of some degree of protection from harm, as important as that is. We also must be concerned with justice and the common good—raising concerns about human and environmental sustainability and the just distribution of nutritious food *and* acknowledging the need for thoughtful regulation that addresses necessary human and environmental protections while pursuing benefit. Such a task might well begin with a good dose of humility.

And so, we approach the "future of food" and the questions we have before us today:

- Should we have genetically modified foods?
- Since we do have genetically modified foods, how ought they be regulated?
- How do we weigh values and risk in biotechnology?
- Is the genetic modification of food necessary to relieve world hunger?

Questions

1. What are some of the benefits and risks associated with genetically modified foods as described by the Food and Agriculture Organization of the United Nation?

Benefits (+)	Risks (–)

2. Do you believe that there should be a mandatory labeling of GM foods?
3. Do you believe that GM foods are safe or unsafe?
4. Use Internet resources to complete the following table by listing five products you consume that are genetically modified. What are the benefits and what are the potential risks?

GM Food Product	Benefits (+)	Potential Risks (–)

5. In India, thousands of farmers who used GM food committed suicide. What was the cause of their extreme actions? How could this tragedy have been prevented?
6. Are GM food regulated at national and international levels?
7. Summarize your learning of new ideas and information presented in this article by completing the following flowchart:

Flow Chart

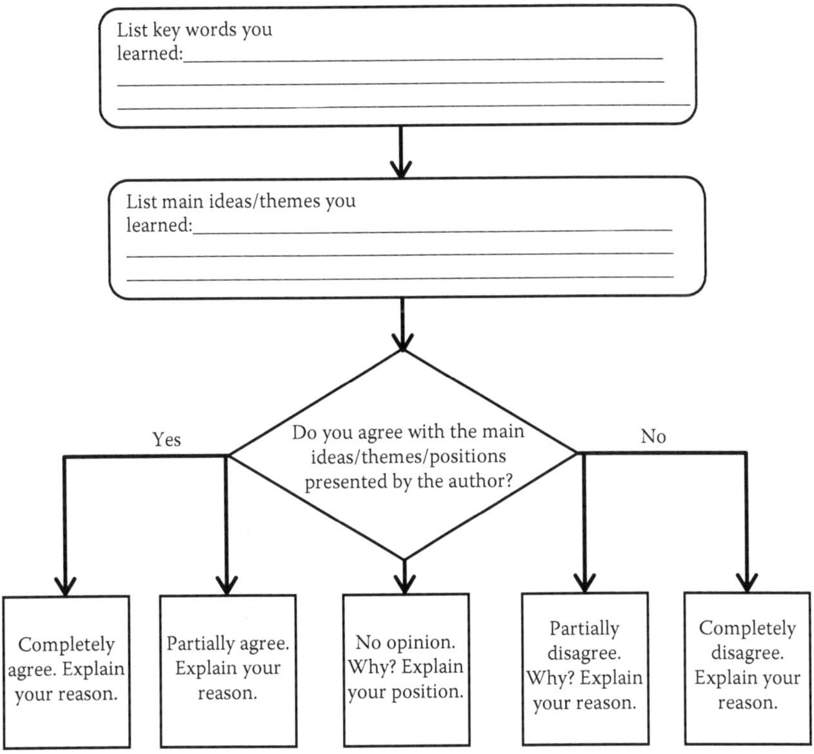

Acknowledgment

Reprinted with permission of Markkula Center for Applied Ethics, Santa Clara University.

Other Resources

Food and Agriculture Organization of the United Nations. *Weighing the GMO arguments*. Available online at: http://www.fao.org/english/newsroom/focus/2003/gmo8.htm

Herrera, Stephan (2005). *Syngenta's gaff embarrasses industry and White House*. Nature Biotechnology, Vol. 23, No. 5, May 2005.

http://www.gm.org/

http://www.centerforfoodsafety.org

References

Associated Press. (2005, March 26). Americans clueless about gene-altered foods. MSNBC.com. Retrieved November 16, 2011, from http://www.msnbc.msn.com/id/7277844/ns/health-genetics/t/americans-cluelessabout-gene-altered-foods/

Dalton, R. (2004). California edges towards farming drug-producing rice. *Nature, 428,* 591.

Stokstad, E. (2004). Monsanto pulls the plug on genetically modified wheat. *Science, 304,* 1088.

12

Titanium Dioxide (TiO_2): A Versatile Semiconducting Material for Environmental and Antibacterial Applications

M. A. Shah

CONTENTS

For a successful technology, reality must take precedence over public relations, for nature cannot be fooled.

Richard P. Feynman

KEY WORDS: Field Emission Scanning Electron Microscope (FESEM), Transmission Electron Microscope (TEM), ultraviolet (UV), hydrophilic, hydrophobic, photonics

Introduction to Titanium Dioxide (TiO_2)

Titanium dioxide (TiO_2) exists in three crystalline forms—rutile, brookite, and anatase—all of which have been prepared synthetically. In each type, the titanium ion coordinates with six oxygen atoms, which in turn are linked to three titanium atoms, and so on. Anatase and rutile are tetragonal, whereas brookite is orthorhombic in structure. Brookite and anatase are unstable forms and are not economically significant (Chen & Mao, 2007).

Titanium dioxide has the highest average refractive index known. For anatase it is 2.55, and for rutile it is 2.76. These high values account for the exceptional light-scattering ability of pigmentary titanium dioxide when dispersed in various media, which in turn yields the high reflectance and hiding power associated with this pigment. Although single-crystal titanium dioxide is transparent as a finely divided powder, it has a very high reflectance, and it is intensely white because its high reflectance is substantially uniform throughout the visible spectrum (Asahi, Morikawa, Ohwaki, Aoki, & Taga, 2001). Pure titanium dioxide possesses by nature an internal crystal structure that yields an innately high refractive index. When particle size and particle size distribution are to be optimized so as to contribute along with its high refractive index to a maximum light scattering, conventional and pigmentary titanium dioxide is obtained (Crap, Huisman, & Reller, 2004). It reflects all the wavelengths of the visible light to the same degree, producing the effect of whiteness to the human eye. All these attributes, together with its opacity, are achieved for an optical particle diameter that is approximately 0.2–0.4 micrometer, which is on the order of half the wavelength of visible light. However, there is another type of titanium dioxide whose median crystal size has been reduced to 0.02 micrometers. These particles—called *ultrafine* or *nanoparticles* of TiO_2—are the subject of this chapter, which also addresses the societal issues of nanotechnology in general.

Preparation Protocols

The history of the nanostructures of titanium dioxide dates to the late 1970s, when the first patent on the preparation of these materials was issued in Japan. It is, in principle, possible to obtain nanoparticles of TiO_2 by simply milling pigmentary TiO_2 to a finer particle size. However, the properties of fine powders in terms of purity, particle size distribution, and particle shape remain highly unsatisfactory. Several wet chemical processes were developed during the 1980s. The different methods for the synthesis of titanium dioxide result in products with different structures (anatase or rutile), crystallinity, and contaminants.

There are two main methods for obtaining TiO_2 for industrial purposes. The first is the so-called sulfate method. The other widespread technique for manufacturing TiO_2 is the vapor-phase oxidation of $TiCl_4$. These two methods have their drawbacks, either in purity or in control of the particle shape, size, and distribution (Shah & Towkeer, 2010; Zhang, Xiao, & Nan, 2010).

On a laboratory scale, nanocrystalline TiO_2 is usually prepared from a solution of titanium salts. Cassaignon, Koelsch, and Jolivet (2007) reported selective synthesis of brookite, anatase, and rutile nanoparticles by the thermohydrolysis of TiO_4 in concentrated aqueous nitric acid, wherein the selectivity was strongly dependent on the medium's acidity. Similarly, using aqueous solutions of TiO_4 and acidic additives, Yin et al. (2002) and Franklyn, Levendis, Demetrius, Coville, and Maaza (2007) confirmed the selective formation of nanoparticulate anatase and rutile nanorods. In addition, Jiu, Isoda, Wang, and Adachi (2006) reported the synthesis of a pure, highly crystalline anatase phase by hydrolyzing titanium isopropoxide (rather than a $TiCl_4$) precursor in the presence of ethylenediamine basic catalyst and cetyltrimethylammonium bromide (CTAB) surfactant. The work of X. L. Li, Peng, Yi, Wang, and Y. D. Li (2006) described the synthesis of TiO_2 nanorods by adding $Ti(OBu)_4$ into a mixed solution of linoleic acid, triethylamine, and cyclohexane and then sealing the solution for two days.

Most of these suggested pathways for the synthesis of TiO_2 nanoparticles involve toxic chemicals that are not easily degraded in the environment. Environmentally friendly chemical synthesis requires alternative solvents such as ionic liquids and water. Water is particularly attractive because it is inexpensive, environmentally benign, and bestowed with many virtues under supercritical conditions.

In our approach/technique, titanium powder—AR (analytical reagent) grade (Ranbaxy, >3 μm)—was used as a source of metal/precursor and purified by ultrasonication in water. In a typical preparation process, 3 mg of titanium powder was added to 30 ml of pure water in a glass vial. The reaction mixture was sonicated for about 20 minutes in a glass vial, transferred into a stainless steel Teflon-lined metallic bomb, and sealed under normal conditions. The closed autoclave was then placed inside a furnace, and the mixture was heated slowly (2°C/min) to 180°C and maintained at this temperature for 24 hours. After cooling, the resulting suspension was centrifuged to retrieve the product, washed, and then finally vacuum-dried for a few hours. The samples were directly transferred to an FESEM (field emission scanning electron microscope) chamber for morphological investigations.

Environmentally Friendly Nanotechnology

There is a growing movement toward so-called green nano that seeks to reduce the toxicity and environmental burden associated with nanosubstances and technologies. Generally, certain material properties are tested as indicators of potential for persistence and for potential to accumulate in the food chain. Several organizations fund research into models that evaluate the behavior of nanomaterials across their life cycle from their creation or extraction from Earth to their post-use fate in the environment, whether by recycling or disposal. There are many issues addressing green technology development.

Chemical synthesis (bottom-up approach) of nanomaterials has been reviewed by many authors, but innumerable improvements and better methods have been reported continually in the last few years. A detailed study has revealed that reported methods involve multistep processes and frequent use of amines and structure-directing chemical groups. Moreover, they involve environmentally harmful chemicals that may be toxic and persistent in the environment. Organic solvents are also similarly problematic because many are toxic and harmful and have raised health and environmental safety issues. The other techniques, based on a top-down approach, are also technically complex and require high temperature, harsh growth conditions, expensive instruments, complicated control processes, and the use or release of excessive and harmful emissions. Thus a simple and a safe approach for industrial-scale mass production and controlled growth of a wide variety of nanomaterials is eagerly awaited and is an open and ongoing challenge to nanotechnologists.

Recently, an interesting and encouraging result obtained by the reaction of titanium metal with water by Shah and Al-Marzouki (2011) provides motivation to synthesize TiO_2 nanostructures without additives or organics at low temperatures. Interestingly, almost uniform-sized nanoparticles were obtained by a simple reaction of titanium powder and deionized water at the very low temperature of 180°C, as shown schematically in Figure 12.1. The typical diameters of the as-grown nanoparticles are ≈25 ± 10 nm, with an average diameter of 15 nm. Figure 12.2 (a,b,c,d) shows the low and high magnification of titanium dioxide nanoparticles prepared by the authors in their laboratory. The morphology is further confirmed by TEM (transmission electron microscope) examination, as shown in Figure 12.3.

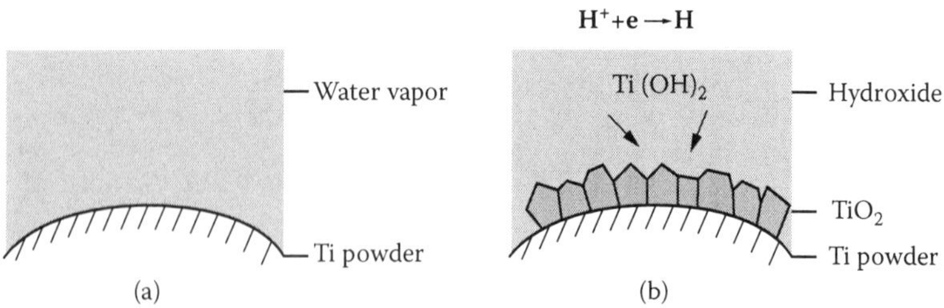

FIGURE 12.1
Schematic illustrations of the growth of (a) Ti particles in the presence of water vapor and (b) layer of $Ti(OH)_2$ formed on the particle surface and subsequent decomposition to TiO_2.

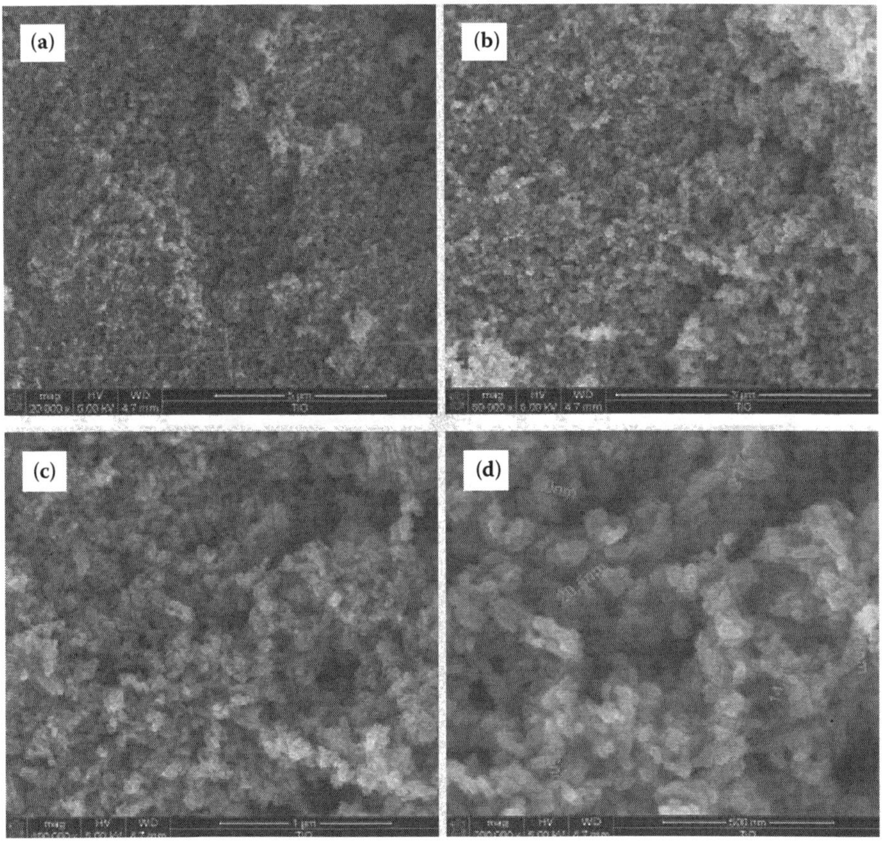

FIGURE 12.2
The typical (a and b) low- and (c and d) high-resolution FESEM images of TiO_2 nanoparticles obtained by the reaction of titanium metal powder with water at 180°C for 24 h.

It is pertinent to mention that new physical and chemical properties emerge when the size of materials becomes smaller and smaller, down to the nanometer scale. Properties also vary as the shape of the shrinking nanomaterials changes. The size- and shape-dependent optical and electronic properties of nanoparticles make an interesting case for environmental scientists to exploit their role in environmental and pollution remediation processes The method described will address the mass production of nanomaterials through economical and safe means. It will also reduce the cost of nanoproducts and eventually lead to advances in the understanding as well as the fabrication of novel high-technology devices in many fields of application from electronics to medicine.

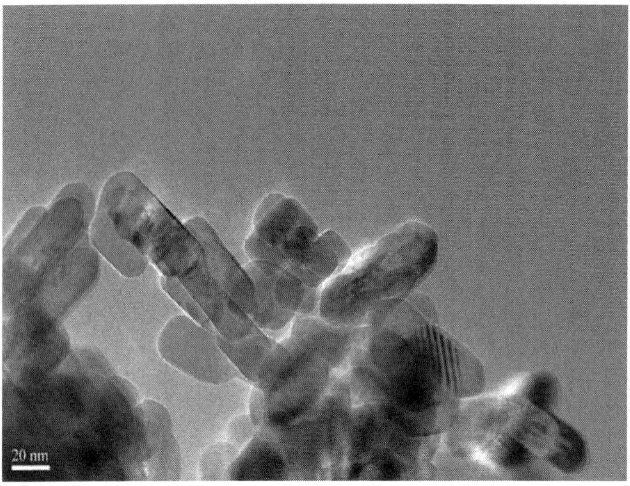

FIGURE 12.3
TEM micrograph of TiO_2 nanoparticles confirms the morphology of the samples.

Environmental Degradation and Its Solution

While we enjoy the comforts and benefits that modern technology has provided us—from composites to computer chips, from drugs to dyes—we are faced with the task of treating wastes generated during manufacturing processes and proper disposal of various products and by-products (Colvin, 2003). Environmental degradation is a serious problem with many sources and causes. Nanoscience and nanotechnologies could play an important role in developing smart materials that can sense and destroy harmful chemical contaminants in our environment. This technology could provide a means to purify air and water resources by utilizing semiconducting nanoparticles. The procedure of preparing TiO_2 nanostructures proposed here is environmentally benign and free from toxic reagents.

Antimicrobial Properties

The ability of a substance to kill or inhibit the growth of microbes or microbiological organism is known as an *antimicrobial property*. Microbes include bacteria, such as *Escherichia coli* and *Salmonella* spp., which cause gastrointestinal illness, as well as *Lactobacillus acidophilus*, the active ingredient in yogurt that improves digestion. Bacteria tend to be in the micron size range, whereas viruses—which are also microbes—are of the nanosize scale. Not all microbes cause disease. Those that do are called *pathogens*. Pathogens can cause disease by infecting the food, water, and air that people eat, drink, and breathe.

The development of nanoscale materials for use as antimicrobials applies to the greater reactivity and surface area of smaller particles for more effective killing of pathogens. Applications include fabric coatings, food packaging, and the surfaces of doorknobs. Studies evaluating the effects of nanoscale silver, fullerenes, titanium dioxide, and carbon nanotubes have identified antimicrobial properties. These types of nanoscale particles are now being tested and applied for use in numerous applications. Thus far, titanium dioxide nanostructures are the most widely developed for antimicrobial use. As a promising photocatalyst, titanium dioxide (TiO_2) nanoparticles are playing a significant role in helping to solve many serious environmental and pollution challenges. It also bears tremendous hope in helping ease the energy crises through effective utilization of solar energy based on photovoltaic and water-splitting devices. TiO_2 particles have been extensively studied for oxidative and reductive transport of organic and inorganic species present as containments in air and water. Technical advances in this area have already led to product development for a variety of day-to-day operations. Commercialization of products such as self-cleaning glasses, disinfectant tiles, and filters for air purification demonstrates the early success of nanodevices for environmental applications. TiO_2 nanoparticles have considerable advantages over other similar photocatalysts due to their chemical stability, nontoxicity, endurance, thin-film transparency, and lower production costs (Hardman, 2006).

The use of TiO_2 has been explored in water treatment to destroy chemicals such as polychlorinated biphenyls, pesticides, and other complex organic contaminants. TiO_2 has been demonstrated to be bactericidal, and one commonly used type of TiO_2 nanostructure has a negative surface charge, resulting in acidic pH in water that creates additional stresses in *E. coli* bacteria. Rincon and Pulgarin (2003) observed a determinative effect on the survival of *E. coli* after photocatalytic exposure: No bacterial growth was observed after UV illumination of a contaminated TiO_2 suspension.

Applications of TiO_2 Nanomaterials

Titanium dioxide nanomaterials are very stable, nontoxic, and cheap. The existing and promising applications of TiO_2 nanomaterials include paint, toothpaste, UV protection, photocatalysis, photovoltaics, sensing, and electrochromics as well as photochromics. TiO_2 nanomaterials normally have an electronic band gap larger than 3.0 eV and high absorption in the UV region. Their benign optical and biological properties allow them to be suitable for UV protection applications (Mor, Varghese, Paulose, Shankar, & Grimes, 2006).

TiO_2 nanomaterials can be imparted with antifogging functions on various glass products, for example, mirrors and eyeglasses having superhydrophilic or superhydrophobic surfaces. Stain-proofing and self-cleaning properties can also be bestowed on many different types of surfaces due to the superhydrophilic or superhydrophobic surfaces. These materials have been used as sensors for various gases and humidity because of their electrical and optical properties, which change upon absorption (Chen & Mao, 2007). TiO_2 can absorb light into the visible light region and convert solar energy into electrical energy for solar-cell applications. One of the many applications of TiO_2 nanomaterials when synthesized with dyes or metal nanoparticles involves the photocatalytic decomposition of various pollutants. TiO_2 photocatalysts can also be used to kill bacteria, as has been carried out with *E. coli* suspensions. The strong oxidizing power of illuminated TiO_2 can be used to kill tumor cells in cancer treatment. The principle of semiconductor photocatalytic reaction is straightforward. Upon absorption of photons with energy larger than the band gap of TiO_2, electrons are excited from the valence band to the conduction band, creating electron-hole pairs.

Societal Issues

The historically unprecedented developments of nanoscience and nanotechnology—in view of their phenomenal expansion and growth and in conjunction with their convergence with information science and molecular biology—confront our society and our natural environment with new challenges. Advances in any discipline inevitably lead to changes in social relations, meanings, and societal patterns. Discoveries at the frontiers of science have the potential to pave the way for radically innovative and integrative approaches, providing new solutions for pressing problems. Scientific discoveries and technological innovations have the power to revolutionize societies over time.

Science, technology, and society are intrinsically interlinked and characterized by mutual interdependence. Myriad factors determine the evolution of society's technological, social, and cultural processes. Among these, the applications of scientific knowledge and associated technological developments are major factors determining social progress and prosperity.

Nanotechnology has already impacted society by virtue of its potential applications, and its impact will increase as commercial products are developed. From the very beginning, society has shaped nanotechnology through visionary ideas, science fiction stories, and innovative research and funding programs. Public relations and public debates have tried to keep up with these interactions between nanotechnology and society. As with all new technologies, the future of nanotechnology will depend on these interactions between human beings with their creative minds and skills, their hopes and fears, and their other values and interests (Birgit & Pardeep, 2006).

Today's world comprises societies that operate over a wide range of technologies—from the pre-industrial to the modern state of the art. Nanotechnology has the promise of both bypassing and bridging many of yesterday's missing technological links between the developed and developing worlds.

Acknowledgments

The author is grateful to Prof. A. Ansari and Prof. K. Siddique for valuable guidance and for making the relevant literature available within a short span of one week during my stay in King Abdul Aziz University, Jeddah.

Glossary

FESEM: field-emission electron microscope.
hydrophilic: water-loving materials.
hydrophobic: water-fearing materials that do not dissolve in water.
photonics: the science of manipulating photons.
TEM: transmission electron microscope.
UV: Ultraviolet.

Questions

1. How can nanomaterials be fabricated economically? How can we determine whether TiO_2 is a biosafe and biocompatible material?
2. Explain how nanoscience may serve as a bridge between human development and economic sustainability?
3. How do biosensors help to detect diseases in advance? Do gas sensors shows better results when nanoparticles of TiO_2 are used?
4. How can nanoparticles be used to protect the environment? Soot particles in the atmosphere are known to be toxic: Is it therefore not obvious that nanoparticles are also toxic?
5. What are the expected benefits of nanotechnologies for citizens and consumers? How could fabric industries benefit impoverished people?

6. What are the major crises the world is facing, and how will nanotechnology address them? Explain how it will bring a revolution in the energy sector.
7. Summarize your learning of new ideas and information presented in this article by completing the following flowchart:

Flow Chart

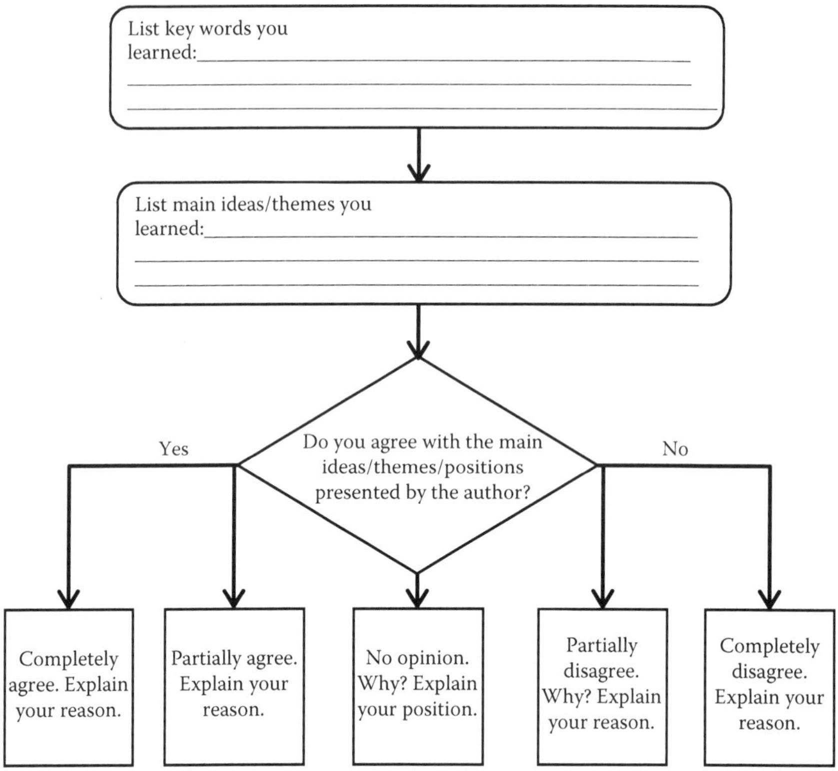

Additional Resources

www.nature.com
www.nanotechproject.org
www.apnf.org
www.nanotechnologyfordummies.com
www.azonano.com

References

Asahi, R., Morikawa, T., Ohwaki, T., Aoki, K., & Taga, Y. (2001). Visible light photocatalysis in nitrogen doped titanium oxides. *Science, 293*, 269–271.

Birgit, R. B., & Pardeep, T. (2006). Societal implications of nanoscience and nanotechnology in developing countries. *Current Science, 90*, 645–658.

Cassaignon, S., Koelsch, M., & Jolivet, J. P. (2007). Selective synthesis of brookite, anatase, and rutile nanoparticles: Thermolysis of $TiCl_4$ in aqueous nitric acid. *J. Mat. Sci., 42*, 6689–6695.

Chen, X., & Mao, S. S. (2007). Titanium dioxide nanomaterials: Synthesis, properties, modifications and applications. *Chem. Rev., 107*, 2891–2959.

Colvin, V. L. (2003). The potential environmental impact of engineered nanomaterials. *Nature Biotechnology, 21*, 1166–1170.

Crap, O., Huisman, C. L., & Reller, A. (2004). Photoinduced reactivity of titanium dioxide. *Prog. Solid State Chem., 32*, 33–177.

Franklyn, P. J., Levendis, D. C., Demetrius, C., Coville, N. J., & Maaza, M. (2007). Phase transformation of hydrothermally synthesized nanoparticles of TiO_2: From anatase to rutile nanorods. *S. Afr. J. Chem., 60*, 71–75.

Hardman, R. (2006). A toxicologic review of quantum dots: Toxicity depends on physicochemical and environmental factors. *Environmental Health Perspectives, 114*, 165–172.

Jiu, J. T, Isoda, S., Wang, F. M, & Adachi, M. (2006). Dye sensitized solar cells based on a single crystalline TiO_2 nanorod film. *J. Phys. Chem. B, 110*, 2087–2092.

Li, X. L., Peng, Q., Yi, J. X., Wang, X., & Li, Y. D. (2006). Near monodisperse TiO_2 nanoparticles and nanorods. *Chem. Eur. J., 12*, 2383–2391.

Mor, G. K., Varghese, O. K., Paulose, M., Shankar, K., & Grimes, C. A. (2006). A review on highly ordered vertically oriented TiO_2 nanotube arrays: Fabrication, materials properties, and solar energy applications. *Solar Energy Materials & Solar Cells, 90*, 2011–2075. Retrieved November 17, 2011, from http://web.mit.edu/kjhuang/www/Literature/Solar%20Cells/Experimental/Oxides/TiO2/A%20review%20on%20highly%20ordered,%20vertically%20oriented%20titania%20nanotube%20arrays.pdf

Rincon, A. G., & Pulgarin, C. (2003). Photocatalytic inactivation of *E. coli*, effect of (continuous-intermittent) light intensity and of (suspended-fixed) TiO_2 concentration. *Applied Catalysis B, Environmental, 44*, 263–284.

Shah, M. A., & Al-Marzouki, F. M. (2011). TiO_2 nanoparticles prepared without harmful organics: A biosafe and economical approach. *Scientia Iranica, 18*, 804–807.

Shah, M. A., & Towkeer, A. (2010). *Principles of nanoscience and nanotechnology*. New Delhi, India: Naroosa Publishing House.

Verianshyah, B., Park, T. J., Lim, J. S., & Lee, Y. W. (2005). *J. Solid Supercritical Fluids, 34*, 51–61.

Yin, H. B., Wada, Y., Kitamura, T., Sumida, T., Hasegawa, Y., & Yanagida, S. (2002). Novel synthesis of phase pure nanoparticulate anatase and rutile TiO_2 using $TiCl_4$ aqueous solution. *J. Mater. Chem., 12*, 378–383.

Zhang, J., Xiao, X., & Nan, J. (2010). Hydrothermal–hydrolysis synthesis and photocatalytic properties of nano TiO_2 with an adjustable crystalline size. *J. Hazardous Materials, 176*, 617–622.

13

What Are the Social Implications of Our Delay in Teaching Nanoscience Education to K–12 Students in the United States?

Judith Light Feather

CONTENTS

KEY WORDS: nanoscience, education, ethical choices, societal implications, K–12 inclusion, new standards, nanomanufacturing centers, workforce training, top-down structure, bottom-up solutions

Ethics: 1. the study of standards of conduct and moral judgment; moral philosophy; 2. a treatise on this study; 3. the system or code of morals of a particular person, religion, group, profession, etc.

Ethical: 1. having to do with ethics or morality; of or confirming to moral standards; 2. conforming to the standards of conduct of a given profession or group.

Introduction

What are the ethical issues associated with education in general? They are the foundation of developing an informed society capable of: making informed decisions, being competitive in a global economy, ensuring education is uniform in all aspects for the nation's citizens (including minorities), and fostering an accurate perspective of the role of the United States in global affairs.

How does science and engineering play a role in society? The advancement of scientific research and engineering leads to innovation and new methods of commercialization, which stimulates our economic development and increases our standing as a global leader.

How important is nanoscience education for the future of our economic and global competitiveness? Nanoscience education introduces a size/scale of inquiry that must be explored for our growth into the Nanomanufacturing Age, as we apply technological discoveries in a competitive global market.

If we are to develop this case study to determine the ethics and social implications of our actions as a country concerning the development and inclusion of nanoscience education, multiple issues within our complex educational matrix must be considered.

1. Will the United States have a viable next generation of K–12 science/engineering standards, accepted by most (all) of the states in this decade? Will those standards incorporate the nanoscale science and engineering necessary to compete globally for the next generation of nanomanufacturing?
2. Education with new standards that include nanoscale science education have been in effect in Taiwan and Australia since 2004 and 2011, respectively. What are the implications for the United States, since we now lag two decades behind those countries, even if we develop our standards in this decade?
3. Revised U.S. standards that include/do not include the nanoscale science education: What would the consequences of each choice be for our society?
4. A global acceptance of Nanoscale Science Education (NSE) standards and Ethics Education for Science and Technology has major implications for Third World impact. What would the implications be for the United States?

Investment in Nanotechnology R&D

Global spending for nanotechnology research and development is escalating exponentially, setting the stage for nanomanufacturing as the next major world enterprise. The advances in microscopy since the 1980s have opened a window into nature that scientists all over the world have embraced, and their governments have supported nanotechnology initiatives to fund research in a race to commercialize enabling technologies that improve our current products. By the end of 2011, the total government funding for nanotechnology research worldwide will be $65 billion, rising to $100 billion by 2014, as shown in Figure 13.1 (Harper, 2011). When figures for corporate research and various other forms of private funding, which are thought to have surpassed government funding figures as far back as 2004, are taken into account, an estimate of nearly a quarter of a trillion dollars will have been invested into nanotechnology by 2015.

Every year, Tim Harper from Cientifica in the United Kingdom undertakes one of the world's most exhaustive searches into the global funding of nanotechnologies to identify not only where the dollars, euros, and yen are being spent, but also to gain a unique insight into the trends shaping tomorrow's

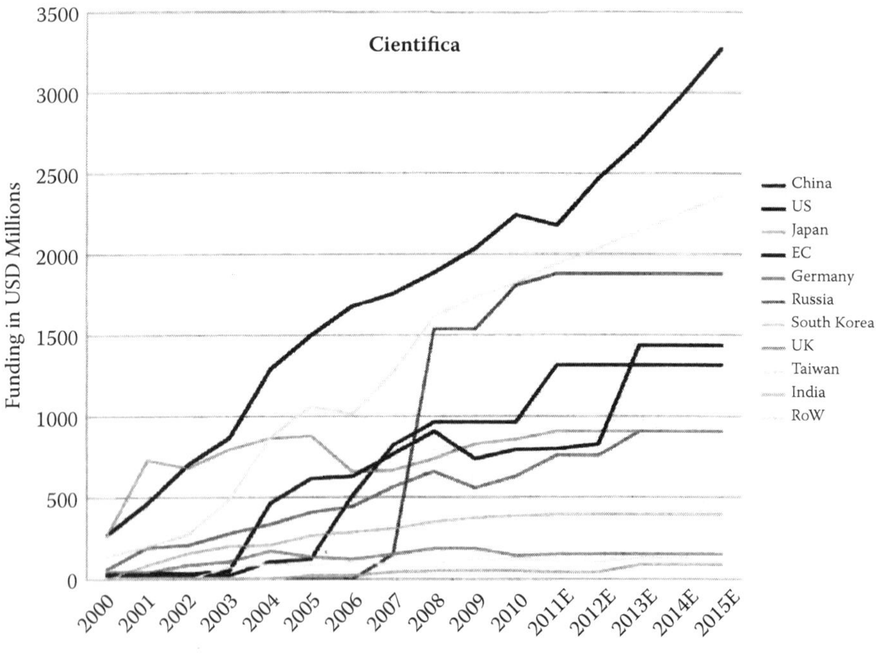

FIGURE 13.1
Funding of nanotechnologies by country. (From Cientifica Ltd., 2011. Reprinted with permission.)

applications. For the 2011 report, they have also integrated data from the World Economic Forum's annual Global Competitiveness Report (Schwab, 2011) to gain insights into how well various countries are able to capitalize on their investment in research. The United States scored the highest. However, on the Quality of Mathematics & Science in Education, the United States was 54th of the 139 countries, with a 4.4 score out of a possible 7 in the ratings.

Ethical Relevance Concerning Our Failing Schools

The conscious effort to change the world by moving out of the Industrial Age of manufacturing into the Nanotechnology Age affects every area of science and the resulting technologies in the immediate future. This statement is based on the National Nanomanufacturing Initiative funding to establish the National Nanomanufacturing Network (NNN),[1] an innovative alliance of academic, government, and industry partners who cooperate to advance nanomanufacturing strength in the United States. The core foundation of the NNN consists of the four National Science Foundation (NSF) nanomanufacturing Nanoscale Science and Engineering Centers (NSECs)—the Center for Hierarchical Manufacturing (CHM), the Center for High-Rate Nanomanufacturing (CHN), the Center for Scalable and Integrated NanoManufacturing (SINAM), and the Center for Nanoscale Chemical-Electrical-Mechanical Manufacturing Systems (Nano-CEMMS)—as well as the Department of Energy (DOE) Center for Integrated Nanotechnologies (CINT) at Sandia National Laboratories, the National Institute of Standards and Technology (NIST), Center for Nanoscale Science and Technology (CNST), and other affiliations. InterNano is the information arm of the NNN—a digital library resource of timely information on nanomanufacturing and a platform for collaboration.

Nanomanufacturing, according to the NNN, is the use of value-added processes to control matter at the nanoscale in one, two, and three dimensions for reproducible commercial-scale production. Areas where nanomanufacturing is being embraced aggressively right now include magnetic data storage, lightweight batteries, and high-efficiency solar cells. The University of Massachusetts at Amherst is a research leader in all three areas of nanomanufacturing education. A small number of community colleges have also been developing workforce training for the predicted 2 million technicians that will be needed by 2015 for these new companies. However, the enrollment has been limited, as high school students and their guidance counselors have not been informed of the relevance associated with the future career paths for nanotechnologies that would require specialized technicians.

In the meantime, K–12 students are still using outdated Industrial Age textbooks, proclaiming that we still don't know what an atom looks like. These students are not receiving adequate instruction in the science, technology, engineering, and mathematics (STEM) courses that will be needed to prepare them for the new nanoscience-related careers. Today's students are further inhibited by the outdated science standards that teachers are required to use in order to "teach to the test" under the No Child Left Behind (NCLB) Act. As I prepared my book, *Nanoscience Education, Workforce Training, and K–12 Resources* (Light Feather & Aznar, 2010), it was obvious that the political stakeholders from the top down were primarily interested in test results for accountability. The most recent report on the key elements of failure relevant to science from the National Academy of Sciences states:

> **Key element: Adequate instructional time.** The NCLB Act has also changed the time for science, technology, engineering, and mathematics instruction in the K–12 curriculum. Particularly in elementary school, the predominant instructional emphasis is on mathematics and English language arts because those subjects are tested annually under the current accountability system. In the 2006–2007 school year, for example, elementary schools (on a nationally representative survey) reported spending an average of 178 minutes per week on science instruction, 323 minutes on mathematics, and 503 minutes on English language arts [Center on Education Policy, 2007]. A closer look at those data revealed that 28 percent of districts reported decreasing their instructional time in science in elementary schools, with an average decrease in those districts of 75 minutes per week. In contrast, 45 percent of districts reported increasing instructional time for mathematics in elementary schools, with an average increase of 89 minutes per week [Center on Education Policy, 2008].
>
> A 2007 study of science education in California paints a starker picture. That survey of nine counties in the San Francisco Bay Area found: '80 percent of K–5th grade multiple-subject teachers who are responsible for teaching science in their classrooms reported spending 60 minutes or less per week on science, with 16 percent of teachers spending no time at all on science' [Dorph et al., 2007]. Those researchers estimate that their results actually overstate the amount of science instruction in the Bay Area because 'teachers who took the time to respond to the survey are more likely to be engaged in science education than those who did not' [National Research Council, 2011b, p. 4]. **Overall, the decrease in time for science education is a concern because some research suggests that interest in science careers may develop in the elementary school years** [Maltese & Tai, 2010]. (National Research Council, 2011b, p. 22)

The conclusions of the report are as follows:

- To make progress in improving STEM education for all students, policy makers at the national, state, and local levels should elevate science to the same level of importance as reading and mathematics.

- Science should be assessed with the same frequency as mathematics and literacy, using a system of assessment that supports learning and understanding. Such a system is not currently available. Therefore, states and national organizations should develop effective systems of assessment that are aligned with the next generation of science standards and that emphasize science practices rather than mere factual recall.
- National and state policy makers should invest in a coherent, focused, and sustained set of supports for STEM teachers to help them teach in effective ways. Teachers in STEM should have options to pursue professional learning that addresses their professional needs through a variety of mechanisms, including peer-to-peer collaboration, professional learning communities, and outreach with universities and other organizations.
- Furthermore, federal agencies should support research that disentangles the effects of school practice from student selection, recognizes the importance of contextual variables, and allows for longitudinal assessments of student outcomes, including the three strategic goals of STEM education and intermediate outcomes.
- Federal funding for STEM-focused schools should be tied to a robust, strategic research agenda. Only knowledge of this sort will allow a full response to the questions that were put to this committee.

Top-Down Management of the Education System

The complexity of our education system is not easily penetrated from the top down. The Institute of Education Sciences (IES)[2] receives $200 million a year to fund cognitive research and post their results on the What Works Clearinghouse (WWC)[3] for educators. This funding was granted by Congress on November 5, 2002, as the Education Sciences Reform Act (ESRA). In conjunction with the National Center for Education Statistics (NCES),[4] this is mandated by Congress to produce the Condition of Education, known as the Nation's Report Card, each year. Table 13.1 shows a minus three in science for fourth grade and a gain of seven points for eighth grade between 1995 and 2007.

The structure of the top-down list of stakeholders who make the decisions on education is shown in Figure 13.2, which illustrates the internal and external influences that add to the complexity of the curriculum development process. The influence is political from the top down, and teachers remain at the bottom, teaching from textbooks that match the state and national tests. The standards necessary to integrate nanoscale, biology, cognitive, and information sciences in the K–12 curriculum is only an advisory function without a national curriculum. Current policies use a uniform measure of accomplishment through standardized testing, and it is becoming obvious that more effective measures need to be developed based on cognitive development and individual learning differences.

TABLE 13.1

Trends in Average Science Scores of Fourth- and Eighth-Grade Students, by Country: 1995–2007

Grade four				Grade eight			
	Average score		Difference[1]		Average score		Difference[1]
Country	1995	2007	2007–1995	Country	1995	2007	2007–1995
Singapore	523	587	63*	Lithuania[2]	464	519	55*
Latvia[2]	486	542	56*	Colombia	365	417	52*
Iran, Islamic Rep. of	380	436	55*	Slovenia	514	538	24*
Slovenia	464	518	54*	Hong Kong SAR[3,4]	510	530	20*
Hong Kong SAR[3]	508	554	46*	England	533	542	8
Hungary	508	536	28*	**United States[4,5]**	**513**	**520**	**7**
England	528	542	14*	Korea, Rep. of	546	553	7*
Australia	521	527	6	Russian Federation	523	530	7
New Zealand	505	504	–1	Hungary	537	539	2
United States[4,5]	**542**	**539**	**–3**	Australia	514	515	1
Japan	553	548	–5*	Cyprus	452	452	#
Netherlands[6]	530	523	–7	Japan	554	554	–1
Austria	538	526	–12*	Iran, Islamic Rep. of	463	459	–4
Scotland	514	500	–14*	Scotland[4]	501	496	–5
Czech Republic	532	515	–17*	Romania	471	462	–9
Norway	504	477	–27*	Singapore	580	567	–13
				Czech Republic	555	539	–16*
				Norway	514	487	–28*
				Sweden	553	511	–42*

Source: International Association for the Evaluation of Educational Achievement (IEA), Trends in International Mathematics and Science Study (TIMSS), 1995 and 2007.

Note: Bulgaria collected data in 1995 and 2007, but due to a structural change in its education system, comparable science data from 1995 are not available. Countries are ordered by the difference between 1995 and 2007 overall average scores. All countries met international sampling and other guidelines in 2007, except as noted. Data are not shown for some countries, because comparable data from previous cycles are not available. The tests for significance take into account the standard error for the reported difference. Thus, a small difference between the United States and one country may be significant while a large difference between the United States and another country may not be significant. Detail may not sum to totals because of rounding. The standard errors of the estimates are shown in tables E-20 and E-21 available at http://nces.ed.gov/pubsearch/pubsinfo.asp?pubid=2009001.

■ Country difference in average scores between 1995 and 2007 is greater than analogous U.S. difference ($p < .05$)

▩ Country difference in average scores between 1995 and 2007 is not measurably different from analogous U.S. difference ($p < .05$)

□ Country difference in average scores between 1995 and 2007 is less than analogous U.S. difference ($p < .05$)

Rounds to zero.

*$p < .05$. Within-country difference between 1995 and 2007 average scores is significant.

1 Difference calculated by subtracting 1995 from 2007 estimate using unrounded numbers.

2 In 2007, National Target Population did not include all of the International Target Population defined by the Trends in International Mathematics and Science Study (TIMSS).

3 Hong Kong is a Special Administrative Region (SAR) of the People's Republic of China.

4 In 2007, met guidelines for sample participation rates only after substitute schools were included.

5 In 2007, National Defined Population covered 90%–95% of National Target Population.

6 In 2007, nearly satisfied guidelines for sample participation rates only after substitute schools were included.

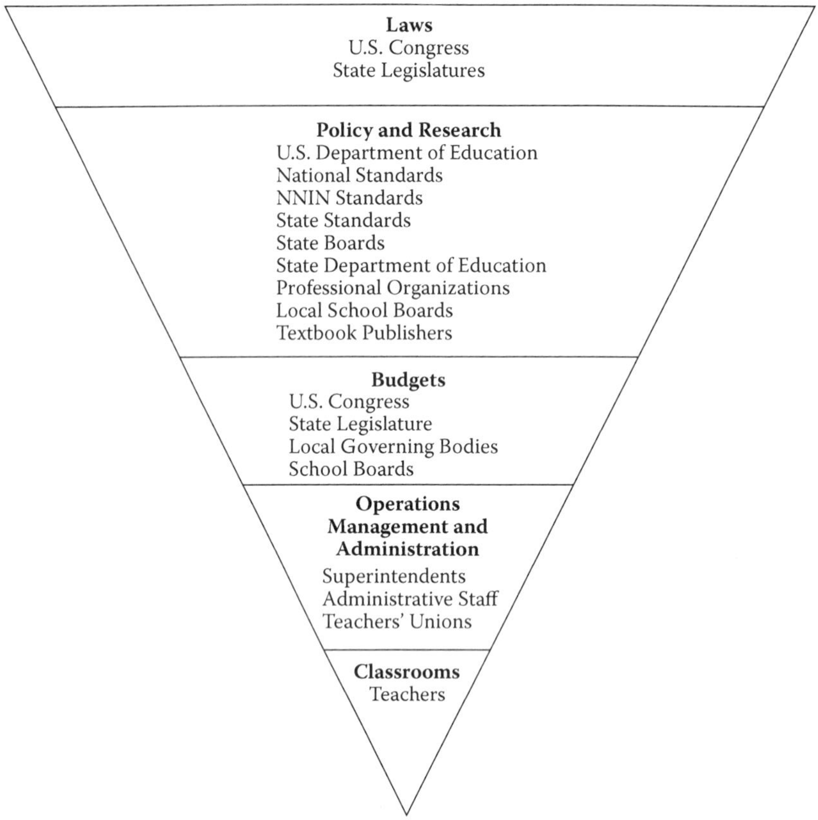

FIGURE 13.2
Structure of top-down list of stakeholders who make the decisions on education. (From *Nanoscience Education, Workforce Training, and K–12 Resources* [p. 21], by J. Light Feather and M. F. Aznar, 2010, Boca Raton, FL: CRC Press. Copyright 2011 by Taylor & Francis Group. Reprinted with permission.)

Questions to consider:

1. Given the future projections and societal impact of nanotechnology, is it ethical for the education stakeholders who make the decisions to ignore the $65 billion that the national nanotechnology initiatives have spent on research and commercialization?
2. Should the new science standards for K–12 include nanoscale science within the matrix?
3. Because standards are not reviewed for 10–15 years, how would that lengthy delay affect the next two generations, which will find themselves at a disadvantage in the new job market in nanomanufacturing?

4. Can you think of solutions that would aid in solving the complexity and cost in reforming K–12 STEM education?
5. Do you consider the delays in addressing the inclusion of nanoscience education—a decision that will affect students' future career choices—to be an ethical issue?

Why All Students Need to Learn About Nanoscale Science

This is the most important issue that needs to be addressed. I have answered this question many times over the years. Many grant reviewers would not fund curriculum development, or online nanolabs that young students could access, because they did not understand that nanoscale science is a size that includes the atoms found within all matter. There is no excuse for withholding new scientific discoveries from our children. This is especially true given that the access provided by the digital world allows continuous updates of all science curricula with e-learning tools without adding to school budgets.

Nanoscale Science: What It Is

At its core, science is the study of nature and how the world works. The advances in microscopy over the past two decades have allowed our scientific communities to peer into the atomic level at the nanoscale, to move and manipulate atoms, and to expand our knowledge base in all branches of science from the microscopic to the cosmos.

Nanoscale science deals with sizes where we can observe the underlying behavior of atoms and particles before they assume their macroscale material properties, including their purpose and movement within the cells of our bodies. If we were to teach our young students this scale of science by allowing them to compare the simulations of the nanoscale behaviors versus everything they see in the macro scale, it would be taught as "the foundation of nature." It is not a separate subject to be added: It is a perspective of size, which is of extreme importance in understanding the patterns and relationships of nature that surround us as a part of our physicality in our everyday lives. Therefore, it is our obligation to introduce this science-based perspective of size to all students with visual elements that exhibit atoms in movement as they transition from the properties of the nanoscale and become the components of material solutions to societal needs. This basic foundation underlies all of physics, chemistry, biology, cosmology, materials science, and engineering. With this knowledge, students can continue to study in depth as they proceed through the higher grade levels. The exposure at the very early grades—with animated visuals that show students the nanoscale size within a skin cell, a leaf, the bark of a tree, stones, the Earth, and even the buckyballs that have been seen in space—will prepare children to understand the scale of size in science and to explore the nano aspect of the physical world as they get older. Done properly with simulations, the visuals can stimulate the wonder of discovery and the desire for lifelong learning that is essential to advancing human progress.

Stakeholders at all levels will need to work together while developing the new science standards for K–12 on the basis of the new NRC framework, which is a work in progress. Providing approaches for teachers/students to expand their knowledge of science as the study of nature will stimulate curiosity in our youngest students and develop a better understanding of how they can relate to the world around them.

Case Study 1: A New Conceptual Framework for K–12 Science Education

In order to begin the process of developing new science standards for K–12 schools, a committee on developing a Conceptual Framework for New K–12 Science Education Standards was appointed by the National Research Council in Washington, DC, in 2010.

"A Framework for K–12 Science Education: Practices, Crosscutting Concepts and Core Ideas," developed by the committee for A Conceptual Framework for New K–12 Science Education Standards (National Research Council, 2011a), has been posted online as a prepublication draft of the book published by National Academies Press, which may be read online or downloaded by chapters as PDF files. The framework offers a structural change for content development as an ecosystem of science, but only mentions nanoscale science in two places. Nonetheless it opens the door for discussion with Achieve, a nonprofit organization that will take the framework to the next step of developing actual standards for new curriculum development.

Curriculum and Instructional Materials

> The framework intentionally does not prescribe a specific curriculum, but it does imply criteria for designing a curriculum and selecting instructional materials. If the framework were to influence what is taught to students, then curriculum policy, the design and development of instructional materials, including technology-based materials and tools, and the processes and criteria by which such materials were developed, selected, and implemented in classrooms would reflect the framework's practices, crosscutting concepts, and disciplinary core ideas.
>
> Enrollment and achievement patterns in schools would reveal whether the vision expressed by the framework applied to all students. For example, if the framework were permeating the system, opportunities for taking challenging science courses would be open to every student, and resources needed to implement a robust standards-based curriculum

> would be allocated in equitable ways. Resources designed to accommodate diverse learners, including those learning English as a second language, would support the focus of the standards on all students having access to opportunities to learn important science and engineering concepts and practices. (National Research Council, 2011a, p. 13-18)

Key questions that were provided to the committee, related to tracing the influence of the framework and standards on the future curriculum, include:

- What curriculum development efforts have been undertaken to provide materials that are well aligned to the framework and new standards?
- Who was engaged in these efforts?
- Were any incentives used to encourage these development efforts, and which of them were most effective?
- How do the new curricula differ from those used in the past, and are teachers prepared to address these differences?
- How has the funding from various federal and state agencies been allocated for curriculum development efforts that are aligned with the framework and standards?
- Is technology to support science learning being marshaled and used effectively to develop technology-based curriculum support materials and tools (e.g., simulations, data access)?
- What has been learned about the effectiveness of the new curriculum with various populations and under different implementation conditions?

Organizational Issues

> Institutional barriers can hamper widespread adoption of framework-based curricula and related approaches to instruction. These barriers include incentive structures, organizational culture, career patterns of teachers and administrators, and financial constraints. This piece of the education reform agenda, which entails both short-term and long-term elements, necessitates uncovering obstacles to system reform and exploring innovative ways to overcome these obstacles. In other words, the components of the system for science education must be coherent, and all of the players must be actively participating. (Briars & Resnick, 2000)

Key questions include:

- What is the process by which the framework is used to craft state-level science standards? Who is involved? How were they chosen?
- How does the capacity of the state and districts to fund education affect the writing of the standards and the development of assessments?

- What is the adoption process for the state science education standards? Is it voluntary or mandatory? What kinds of incentives or support are provided to districts to facilitate this adoption?
- To what extent does the state department of education provide funding for adopting new framework-aligned science curricula and professional development programs for teachers and administrators?

Scenario for Case Study 1

The dilemma is intensified, as many states have recently cut their education budgets and laid off many teachers. Failing schools are being closed, and the remaining teachers will have to handle more students with less time to adopt the new curriculum standards. The federal budget for states has also been cut, and the governors would like to accept the new standards, but are in a quandary concerning development and implementation of new curricula, along with the ability to meet federal laws on accountability and assessments. The Center on Budget Policy and Priorities has released a new report available for review (Williams, Leachman, & Johnson, 2011).

Of the 47 states with newly enacted budgets, 38 or more states are making deep, identifiable cuts in K–12 education, higher education, health care, and other key areas in their budgets for fiscal year 2012. Even as states face rising numbers of children enrolled in public schools, students enrolled in universities, and seniors eligible for services, the vast majority of states (37 of 44 states for which data are available) plan to spend less on services in 2012 than they spent in 2008—in some cases, much less.

Washington's budget cuts an amount equal to $1,100 per student in K–12 funds for reducing class size, extended learning time, and teachers' professional development.

Questions

Consider the following ethical questions and social implications:

1. Take a look at your state budget cuts for education in the report and develop solutions that would be ethical.
2. Can teachers learn the material on their own from all the free online resources developed by NSF (Light Feather & Aznar, 2010, pp. 123–295), or are additional resources necessary?
3. Should teachers/students have to wait for all the stakeholders to make the decisions on what they can teach from the top-down structure of education stakeholders? Can science actually be taught as inquiry, and can the students learn with the teachers, based on inquiry rather then being left behind?
4. List the long-term and short-term effects of doing nothing to change K–12 science.

Case Study 2: Benchmarks for Inclusion of Nanoscale Science in New Standards

Meanwhile, the National Science Foundation (NSF) has sponsored two workshops for stakeholders that have been developing benchmarks to include nanoscale science into this concept of the framework's standards.

A new report has just been released entitled "NSF Workshop Report: International Benchmark Workshop on K–12 Nanoscale Science and Engineering Education (NSEE)," which outlines the important standards that could be included and identifies the stakeholders that attended the meetings (Murday, 2011). The solutions are workable, but will require a concerted effort by all the political stakeholders before the standards are finalized and sent forward to the governors for state acceptance.

Dr. James Murday, editor of the reports, has met with the engineering adviser and the science adviser of Achieve,[5] a nonprofit organization that will transform the Conceptual Framework for New K–12 Science Education Standards into the actual standards for the states to approve and accept. Dr. Murday had organized the two stakeholders meetings for the National Science Foundation (NSF) to discuss developing standards/benchmarks for inclusion of nanoscale science education (NSE) in our K–12 classrooms. Both reports from those meetings are available for all stakeholders and education experts to review at: http://www.nsf.gov/crssprgm/nano/reports/nsfnnireports.jsp (accessed December 13, 2011).

Inherent Risks of Current Complacency

Since nanoscale science and engineering is relatively new, there is high risk NSE will not achieve the presence in the pending K–12 Next Generation of Science SOL warranted by its rapidly growing impact. Despite its importance to twenty-first-century science, engineering, and technology, the nanoscale is explicitly mentioned in the NRC draft Framework for Science Education report only as a topic for grades 9–12 engineering and technology: "Using Tools and Materials." Yet nanoscale science/engineering knowledge and nano-enabled technologies will be pervasive by the time students presently in the K–12 grades enter the adult world.

This report was focused on an informed citizenry, that is, "STEM for All"—what an informed citizen should know when graduating high school to enable career and college choices and to make cognizant decisions on societal issues. As pointed out, it will also be necessary to develop standards/curricula for those individuals who will populate the science and engineering innovation workforce—"All STEM for Some." Advanced placement tracks and magnet schools, although important for these students, are not specifically addressed in this report because of limited time and a deliberate focus to impact the broadest possible cross section of students.

As benchmarks for the U.S. status, we pose three important questions:

1. Can K–12 students learn NSE?
2. Are nanoscale concepts important to science education?
3. Do we have good content ready for the classrooms?

Can K–12 Students Learn NSE?

The simple answer is an unqualified yes; if they can learn science, they can learn NSE. But the level of understanding depends on many independent variables, some of which include: the student's knowledge of math, science, and English; the way the assessment questions are given; and the ability of the teacher to teach the basic science concepts (in the textbooks) and at the nanoscale (something not in the standard textbooks). Each of the three research projects mentioned here reflects some of the variables, but not all of them.

Are Nanoscale Concepts Important to Science Education?

As highlighted in this workshop report, nanoscale concepts are essential to K–12 science education. However, the current mandate for teachers is to teach what is in the standards. Thus, there must be a strong impetus to include nanoscience in the K–12 Next Generation of Science SOL. From the perspective of those who are developing nanoscale instructional modules, nanoscale concepts will enhance the learning of basic science concepts that are currently mandated. That is to say, most of the present standards cover nanoscale concepts only implicitly, at best, and need to be revised.

NSE is a perfect approach toward meeting the current national priority to improve STEM education, since it illustrates the integration of STEM subjects. Science and engineering at the nanoscale does not separate physical sciences from biological sciences; engineering and math are very much an integral part; and there is a clear societal impact. This approach is termed *horizontal integration of STEM*. Thus, there is an opportunity for designing nanoscale modules to fill this need.

Do We Have Good NSE Content for the Classrooms?

Some good content in NSE has been developed (see Appendix A of the NSF workshop report, pp. 61–71). Most of these modules are for supplementary purposes, since the teachers are not currently required to teach nano and thus view the topic as something extra or supplemental. The present K–12 NSE modules must be revised and extended to address deficiencies such as: They do not adequately cover all of the big ideas, do not yet address learning progressions, nor have they been sufficiently vetted by K–12 teachers. Good instructional materials are costly and time consuming to develop. There are many parameters to take into consideration: level of students, connection to science classes, safety (for experiments), cost (affordability), linkage to standards and learning goals, student interest, relevance to society or job opportunity, good assessment process, ease of use by teachers, building student knowledge base and self-esteem, teaching teamwork and leadership, taking the latest research into the classrooms, and real design experience based on sound science concepts and understanding. In addition, the instructional materials need to be field tested in different classrooms in different regions (states) of the country.

Recommendations From the Report on Standards of Learning (SOL)

Recommendation 1: Appropriately introduce the nanoscale into the next-generation science SOL. Indexed to the NRC Draft Framework, the many examples in Section II of this report show how Nano Science Education (NSE) ideas might be incorporated into the new science standards for college and career readiness. A few illustrations are:

Life Sciences 1A: Structure and Function

Standard: Students know that nanoscale molecular machines (largely proteins) are responsible for biological cell mobility (cilia and flagella), muscle movement (actin/myosin), cellular metabolic transport, and controlled flow in and out of cells via ion channels in cell membranes. Students also know that nanostructures play an important role in cell structure (cytoskeleton), adhesion/traction (gecko), and self-cleaning surfaces (lotus leaf).

Physical Sciences 1A: Atomic Structure of Matter

Standard: Students know examples of structures at the nuclear, atomic, nano, micro, and macro scales. They also know the concept of surfaces and interfaces in relation to those structures.

Physical Sciences 1B: Properties of Matter

Standard: Students know that material properties change as scale goes from micro to nano; that three different reasons can evoke those changes (surface/interface effects, collective effects, quantum effects); and what specific properties may change (electrical, physical, biological, etc.)

Physical Sciences 2C: Forces in the Transformations of Matter

Standard: Students know that surfaces, interfaces, and nanostructures behave differently than do bulk materials due to an environment that is subject to different forces.

Physical Sciences 3A: Descriptions of Energy

Standard: Students know that atomic, molecular, and nanostructure self-assembly is an example of a total energy minimization process and that boundary conditions can be imposed to direct that assembly.

Engineering and Technology 1B: Nature of Technology

Standard: Students know that miniaturization into the nanoscale enables the continuing information technology revolution, and opens the possibility for close coupling of biotic and abiotic systems. It also underpins many of the potential solutions to sustainable energy and to medicine and health—both considered major economic engines for the coming decades.

Engineering and Technology 3: Technological Systems

Standard: Students examine nano-enabled technologies as state-of-art paradigms of technological systems.

Engineering and Technology 4A: Interactions of Technology and Society

Standard: Students are informed on the benefits/risks associated with nano-enabled technologies.

Recommendation 2: Establish a formal mechanism to enable NSE subject-matter experts to work with K–12 educators and workforce experts toward the continuing evolution of NSE content in standards, curriculum, and assessment. International participation should be involved.

Recommendations From the Report on K–12 Curricula and Teaching Aids

Findings: There are many websites (such as those listed in Appendices A, pp. 61–71, and C, p. 76, of the NSF workshop report) with materials that address curricula supplements, teaching aids, and science and engineering fair projects. In particular, the NSF-funded efforts have been very productive at developing innovative approaches to K–12 nanoeducation. However, the materials are widely dispersed, are of nonuniform format, have varying degrees of refinement, and are not integrated into a learning progression. Taiwan and Australia have a better-developed NSE learning progression. As the new Framework for K–12 Science Education and the associated standards are adopted, it is imperative that we move from stand-alone, one-off teaching-aid modules to a coherent, progressive sequence. In his 2011 State of the Union speech, President Obama spoke of connecting every part of America to the digital age—a student who can take classes with a digital textbook. With its rapid progress, NSE is a perfect topic to utilize digital textbooks, where continual updates are more readily implemented.

Recommendation 1: Create a central website that provides a registry for NSE materials. The website materials need to be of high quality in a format readily usable by K–12 teachers, carefully indexed to the Next Generation of Science SOL, and readily accessible from the National Science Teachers Association (NSTA) website.

Recommendation 2: Additional well-designed, highly interactive, media-rich, online learning tools should continue to be developed with attention to utilizing the new approaches being widely utilized by youth.

Recommendation 3: A National NanoFlexbook with free electronic modules for teachers and students should be created. A national competition might be used to stimulate additional high-quality "flexbook" materials to provide a coherent, vetted learning progression.

Recommendations From the Report on K–12 Teacher Education and Training

Findings: There will be growing inclusion of nanoscale science, engineering, and technology into standards of learning. There are also growing learning resources for K–12 audiences that address nanoscale science, engineering, and technology. Teachers will need to be trained to use these resources.

Recommendation 1: Provide funding for teacher professional development in NSE topics, which are not yet included in the traditional venues. That professional development might include the creation of new teaching modules as part of the effort.

Recommendation 2: The various NanoCenters can be a vital resource to provide materials, training, and information. They should be encouraged to be more proactive toward K–12 teacher training, particularly in their states.

Recommendation 3: A recurring international workshop should be held for K–12 NSE educators to share experiences and ideas. These workshops must include K–12 teachers, college/university professors, science and engineering professional societies representation, and leaders from state and federal education agencies.

By 2020, when present K–12 students are finishing their primary/secondary education, the impact of NSE will be pervasive. The Next Generation of Science SOL efforts present an opportunity, likely a once-in-a-decade opportunity, to influence an important, widely used document that will shape K–12 STEM education.

It is crucial to insert information about nanoscale science, engineering, and technology that will be a major factor during the adult life of these students. NSE is relatively new and evolving rapidly; it is premature to consider it as a discipline of its own. In fact, as NSE is pervasive to all areas of science, both conceptually and in application, it may be best addressed by embedding it throughout all science courses. The most effective approach to inserting NSE into the standards will be to adapt the recent NRC recommendations for K–12 engineering education by infusion (embed relevant learning goals for NSE into the pertinent science and engineering standards) and by integration (mapping) of "big ideas" in NSE into the pertinent standards.

Scenario for Case Study 2

The efforts to work with Achieve in the next twelve months could be successful for inclusion of new standards that include NSE for grades K–12. If successful, the new standards will be presented to the governors for acceptance. The development of the curriculum based on the new standards will be a state-by-state effort, allowing for another twelve–twenty-four months for completion. We will be well into 2015 before teachers in the classrooms are exposed to the new materials, and if teacher training is involved, we may be extended into another two-year period. At the earliest, students will start learning NSE in grades K–12 between 2018 and 2020. It will take another ten years of assessment of the materials before updates or changes can be agreed upon by the stakeholders for new materials to be expanded for inclusion. This would bring us to 2030, unless schools change to digital textbooks that can be updated with the new scientific discoveries as they manifest.

Questions

1. How can the United States compete in a global market if our students are not learning the most advanced scientific information on the planet in a timely manner?
2. What are the social implications caused by the delay and difficulties involved in updating the U.S. curriculum?
3. The United States does not have a national curriculum mandate, although the testing and assessments are mandated nationally. What are the social implications of this dichotomy in education?

Case Study 3: Global Comparative Efforts to Include NSE for K–12

Taiwan

Taiwan began its National Nanotechnology Program (NNP) in 2003, encompassing four sectors : industrialization of nanotechnology, academic research, core facility sharing, and human resource development.

The aim through the establishment of common core facilities and education programs is to achieve academic excellence in nano research, to create innovative industrial applications, and to speed up the commercialization of nano-enabled technology. There were ten government agencies , including the Ministry of Economic Affairs, the National Science Council, the Ministry of Education, the Department of Health, and the Environmental Protection Agency, that sponsored Phase I of this program with a total budget of about US$620 million for a period of six years (2003 to 2008). Beginning in 2009, Phase II of the NNP was launched for another six years.

The ultimate goals of the NNP human resource development program (HRDP) are:

- Cultivating leaders and researchers for nanoscience and technology
- Nurturing experts and entrepreneurs for nanoindustries and business
- Educating the general public to understand and support nanotechnology
- Imbuing a lifelong learning subject and education environment

The human resource development program can be classified into three categories, as shown in Figure 13.3. To carry out the human resource development program efficiently and to spend the budget effectively, a program office was established to take charge of the planning, coordinating, administrating,

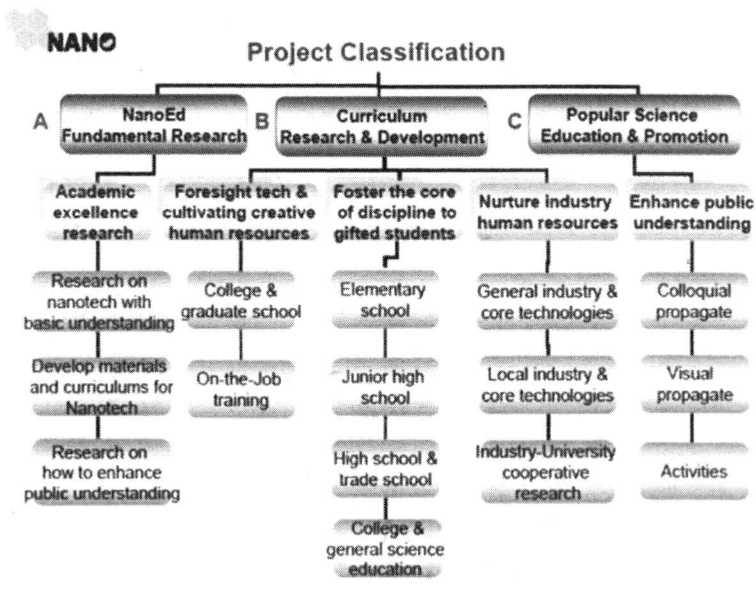

FIGURE 13.3
Categories for human resource development in Taiwan NNP.

and auditing work. Ten regional centers along with an e-learning platform were created and sponsored across the country. Five of them were devoted to advanced education, that is, college level and beyond, the others for K–12 education. It is the first time in Taiwan that advanced science and technology has been introduced to K–12 education through a national human resource development program. Unlike other countries, Taiwan extended its nanoeducation into K–12 from the very beginning of its NNP in 2003.

It has the most mature nanoeducation effort, with the goals of:

- Cultivating K–12 seed teachers by offering lectures, workshops, and lab training courses
- Developing universal K–12 teaching and learning materials on nanotechnology
- Promoting international academic collaboration on K–12 nanotechnology education
- Collaborating with the National Taiwan Science Educations Center, the National Museum of Natural Science, and the National Science and Technology Museum to promote nanotechnology education for the general public

The range of universities, schools, and K–12 teachers engaged in the nanoeducation effort is shown in Table 13.2. In addition to the previously mentioned activities, there are annual K–12 teacher workshops, nano innovation contests, and the Taiwan nano exhibition. Seed teachers with science background from K–12 schools were recruited to nearby universities or colleges for training in the form of lectures, hands-on experiments, and lab training courses on a regular basis for a period of six months to one year.

TABLE 13.2

K–12 Nanotechnology Education Statistics

NANO

K-12 Nanotechnology Education

K-12 Nanotechnology Education	
Universities (2003-2009)	**401**
leading schools (2003-2009)	**364**
K-12 active seed teachers (2003-2009)	**1,082**
K-12 potential seed teachers (2003-2009)	**8,270**
Activities 2003-2009	
Conferences	**235** (86,765 participants)
Seminars, competitions, exhibitions, workshops, speeches	**2,245** (26,5797 participants)

It is important to note that Taiwan, like the United States, did not have a national curriculum mandate before this program was initiated. However, the nanoscale size of scientific discovery was deemed important enough for the future of their country to bring all the stakeholders together for cooperation to develop the curriculum together, enlisting the K–12 seed teachers from the very beginning of the program.

The resulting teaching materials developed by the seed teachers are reaching all K–12 students. With this nano knowledge, the K–12 seed teachers were encouraged to develop a wide variety of teaching materials, experimental kits, animated cartoons, etc., suitable for K–12 students. More than 1,600 books, cartoons, teaching plans, lecture notes, teaching kits, and references have been developed. Many of the K–12 NanoEd materials, such as *A Fantastic Journey of Nana and Nono, NM Magic House*, and *The Wonderland of Nanotechnology* have been translated into other languages and have gained great popularity. This effort in K–12 NSE education is now in the process of being evaluated.

Australia

The Melbourne Declaration on Educational Goals for Young Australians was signed by state, territory, and commonwealth ministers for education in 2008. The agreement highlighted objectives for Australian education, including the development of a national Australian curriculum. Previously the curricula were determined by each state and territory. For nanoeducation, this posed a challenge in that some states or territories included nanotechnology in the curriculum and others did not. However, nanotechnology has been specified under the new curriculum. The Australian Curriculum for science years K–10 was released in December 2010; it details what students are expected to learn for each year and to what competency.

Assessment of the curriculum is still determined by the state and territories, as is how to apply the curriculum in their classrooms. Currently, state and territory education departments are assessing existing curricula and education resources while testing the new curriculum in select schools.

On a national level, the Department of Education, Employment and Workplace Relations (DEEWR) and the Ministerial Council for Education, Early Childhood Development and Youth Affairs (MCEECDYA) has engaged Education Services Australia (ESA), a not-for-profit government corporation, to develop two tools to assist Australian teachers with the new curriculum: Scootle[6] and Curriculum Connect.[7]

Both resources are available to teachers and educators through their local education department, educational organization, or association, or by registering with the ESA. The websites are not available outside Australia and New Zealand. Both sites are portals accessing a pool of digital resources for teachers to use. Scootle allows teachers to seek out resources by search terms or resource content. Curriculum Connect, which is still in trial format, showcases resources suitable for each component of the Australian curriculum.

In 2009, the Department of Innovation, Industry Science and Research (DIISR) also released the innovation agenda, "Powering Ideas: An Innovation Agenda for the 21st Century." Part of the agenda was the creation of the National Enabling Technologies Strategy (NETS). Its purpose is to provide a responsible framework for the development of enabling technologies, which is defined as nanotechnology, biotechnology, and other emerging technologies.

A component of NETS is the Public Awareness and Community Engagement section (NETS-PACE). NETS-PACE provides factual information to the public while seeking feedback on public attitudes toward these technologies. To this end, NETSPACE undertakes Web 2.0 activities, community events and surveys, and material publication and also develops education resources for K–12 students.

TechNyou science education resources (http://education.technyou.edu.au/) provides high school science teachers education resources relevant to the Australian Curriculum covering the fields of nanotechnology and biotechnology. The website will present the resource in two formats. Format 1, shown in Figure 13.4, displays how the resource relates to the various applications of nanotechnology and biotechnology and where the two fields overlap. Format 2, shown in Figure 13.5, directly relates the resource to the requirements of the National Curriculum.

Each format has different benefits. Format 1 shows the diversity of applications of the fields, while Format 2 allows opportunity to reveal how these technologies relate to more traditional disciplines, such as physics and chemistry's relevance to nanotechnology. The different displays also cater to different audience needs. Format 1 is for teachers who wish to explore the technologies and applications; Format 2 is for teachers who simply wish to find relevant material for their classroom plans (see Figure 13.5).

The following attributes were deemed necessary in all new resources:

- Highly editable and downloadable material with multiple entry points to the website
- Inquiry-based learning to develop critical thinking, and a combination of pracs and hands-on learning with virtual labs for high science using real university data
- Games with different difficulty levels and savable progress
- Contemporary issues, such as water purification, addressed using current science and showcasing scientists and careers
- Teacher professional development (TPD) workshops to support the resource using the train-the-trainer model with the assistance of local STAs and education departments

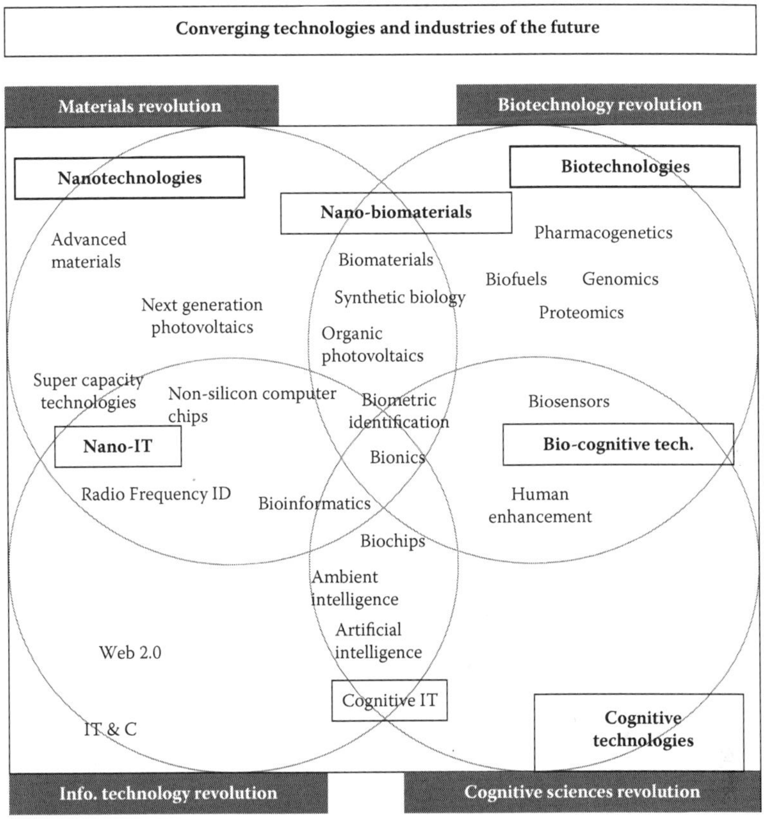

FIGURE 13.4
Format 1 of nanotechnology and biotechnology resources.

The assessments and testing were accomplished in 2011 and the resources and teacher professional development program can be accessed at the new website: (http://education.technyou.edu.au/).

Europe

The European Commission, in its document "Nanosciences and Nanotechnologies (N&N): An action plan for Europe 2005–2009," calls upon member states to "foster interdisciplinary training and education for R&D in N&N, focusing on physics, chemistry, biology, toxicology and ecotoxicology, and engineering, but also including entrepreneurial studies, risk assessment, and social and human sciences where appropriate" (European Commission, 2005). In line with this, the GENNESYS initiative has devised a new education scheme and training structure and made a distinct proposal of how to establish a European education in nanomaterials science, which involves all sectors engaged in GENNESYS.

Year 10 Content descriptions		
Science Inquiry Skills	**Science as a Human Endeavor**	**Science Understanding**
1. Questioning and predicting Research information from a variety of sources to formulate scientific questions and develop testable hypotheses	**1. Nature and history of science** ♥ ♡ Ideas of the world change as scientific theories and models develop	**1. Evolution** Evolution by natural selection and the diversity of plants and animals
2. Investigation methods Choose the most appropriate investigation type for the question including experimental investigations involving repeat trials and replicates and multiple variables	**2. Influence of science** ♥ ♡ Science provides reliable knowledge and enables valid predictions and conclusions to inform choices	**2. DNA** ♥ The structure and function of DNA, genes and chromosomes
3. Using equipment Select and use specialized equipment and materials that are suitable for the investigation	**3. Contribution of scientists** ♥ ♡ Scientists are recognized by society in various ways for their contribution to human understanding	**3. Genetics** ♥ The role of genes in determining patterns of inheritance and the chemical processes in cells
4. Managing risk Conduct a risk assessment of a practical activity	**4. Collaboration in science** Science research commonly involves teams of scientists with expertise from a diversity of specializations	**4. The Universe** The evidence supporting the big bang theory and the major processes that have produced galaxies and planetary systems
5. Observing and measuring Collect data in a consistent efficient and ethical manner including methods that use ICT	**5. Science and culture** The knowledge of a cultural group can contribute to scientific understandings in areas such a agriculture, sustainability and technological design	**5. Plate tectonics** Plate tectonics explains global patterns of geologic activity, continental movement, and the characteristics of the Australian continent
6. Analyzing results Represent and analyse data appropriately including using simple statistical methods and ICT	**6. Science careers** Science engineering and technology are interdependent in the work of many careers and industries	**6. Forces and motion** ♡ Forces, motion and conservation of energy, their interactions and how they can be described qualitatively and quantitatively
7. Developing explanations Draw conclusions that are consistent with the evidence and critique these conclusions with reference to scientific concepts		**7. Matter and energy** ♡ Large and small scale physical systems rely on dynamic interactions between matter and energy
8. Communicating Communicate scientific ideas and information for a particular audience and purpose including making evidence-based arguments		**8. Atoms** ♡ The periodic table as an organizer based on the electronic structure of elements
9. Reflecting on methods		**9. Chemical change** ♡ Chemical reactions can be represented by symbolic equations and are affected by factors such as temperature, catalysts

♥ = Biotechnology
♡ = Nanotechnology

FIGURE 13.5
Format 2 of nanotechnology and biotechnology resources.

These initiatives could—together with the Bologna Process,[8] which has facilitated cooperation between and comparison of European Higher Educations—form a common European strategy for the education of future N&N scientists and engineers. However, so far only national and/or local initiatives exist, and they range from single university courses to full five-year science or engineering programs. True interdisciplinarity is reflected in those educational programs where N&N have been built-in from the beginning,[9,10] and where courses from different subjects are fully concerted or integrated. These pioneering programs could preferably act as models for a more general initiative on creating a European scheme for higher nanoeducation.

To ensure some sort of impact, most European Union (EU) initiatives on education aim at reforming the uniform higher education area, or at highlighting educational needs of the society in very general terms. Examples of the latter are the Key Competences Framework (European Commission, 2007), the European Qualifications Framework (European Commission, 2011b), and the project New Skills for New Jobs (European Commission, 2010). They all intend to improve the educational systems within Europe so as to better match the supply of skills to the needs of the labor market. In the latter of these projects, there is a focus on development of the vocational education sector.

NSE for K–12 in Europe Would Involve Twenty-Three Different Languages

When it comes to introducing NSE at a K–12 level, the development of common standards should be a possible way forward. However, difficulties occur in terms of developing and adopting common core (science) standards within the EU. Not only will it have to be formulated in twenty-three different languages, the work will also involve authorities and organizations at a national level or, in some European countries, even at a local level. Since no two European countries have similar organizations of their K–12 education, this is not likely to happen.

One of the underlying reasons for differences in K–12 education is the variation in educational attainment within the population of Europe. All European countries have a common goal in raising the educational attainment in general, but the strategies for doing this necessarily differ (European Commission, 2009). One should also keep in mind other significant demographic and historical differences, which always will be reflected in a lack of clarity in official messages delivered by the EU. European initiatives on education, for example, the "Strategic Framework for Education and Training" (European Commission, 2011a) thus mainly result in descriptive policy documents and statistics. When looking for "priority areas" within the European Commission's work on education and training, one finds priorities such as teacher education, key competences, language learning, information and communication technologies, math, science and technology, active citizenship and social cohesion. All this can be generally agreed upon among the states. However, narrowing subject areas by identifying nanoscience and nanotechnology as priority areas in K–12 education probably would lead to endless discussions between the member states, thus paralyzing further work.

Several European countries are in the process of developing new (science) K–12 curricula. These show hardly any sign of NSE, and from the writings one can conclude that the degree of interdisciplinarity in the science curricula (still) decreases when moving from kindergarten toward year 12.

Teachers' Education

That interdisciplinarity is more prominent in the first years in school, which is in line with the way most European teachers' educations are structured. Science for the teachers of the early grades transforms into biology, chemistry, and physics for the grade 7–12 teachers. It would require a fundamental change of most European teachers' educations to fully include the engineering perspectives of STEM education and to promote interdisciplinarity in grades 7–12. Since the timescale for such changes is long, a more viable approach would be to strongly invest in further training for teachers.

This strategy has been adopted, for example, in Sweden, where the equivalent of US$3,000 per teacher is being spent on further education between the years 2007 and 2014. However, the money is primarily invested in developing the teaching of mathematics, Swedish (first language), and English (second language), since these subjects in Sweden are considered key subjects.

In 2007, the McKinsey report on the world's best performing schools concluded: "Above all, the top performing systems demonstrate that the quality of an education system depends ultimately on the quality of its teachers" (McKinsey, 2007). With this in mind, it is surprising that so few European countries invest in reforming and developing their teachers' education to make it up to date and an attractive choice for young people.

Resources Outside School

Within Europe there are only a few official EU Web resources presenting NSE for young people. Timefornano[12] is a website financed by the European Commission's 7th Framework Programme, FP7, and run by a consortium of European Science Centers. The content is a mix of pure information, pictures, competitions, and quizzes. Which age group the site is targeting remains to be discovered. But to reach younger kids in the smaller European countries, the content has to be translated into more than eight languages. Nanoyou,[13] the most extensive Web resource, is an initiative financed within the FP7 as well, and aiming at young people in the 11–25 age group. Partners in this initiative include the nanoscience centers in Aarhus (DK), Barcelona (E), Cambridge (UK), and Grenoble (F), which all are involved in higher education on NSE. The website is one arena where NSE students can act as ambassadors for NSE in the K–12 age group.

Future Opportunities for Interaction

The platform Nanoyou could be developed to become the arena where European NSE students meet teachers, pupils, and the public. But the critical mass is yet to be reached—and overcome. Thus far, only about fifty European schools have become partner schools of Nanoyou. The platform also lacks connections to industry. Such cooperation, showing applications—examples from reality—would appeal to all students.

Scandinavia NSE Programs

The project Nano Connect Scandinavia[14] is a resource aiming at supporting networks and spreading knowledge to "tie universities, businesses, scientists and students together, creating a world class nano cluster in south-western Scandinavia." It is mainly used by students of NSE, especially the newly graduated ones looking for jobs within the nanoindustry of Scandinavia.

The Scandinavian NSE programs, which truly focus on interdisciplinarity, for example, the ones at the universities of Lund (SE) and Aarhus (DK), so far seem to be the ones most successful in providing the developing nanoindustry with a highly skilled workforce. The success of these university programs could be used to further promote interaction between students, K–12 teachers, and industry. Nano Connect Scandinavia is a possible platform for this interaction, but to reach a greater impact, a European arena with long sustainability is needed. This necessarily has to be influenced by interdisciplinarity and a true bottom-up perspective. The use of ideas and experiences by our students is crucial.

Scenario for Case Study 3

Taiwan has been the earliest adopter of developing NSE educational programs for K–12, with the most advanced curriculum development that has also included seed teachers since 2003. Australia also created a national curriculum that started in 2008 and reached the classrooms in 2011. These two countries are far ahead of the United States, which will not be able to introduce a national curriculum. The steps being taken to include NSE in the current changes in the science of learning standards are our only avenue to eventually introduce NSE in K–12. The delays of another decade while states prepare the curriculum, set up teacher training, and finally perform assessments will have many social implications in a global competitive market.

Questions

1. Develop a strategy based on the full report of how you would debate these issues with stakeholders to prevent catastrophic failure for global competition.
2. Our Industrial Age manufacturing has already been outsourced in other countries. Look at the report on international programs for Taiwan, Australia, New Zealand, Europe, and the Scandinavian countries. Can we afford to keep our students undereducated so that the nanomanufacturing that is already in stages of development will also be outsourced, while we try to train workforce technicians by 2015?

Case Study 4: Global Acceptance of NSE Standards, With Implications for Third World Impact

Introduction

Nanotechnology is one of the most rapidly developing fields of technology, with many promising applications in medicine, energy, manufacturing, and communications.[15] Like any new technology, it raises ethical issues. The possible benefits and harms of nanotechnology, as well as its implications for international relations in science and technology policies, are under debate. UNESCO is taking initiatives to map the ethical dimensions of nanotechnology from a global perspective in an effort to explore the implications for its member states and possible actions for the organization.

A first examination of the ethical dimensions was done in a special session on nanotechnologies during the third Ordinary Session of COMEST in Rio de Janeiro in December 2003. It was also an issue presented in a number of debates during the fourth Ordinary Session of COMEST in Bangkok in March 2005. Following these meetings, the Division of Ethics and Technology of UNESCO then decided to establish a group of experts to explore the ethical issues in relation to nanotechnologies.

As an output of the early steps of the working methodology, a characterization of nanotechnologies by its interdisciplinary approach emerged as a possible area of consensus. Proposals based on the discussion among ethics experts can be divided into three actions: awareness-raising, education, and policies for research and development. Before discussing these, we will address some central features of nanotechnology.

Some Central Features of Nanotechnologies

Interdisciplinary and Cross-Disciplinary Dimensions

One way to characterize nanotechnologies is to refer to their interdisciplinary dimension. There is not one particular technology that should be called nanotechnology, but the word rather embeds several technologies and sciences. From the scientific point of view, nanotechnologies question the usual distinction between scientific disciplines. Furthermore, the distinction between science and technology is very difficult to make in this field (scientists and engineers have to work in the same teams). Finally, the very boundaries of science and the humanities are blurred as the nanoscale is being explored. Constructive interactions between science and the humanities must be raised and reinforced to avoid mutual distortions and misunderstandings.

Nanotechnologies as an Opportunity

Nanotechnologies are an enabling technology. Therefore nanotechnologies require a holistic approach, which implies a truly interdisciplinary dialogue.

This applies to any of the proposed actions proposed here: debate, education, research, and policy. Reciprocally, the development of nanotechnologies may be an opportunity to develop interdisciplinary cooperation, helping to address one of the most basic demands of ethics of science and technology.

Need for Awareness-Raising and Debate on Nanotechnologies: Some Specific Issues

Science Fiction, Apocalyptic, and Utopian Scenarios to Be Dismissed

Having an early, informed, and interdisciplinary public debate is required for nanotechnologies. The failure to have this kind of debate is to a large extent the cause of the backlash and public defiance encountered, for example, with genetically modified organisms. The case is even reinforced by the manifest influence of science fiction in the current debate and public opinion about nanotechnologies. The shaping of the debate by science fiction considerations would be unfortunate (for example, it should be made widely known that the apocalyptic visions of the "gray goo" scenario or the utopian views of nanomanufacturing are unrealistic in the light of current scientific knowledge). Following are some of the issues that could be brought in such early, informed, interdisciplinary public debates. Issues mentioned in a subsequent section ("Need for Research and Development Policies") equally require public, informed, and interdisciplinary public debate.

Environmental Impact and Health Issues

An early, informed, and interdisciplinary public debate could take place on the environmental impact and health issues in order to maximize the benefits that can be expected from nanotechnologies. The opportunities and risks of nanotechnologies in products and applications that involve human contact or that may affect the environment must be weighed.

The difficulty is that the potential toxicity of nanoengineered particles is subject to scientific uncertainty in a very fundamental way. Indeed, the very definition of the toxicity of these particles is problematic. (The definition is clear; the next step would be to define what mechanism is causing the toxicity to proceed with testing.) Furthermore, there are no clear views on how this toxicity, if defined, could be scientifically and indisputably tested. Finally, there are no scientific studies on the toxicity of many particles (as the recent debate in the European Parliament emphasized). One of the issues could be that such a toxicity may be slow to manifest itself, as was the case for asbestos. Therefore, the question of the applicability of the precautionary principle would need to be studied and discussed, and scientific uncertainty should not lead to skipping the necessary debate. (See also the subsequent discussion in the subsections "Scientific and Technical Knowledge" and "Ethical Research and Ethics in Connection with Legal Issues.") In this connection, issues of risk analysis and standardization require in-depth ethical, and not only scientific, consideration.

Nanomedicine Issues

Nanotechnologies applied to medicine give rise to several issues that need to be discussed in an early, informed, and interdisciplinary public debate. The easy availability of new diagnostic methods is one issue (e.g., susceptibility to diseases could be measured, with implications toward insurance coverage). There is also another set of issues related to prospects of enhancement of the human body (e.g., What is a genuine part of the body? What is an enhancement and who defines it?).

Privacy and Confidentiality

An early, informed, and interdisciplinary public debate could take place in regard to basic ethical and legal notions of privacy and confidentiality. Indeed, nanotechnologies allow for unprecedented surveillance devices (ranging from nanocameras to nanotracers locatable by GPS), and the question of the acceptability and conditions of use of such devices would need to be considered.

Need for Ethics Education

General Need for Ethics Education Reinforced in Nanotechnologies

Public engagement and educational strategies for nanotechnologies are deemed necessary. The need for adequate ethics education of scientists is commonly emphasized. It is a consequence of the ethical demands for interdisciplinarity and a holistic view in the broadest sense. The interdisciplinarity of nanotechnologies therefore strengthens the need for both formal and informal ethics education for scientists and engineers involved in nanotechnologies and for the public at large. Incidentally, it also strengthens the need for science education for professionals in social and human fields involved in nanotechnologies ethics. The role of nongovernmental organizations (NGOs) in the decision-making process is central for raising awareness and monitoring the issues. Governments should also be involved in public debate on nanotechnologies, as they play a determining role for heavy investments and for possible legal actions. Identifying interested groups is also particularly important, and different professional viewpoints need to be factored in. Attitudes toward nanoethics education may differ according to cultural background. To the extent possible, education to encourage critical thinking should be favored.

Specific Additions to Content of the Programs

The demand for ethics education for scientists and engineers is addressed by UNESCO through its Ethics Education Program (EEP). In this framework, the consideration of nanotechnologies will have implications for the content of ethics education. (See, for example, the issues mentioned in two previous sections, "Some Central Features of Nanotechnologies" and "Need for Awareness-Raising and Debate on Nanotechnologies: Some Specific Issues," e.g., the questioning of the usual physical and philosophical rules when it comes to the nano scale; the implementation of the precautionary principle.) The challenges that cultural diversity represents for the development of a core program should also be taken into account.

Guidelines

In addition to a general core program on ethics education that could be adopted for different regions, specific ethical guidelines could also be elaborated as an indicative basis (voluntary compliance guidelines) and be incorporated in the education programs (see also the discussion in a subsequent subsection, "Voluntary Guidelines") along with other UNESCO-originated education materials. The elaboration of such guidelines would require an extensive consultation process and could take place in the framework of the ongoing reflection of UNESCO on science ethics.

The objective of the guidelines is to provide practical guidance not only to individual researchers, but also to UNESCO member states in their implementation of the progressive realization of the ethical recommendations regarding nanotechnologies, especially in the national and regional context. The guidelines represent a first attempt by UNESCO to propose a harmonization of ethical principles related to nanotechnologies and to recommend actions to be undertaken for research and applications in this field.

Need for Research and Development Policies

Scientific and Technical Knowledge

There is a lack of knowledge in many issues pertaining to nanotechnologies, which demonstrates the need for more scientific research. This lack of knowledge should be acknowledged so that further research can address the many issues pertaining to nanotechnologies (e.g., environmental and health impact, fundamental properties of nanoparticles). This research (as mentioned previously in the subsection on "Interdisciplinary and Cross-Disciplinary Dimensions") should be interdisciplinary.

Social Sciences Research to Guide Policy

Strengthening current social science research would also be necessary to determine the social and economic impact of nanotechnologies in order to guide appropriate research, industrial, and development policies (see the discussion in the subsequent subsection on "Nanotechnologies and Development"). This research would need to be interdisciplinary (see the previous discussion in the subsection on "Interdisciplinary and Cross-Disciplinary Dimensions" for issues that should be addressed).

Ethical Research and Ethics in Connection With Legal Issues

Research in ethics needs to be developed in association with nanotechnologies. Ethical considerations are still insufficient within the vast financial effort devoted to nanotechnologies. Ethicists should be encouraged to address nanotechnologies, and nanotechnologies teams should endeavor to be in close interaction with scientists.

Nanotechnologies and Development

Even being interdisciplinary, scientific research alone cannot solve value problems regarding nanotechnology. Social science research (see the previous subsection on "Social Sciences Research to Guide Policy"), as well as more debate and awareness are needed (see the previous section on "Need for Awareness-Raising and Debate on Nanotechnologies: Some Specific Issues"). Therefore UNESCO could assist countries in identifying technologies that are most appropriate and relevant for development. It is necessary to distinguish the actions to be undertaken at an international level and issues that need to be addressed from a local perspective. Countries whose national resources may be replaced by nanoengineered materials, for example, should rather look for a best use of their resources.

In this debate, the following topics should be considered:

- The utility of particular nanotechnologies for development
- The comparative advantages of a given nanotechnology for a given country
- The ability to turn research effort into applications that are useful for development
- The possible environmental risk: risk assessment and management
- The impact of the intellectual property regime
- Sharing of benefits[16]
- International cooperation[17]

Voluntary Guidelines

Early assessment of the ethical and social implications of nanotechnologies raises opportunities to develop a normative framework in this field. Therefore, voluntary guidelines on science ethics and nanotechnologies, as mentioned in the previous discussion in the subsection on "Guidelines" (particularly in regard to safety issues), could be elaborated on in a consultative process and proposed as an indicative ethical framework for countries, corporations, or scientific organizations. Such guidelines could also inspire national regulations.

Institutionalization

Creation of national commissions or committees on the ethics of science and technology to deal not only with nanotechnologies, but also with other emerging technologies, should be also envisaged.

Ethics Education Program: During the 32nd UNESCO General Conference (2003), member states expressed the need to initiate and support teaching programs in ethics, not only in bioethics, but in all scientific and professional education. On the basis of these recommendations and statements, UNESCO initiated the Ethics Education Program in 2004 (UNESCO, n.d.). The overall objective of UNESCO activities in this program is to reinforce and increase the capacities of member states in the area of ethics education. This is a long-term objective. For the biennium 2004–2005, the activities in the area of ethics teaching have primarily focused on Eastern and Central Europe. During 2006–2007, priority has been given to Southeastern Europe and part of the Arab region (Gulf region).

New reports and studies: NSF and WTEC (World Technology Evaluation Center) have published a report outlining the research directions for societal needs in 2020 (Roco, Mirkin, & Hersam, 2010). A new study, *Current Outlook: International Societal Needs for Nanotechnologies*, includes opinions of leading experts from over thirty-five countries and brainstorming meetings hosted by the World Technology Evaluation Center[18] (WTEC) in 2010 in Chicago, Hamburg, Tokyo, Singapore, and Arlington, Virginia. From a position in 2000 where "science leaps ahead, ethics lags behind" (Mnyusiwalla et al., 2003), we are in the process of achieving a more appropriate balance between science and ethics internationally.

A new resource for intellectual exchange: The Society for the Study of Nanoscience and Emerging Technologies (S.NET)[19] is poised to become an international organization promoting open intellectual exchange toward the advancement of knowledge and understanding of nanotechnologies in society. S.NET represents diverse communities, viewpoints, and methodologies in the social sciences and humanities. It welcomes contributions from scientists and engineers that advance the critical reflection of nanotechnologies and related developments.

Coordination and Harmonization

The call for international coordination and harmonization is loud and clear with working parties in various venues. The initial emphasis with regard to the ethical and societal dimensions concerns specific cultural values and citizens' attitudes at the international and European Union levels. Recognition of these differences is an important prerequisite for the international diffusion of nanotechnologies products and processes.

There is a two-pronged approach related to the ethical, legal, and societal issues (ELSI):

1. The proactive and anticipatory approach consists of imagining potential or likely future applications of nanotechnologies in society and then to appreciate and evaluate their impacts. Prospects of human enhancement are currently proving divisive.
2. The other approach attempts to understand nanotechnologies as a sociopolitical project. This involves tracking the trajectories that are continued and intensified.
   - Assessing the visionary dreams and societal expectations that drive the research funding
   - Tracking the funded research to detect disruptive results

The perceived divisions of moral labor (ethical considerations are delegated to advisory committees) has been and will continue to be challenged, for example, by the Code of Conduct for Nanotechnology Research proposed by the European Commission.

A society that observes itself in an experimental mode must repeatedly ask itself, "How are we doing?" Since answering this question involves judgments, interventions, and calls for action, this goes beyond the role of nanotechnology observatories as they are currently perceived. This will open the door for new requirements to include social science and humanitarian studies.

Scenario for Case Study 4

Picture your future if the United States does not include nanoscale science or ethics education for science and technology in the education standards now. Teachers are already being trained all over the world in understanding bioethics and how nano-biotechnology can change the future. The international communities understand the importance of nanoscience education, including the ethical and societal implications of all the emerging technologies of this decade. How and what could you change now that would guarantee our innovative competency as a global leader for this decade?

Questions

Questions and issues to consider if these efforts to develop new science benchmarks/standards that include nanoscience education are unsuccessful:

1. Do we have to wait for top-down congressional laws to change our methodology in the classrooms of America?
2. What do we want to keep in our classrooms that we are already doing well?
3. What do our students need to know and be able to do in the twenty-first century?
4. Can local communities change the current system from the bottom up?
5. Can teachers learn the material on their own from the free resources developed by NSF funding?
6. If science is based on inquiry, why can it not be taught as inquiry?
7. If a teacher does not know the answer to a question a student might ask, why not turn it into inquiry, working together to search the Internet for resources that answer the question?
8. Many of the resources developed for early grade levels are also available on the Internet as informal science, along with museums participating in this area of nanoscience education.
9. Teachers are not expected to be scientists, so why not try some of the nanoscience games[20] with students to get started?
10. Rethink education from the bottom up and learn together.
11. Should education regarding ethics and social implications be included in primary, secondary, MBA, and humanities courses? A new board game—NanoVenture: The Nanotechnology Board Game,[21] which explores the connections between nanoscience, nanotechnology, and society—may be a useful tool.
12. Summarize your learning of new ideas and information presented in this chapter by completing the following flowchart:

Flow Chart

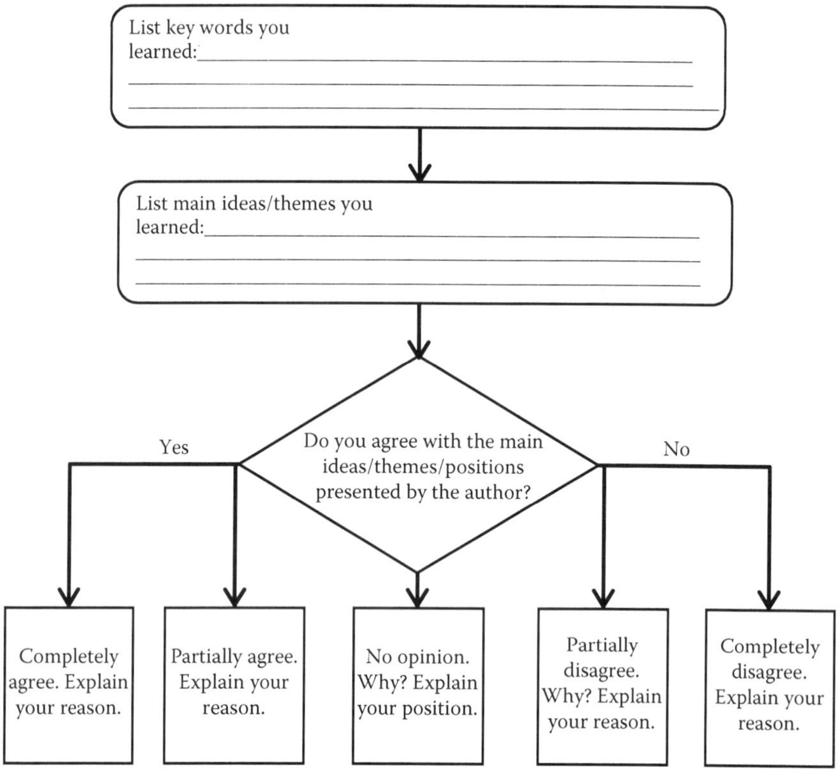

Notes

1. National Nanomanufacturing Network. Retrieved December 14, 2011, from http://www.internano.org/content/view/66/159/
2. Institute of Education Science (IES). Retrieved December 14, 2011, from http://ies.ed.gov/aboutus/
3. What Works Clearinghouse (WWC). Retrieved December 14, 2011, from http://ies.ed.gov/ncee/wwc
4. National Center for Education Statistics (NCES). Retrieved December 14, 2011, from http://nces.ed.gov/timss/table07_4.asp
5. Achieve. Retrieved December 13, 2011, from http://achieve.org/
6. Scootle. Retrieved December 14, 2011, from http://www.scootle.edu.au/ec/p/home
7. Australian Curriculum Connect. Retrieved December 14, 2011, from http://www.esa.edu.au/projects/australian-curriculum-connect
8. The official Bologna process website 2010–2012. Retrieved December 14, 2011, from http://www.ehea.info

9. Interdisciplinary Nanoscience Center, Aarhus University, Denmark. Nanoscience curriculum. Retrieved December 14, 2011, from http://inano.au.dk/education/nanoscience-curriculum-english-version/
10. Lund University, Sweden. Engineering Nanoscience Masters program. Retrieved December 14, 2011, from http://www.lth.se/english/education/programmes/master_engineering/engineering_nanoscience
11. International Congress on Nanotechnology and Research Infrastructures, Barcelona, Spain, 26–28 May 2010. Retrieved December 14, 2011, from http://www.gennesys2010.eu
12. Time for nano. Retrieved December 14, 2011, from http://www.timefornano.eu
13. Nanoyou. Retrieved December 14, 2011, from http://nanoyou.eu
14. Nano Connect Scandinavia. Retrieved December 14, 2011, from http://www.oresund.org/nanoconnect
15. The following discussion is an outline of a UNESCO policy advice on nanotechnologies and ethics. The working document was drafted by the Division of Ethics of Science and Technology and resulted from the second meeting of the group of experts on the ethics of nanotechnologies, UNESCO HQs, 6–7 December 2005. The final copy is available on the Internet. Retrieved December 13, 2011, from http://unesdoc.unesco.org/images/0015/001521/152146e.pdf
16. Similarly to the disposals of the Universal Declaration on Bioethics and Human Rights: Benefits from advances in nanotechnology shall be made available to all, with due regard for the dignity and human rights of each individual.
17. Similarly to the disposals of the Universal Declaration on Bioethics and Human Rights: In the framework of international cooperation with developing countries, states should seek to encourage measures enabling: (1) assessment of the risks and benefits pertaining to nanotechnologies to be carried out and abuse to be prevented; (2) the capacity of developing countries to carry out research on nanotechnologies, taking into consideration their specific problems, to be developed and strengthened; (3) developing countries to benefit from the achievements of scientific and technological research so that their use in favor of economic and social progress can be to the benefit of all; (4) the free exchange of scientific knowledge and information in the areas of nanotechnology to be promoted.
18. World Technology Evaluation Center. Retrieved December 14, 2011, from http://wtec.org/
19. Society for the Study of Nanoscience and Emerging Technologies. Retrieved December 14, 2011, from http://www.thesnet.net/TheSociety.html
20. Nanomission Game. Retrieved December 14, 2011, from http://nanomission.org/
21. NanoVenture Game. Retrieved December 14, 2011, from http://mrsec.wisc.edu/Edetc/supplies/nanoventure/index.html

References

Briars, D. J., & Resnick, L. B. (2000). *Standards, assessments—and what else? The essential elements of standards-based school improvement*. CSE Technical Report 528. Los Angeles, CA: Center for the Study of Evaluation. Retrieved December 14, 2011, from http://www.cse.ucla.edu/products/Reports/TECH528.pdf

Center on Education Policy. (2007). *Choices, changes, and challenges: Curriculum and instruction in the NCLB era*. Washington, DC: Author.

Center on Education Policy. (2008). *Instructional time in elementary schools: A closer look at changes for specific subjects*. Washington, DC: Author.

Dorph, R., Goldstein, D., Lee, S., Lepori, K., Schneider, S., & Venkatesan, S. (2007). *The status of science education in the Bay Area: Research brief*. Berkeley, CA: Lawrence Hall of Science, University of California, Berkeley.

European Commission. (2005). *Nanosciences and nanotechnologies: An action plan for Europe 2005–2009*. Retrieved December 14, 2011, from http://ec.europa.eu/research/industrial_technologies/pdf/nano_action_plan_en.pdf

European Commission. (2007). *Key competences for lifelong learning: A European framework*. Retrieved December 14, 2011, from http://ec.europa.eu/dgs/education_culture/publ/pdf/ll-learning/keycomp_en.pdf

European Commission. (2009). *Progress towards the Lisbon objectives in education and training*. Retrieved December 14, 2011, from http://ec.europa.eu/education/lifelong-learning-policy/doc/report09/report_en.pdf

European Commission. (2010). *New skills for new jobs: Action Now: A report by the expert group on new skills for new jobs prepared for the European Commission*. Retrieved December 14, 2011, from http://ec.europa.eu/social/main.jsp?catId=822&langId=en

European Commission. (2011a, February). *Strategic framework for education and training*. Retrieved December 14, 2011, from http://ec.europa.eu/education/lifelong-learning-policy/doc28_en.htm

European Commission. (2011b, November). *The European qualifications framework*. Retrieved December 14, 2011, from http://ec.europa.eu/education/lifelong-learning-policy/doc44_en.htm

Feather, J. L., & Aznar, M. F. (2010). *Nanoscience education, workforce training, and K–12 resources*. Boca Raton, FL: CRC Press.

Harper, T. (2011 July). *Global funding of nanotechnologies and its impact*. White paper. Retrieved December 13, 2011, from www.cientifica.com

Maltese, A. V., & Tai, R. H. (2010). Eyeballs in the fridge: Sources of early interest in science. *International Journal of Science Education, 32*(5), 669–685.

McKinsey & Co. (2007). *How the world's best performing school systems come out on top*. Retrieved December 14, 2011, from http://www.mckinsey.com/App_Media/Reports/SSO/Worlds_School_Systems_Final.pdf

Murday, J. S. (Ed.). (2011). *NSF workshop report: International Benchmark Workshop on K–12 Nanoscale Science and Engineering Education (NSEE)*. National Science Foundation, Washington, DC, 6–7 December 2010. Retrieved December 13, 2011, from http://www.nsf.gov/crssprgm/nano/reports/nsfnnireports.jsp

Mnyusiwalla et al. (2003). http://www.wtec.org/nano2/Nanotechnology_Research_Directions_to_2020/chapter13.pdf

National Research Council. (2011a). *A framework for K–12 science education: Practices, crosscutting concepts, and core ideas*. Committee on Conceptual Framework for the New K–12 Science Education Standards, National Research Council. Washington, DC: National Academies Press. Retrieved December 14, 2011, from http://www.nap.edu/catalog.php?record_id=13165

National Research Council. (2011b). *Successful K–12 STEM education: Identifying effective approaches in science, technology, engineering, and mathematics*. Committee on Highly Successful Science Programs for K–12 Science Education, Board on Science Education and Board on Testing and Assessment, Division of

Behavioral and Social Sciences and Education. Washington, DC: The National Academies Press. Retrieved November 19, 2011, from http://www.scribd.com/doc/65547927/Successful-K-12-STEM-Education-NRC

Roco, M. C., Mirkin, C. A., & Hersam, M. C. (Eds.). (2010). *Nanotechnology research directions for societal needs in 2020: Retrospective and outlook*. Berlin and Boston: Springer. Retrieved December 14, 2011, from http://www.wtec.org/nano2/Nanotechnology_Research_Directions_to_2020/

Schwab, K. (2011). *Global competitiveness report 2011–2012*. Geneva, Switzerland: World Economic Forum. Retrieved December 14, 2011, from http://www.weforum.org/issues/global-competitiveness

UNESCO. (n.d.). *Ethics Education Program*. Retrieved December 14, 2011, from http://www.unesco.org/new/en/social-and-human-sciences/themes/ethics-education-programme/

Williams, E., Leachman, M., & Johnson, N. (2011). *State budget cuts in the new fiscal year are unnecessarily harmful*. Center on Budget and Policy Priorities. Retrieved December 14, 2011, from http://www.cbpp.org/cms/index.cfm?fa=view&id=3550

14

Nanotechnology Research and Science, Technology, and Society Education at Bilkent University

Haldun M. Ozaktas

CONTENTS

KEY WORDS: nanoscience; nanotechnology; research centers; research funding; engineering ethics; science, technology, and society; community research; Bilkent University; Turkey

Nanotechnology Research Activities at Bilkent University and in Turkey

Bilkent University was established in Ankara in 1984 as the first private non-profit university in Turkey. It was able to quickly attract high-quality faculty and students, and shortly became one of the leading universities of Turkey. It presently ranks first in Turkey, 32nd in Europe, and 112th worldwide in the Times Higher Education rankings. Presently it has nine faculties, several vocational schools, and around 12,000 students (Bilkent University, 2011a).

The university has particular strength in physical sciences and engineering. It is the home of several nanotechnology research centers.

The Advanced Research Laboratories (ARL) was built in 1992 and equipped with a grant from the Ministry of Defence. ARL is a multidisciplinary laboratory with researchers from different backgrounds and is open to researchers throughout Turkey. Photonic gap materials, metamaterials, compound

FIGURE 14.1 (See color insert)
An interior from the ARL facility at Bilkent University, Ankara.

quantum dots, and group four nanocrystals are just a few examples of the materials under study. The lab is also actively engaged in studies of photodetectors, lasers, plasmonics, and integrated optical devices (ARL, 2011). Figure 14.1 depicts the ARL facility at Bilkent University.

The Nanotechnology Research Center (NANOTAM) is a research center serving researchers from many departments of Bilkent University as well as other universities. Established in 2003, the center has 4,200 m^2 of office and lab space, including clean rooms. Research areas in nanotechnology include nanophotonic crystals, photonic metamaterials, nanomaterials, nanoelectronics, nanoplasmonics, and nanotechnology-based solar cells (NANOTAM, 2011). Figure 14.2 depicts the NANOTAM facility at Bilkent University.

UNAM, the National Nanotechnology Research Center, was established in 2006, initially with funds primarily from the State Planning Agency. The center has 9,000 m^2 of office and lab space, including clean rooms, serving researchers from within and outside Bilkent University. Research areas include nanobiotechnology, nanotextiles, nanochemistry, nanoelectronics, nanophotonics, and energy (UNAM, 2011). Figure 14.3 depicts the UNAM facility at Bilkent University.

Many other nanotechnology research centers exist in Turkey. Among these, we mention the Central Laboratory at the Middle East Technical University (METU), which is home to the Micro and Nanotechnology Program at the same university. Major research subjects include nanophotonics, optoelectronics, nanoelectronics, nanomagnetism, nanomaterials, nanofabrication,

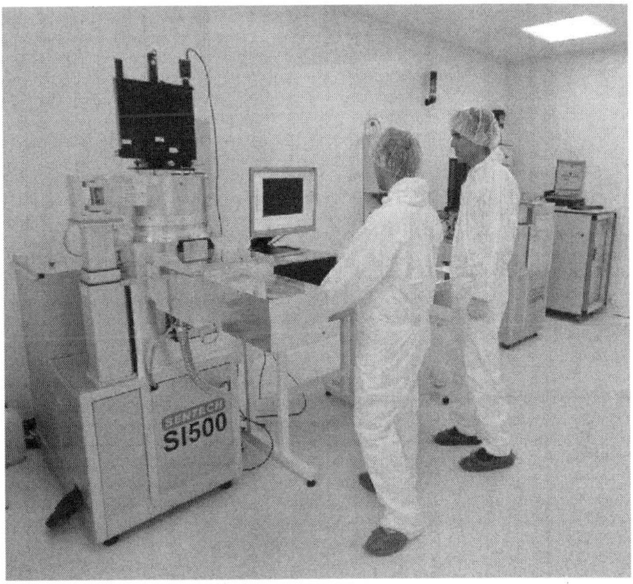

FIGURE 14.2 (See color insert)
An interior from the NANOTAM facility at Bilkent University, Ankara.

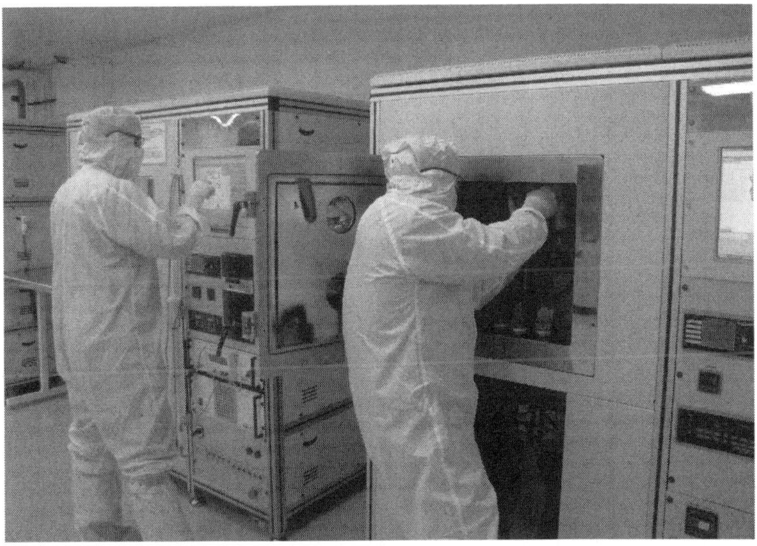

FIGURE 14.3 (See color insert)
An interior from the UNAM facility at Bilkent University, Ankara.

and nanobiotechnology (METU, 2011a, 2011b). Also at Middle East Technical University is MEMS, the Micro-Electro-Mechanical Systems Research and Application Center. While not specifically dedicated to nanotechnology research, MEMS heavily serves the nanoscience and nanotechnology community from many institutions with its extensive facilities (METU-MEMS, 2011). The history of both centers goes back about twenty years.

SUNUM, the Sabanci University Nanotechnology Research and Application Center, supports research programs spanning biology, electronics, physics, chemistry, materials, and mechatronics. Research areas include advanced materials, nanobiotechnology, nanomedicine, nanoelectronics, nanomechanics, nanooptics, and alternative energy. The center, which started operations in 2011, has a 7,140-m^2 building including clean rooms (SUNUM, 2011).

A common feature of many (though not all) nanotechnology centers in Turkey is that they have been primarily established within the last ten to fifteen years within universities, through State Planning Agency (DPT) infrastructure grants. These also include the Bionanotechnology Research Center at Fatih University, Istanbul; the Nanomedicine Research Center at Gazi University, Ankara; the Nanotechnology Research Centers at Gebze Institute of Technology, Gebze; and Istanbul Technical University, Istanbul, as well as SUNUM and UNAM already mentioned (DPT, 2010).

It should be noted that the establishment of dedicated research centers through State Planning Agency grants is not unique to nanoscience and nanotechnology, but seems to be a particular pattern resulting from the lack of more diverse sources of funding. The Scientific and Technological Research Council of Turkey (TUBITAK), the other major funding body, mostly funds specific research projects, leaving the State Planning Agency as the major grantor for the establishment of infrastructure in the form of research centers. (However, nanotechnology research is also undertaken in some centers connected to TUBITAK [Wikipedia, n.d.] The Department of Defence, which is a potential source of funding for high-tech research, does not yet seem to have played a major role in funding nanotechnology research in Turkey. It may also be worth noting that neither funding bodies nor public policy bodies have come forth with very specific targets, goals, or policies regarding nanotechnology research, beyond establishing it as a priority area.

Overall, it could be said that the state of nanotechnology research in Turkey is similar to that of other high-tech research areas. While Turkey does not have a large volume of research and investment in this area, the existence of a number of particularly strong centers is noteworthy.

A number of major corporations in Turkey undertake nanotechnology research and development, but their numbers are limited. Large companies such as Arcelik (mostly home appliances), Dyo (paints), and Aselsan (electronics, especially military) have research and development operations on nanomaterials. Overall, the interest of industry in nanotechnology is not at a very high level, with more watching than action. Most companies do not yet feel sufficient competitive pressure to become heavily involved in nanotechnology. Likewise, interaction between academia and industry is not very strong. Neither is the number of startups in research parks doing nanoscience or nanotechnology very high.

Ethical and Social Issues Education

Most practicing researchers, in all areas and in all countries, do not have any direct interest in ethical and social issues related to the research they are undertaking and the technology they are developing. While most researchers are conscientious about doing their research in an ethical manner (research ethics), they are not directly involved in discussions about ethical and social issues regarding the applications of the technology. One reason for this is insufficient exposure to the general background and conceptual and intellectual tools that would support such involvement.

Such exposure is not only important for researchers, but also for practicing engineers and professionals. Based on this understanding, ethical and social issues education is offered to and compulsory for all undergraduate engineering majors at Bilkent University. The main vehicle to this end is the course Science, Technology, and Society (Bilkent University, 2011b), which was initiated in 1995. The course is now taken by about 500 students every year. It is a course of wide breadth covering a diversity of topics and teaching methods, and is probably the first course in Turkey to deal with the ethical, social, cultural, political, economic, legal, environment and sustainability, health and safety, and reliability dimensions of science, technology, and engineering.

During the early 1990s, gaining ABET (Accreditation Board for Engineering and Technology) (ABET, 2011) accreditation became a goal for elite universities in Turkey. Until that time, the need for incorporating ethical or social dimensions courses in engineering programs had not been felt by most institutions in Turkey. The present course was established at that time, and since then it has been the key course in meeting ABET requirements on ethical and social dimensions.

The course is broadly designed and covers a very wide array of subjects. The topics and contents have changed significantly over the years. The current syllabus has a current-issues focus and emphasizes subjects students would be familiar with from the media with the hope that they will be able to more easily relate to such topics:

- Ethics and technology
- Political positions and science and technology issues

- Energy
- Ecology
- Population
- Health and technology
- Technology and the third world
- Future technologies
- Intellectual property
- Copyrights, patents, and trade secrets
- Philosophical aspects of artificial intelligence
- Technological disasters
- Responsibilities of scientists and engineers
- Technology decisions and the democratic process

Additional topics are covered through guest seminars and video presentations. The texts that are presently used are *Technology and Society: A Bridge to the 21st Century* (Hjorth, Eichler, Khan, & Morello, 2003) and *Taking Sides: Clashing Views in Science, Technology, and Society* (Easton, 2008).

Many science and engineering departments that seek to meet educational or accreditational objectives choose to implement courses such as "Science and Ethics" or "Engineering Ethics." These courses usually consist of an introduction to ethical theories followed by case studies. While there is nothing wrong with this approach per se, such courses usually do not cover the diverse manifestations of the interaction of science, technology, and society in a more multidisciplinary fashion. Thus the choice at Bilkent University has been in favor of a "Science, Technology, and Society" studies approach. Further discussion of the philosophy and rationale behind the course design may be found in Ozaktas, 2012.

In its early years, the course was offered to about 150 or so computer-electrical-industrial engineering students. However with a threefold increase in admissions, the course now has nearly 500 students, and with the addition of a new Mechanical Engineering Department will exceed this number. This increase has necessitated the elimination of term papers/projects and small group discussions and the adoption of teaching techniques suitable for large classes. These include assigning debate topics to groups of students, who are expected to compose arguments for or against certain positions on a controversial topic. Selected opposing groups debate on stage in front of the whole class. The large audience also requires considerable stage performance and techniques of mass voting and collective opinion forming to involve the class as much as possible. The large number of students is far from ideal, but teaching it to smaller classes would entail a significant cost, which at present excludes this option.

Various novel project concepts have been developed throughout the history of the course, including Web-based projects and the incorporation of the concept of "community research," which in contrast to "high technology" and "big science" (research heavily funded by governments and corporations and practiced by professional scientists), involves communities in the identification of their problems (such as public health and social issues) and encourages their participation in applying the results to directly improve their lives.

We all know the close relationships between sciences, technology, industry, and the economy. When people speak of science today, what comes to mind are usually big and expensive projects that have captured the imagination: space exploration, chemicals and pharmaceuticals, electronics and information technology, and the human genome project. As greater and greater wealth is being poured into such research, the discontent felt by ordinary people increases. They are at the same time impressed and disillusioned by the exponential progress, which despite the many conveniences it brings, somehow falls short of solving what they feel to be their real and immediate problems, and improving their conditions. Community research projects aim to directly address these needs and problems, and can potentially have empowering benefits for communities. For additional information about community research and related concepts such as grassroots science and science shops, we refer the reader to an article by Hansen (2005) and the websites of the Loka Institute (2011) and Living Knowledge (2011).

It is hoped that such courses dedicated to the ethical and social issues surrounding science and technology, along with general education courses in the social sciences and humanities, will provide scientists and engineers with the intellectual foundations that will make them more interested in and better able to deal with ethical and social issues pertaining to their area of research or practice. Interest and involvement of researchers and practitioners in these aspects of their subjects are also important for more meaningful discussion of these issues to take place among the general public. This, in turn, would hopefully lead to more democratic decision making regarding science and technology policies.

Questions

1. Compare and contrast the research areas of the various nanotechnology research centers in Turkey.
2. What are the common features of most nanotechnology centers in Turkey?
3. How is nanotechnology research funded in Turkey?
4. Comment on the interest of industry in nanotechnology in Turkey.
5. Why are most nanotechnology researchers not very interested in ethical and social issues relating to nanotechnology?

6. State three important features of the course Science, Technology, and Society at Bilkent University.
7. Why was a "Science, Technology, and Society" approach preferred over an "Engineering Ethics" approach at Bilkent?
8. What are some of the benefits of community-research-type projects in education?
9. Summarize your learning of new ideas and information presented in this chapter by completing the following flowchart:

Flow Chart

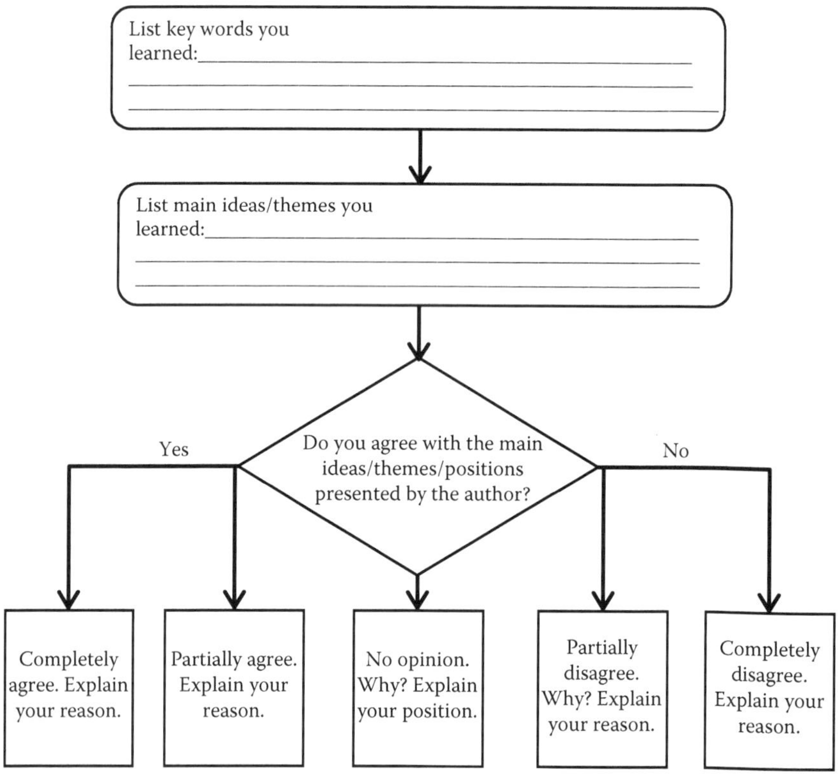

Acknowledgment

The author acknowledges partial support of the Turkish Academy of Sciences.

References

ABET (Accreditation Board for Engineering and Technology). (2011). Home page. Retrieved September 28, 2011, from http://www.abet.org

ARL (Advanced Research Laboratories). (2011). Home page. Retrieved September 28, 2011, from http://arl.bilkent.edu.tr

Bilkent University, Ankara. (2011a). Home page. Retrieved September 28, 2011, from http://www.bilkent.edu.tr

Bilkent University, Ankara. (2011b). Course syllabus for GE 301: Science, Technology, and Society. Bilkent University, Ankara, Turkey. Retrieved September 28, 2011, from http://www.ee.bilkent.edu.tr/~ge301

DPT (State Planning Agency, Ankara). (2010, December). University and Public Institutions Research Centers (in Turkish). Retrieved September 28, 2011, from http://www.dpt.gov.tr/DocObjects/View/10120/Universite_ve_Kamu_Kurumlari_Arastirma_Merkezleri.pdf

Easton, T. (2008). *Taking sides: Clashing views in science, technology, and society* (8th ed.). New York, NY: McGraw-Hill.

Hansen, T. B. (2005). Grassroots science: An ISYP Ideal? *ISYP Journal on Science and World Affairs, 1*, 61–72.

Hjorth, L. S., Eichler, B. A., Khan, A. S., & Morello, J. A. (2003). *Technology and society: A bridge to the 21st century* (2nd ed.). Upper Saddle River, NJ: Prentice-Hall.

Living Knowledge: The International Science Shop Network. (2011). Home page. Retrieved September 28, 2011, from http://www.scienceshops.org

Loka Institute. (2011). Home page. Retrieved September 28, 2011, from http://www.loka.org

METU (Middle East Technical University). (2011a). Home page for Central Laboratory. Retrieved September 28, 2011, from http://www.centrallab.metu.edu.tr

METU (Middle East Technical University). (2011b). Micro and Nanotechnology Program. Retrieved September 28, 2011, from http://www.mnt.metu.edu.tr

METU-MEMS (Micro-Electro-Mechanical Systems Research and Application Center, Middle East Technical University). (2011). Home page. Retrieved September 28, 2011, from http://mems.metu.edu.tr

NANOTAM (Nanotechnology Research Center). (2011). Home page. Retrieved September 28, 2011, from http://www.nanotechnology.bilkent.edu.tr

Ozaktas, H. M. (2012). Teaching science, technology, and society to engineering students: A sixteen-year journey. *Science and Engineering Ethics*, in print.

SUNUM (Sabanci University Nanotechnology Research and Application Center). (2011). Home page. Retrieved November, 2011, from http://sunum.sabanciuniv.edu

UNAM (National Nanotechnology Research Center). (2011). Home page. Retrieved September 28, 2011, from http://nano.org.tr

Wikipedia. (n.d.). Nanotechnology (in Turkish). Retrieved September 28, 2011, from http://tr.wikipedia.org/wiki/Nanoteknoloji

15

Application of Simulation for Emulating Nanotechnology and Emerging Technologies in Education

Ahmed S. Khan

CONTENTS

Introduction

Today, with many colleges offering distance-education programs in engineering, technology, and other scientific disciplines, the demand for truly interactive simulation-based online labs is on the rise. An interactive laboratory is capable of partially substituting for or extending conventional hands-on laboratories. Virtual laboratories have great potential for facilitating the active learning mode and transforming "e-learning by reading and watching" into a more efficient "learning by doing." They also provide an efficient and effective way of teaching complex and dynamic engineering systems. A simulation-based teaching environment enables students to acquire experience and evaluate their previous results (Cherner, Karim, Khan, Rubanchik, & Mullett, 2008).

Phenomenal interest in new and emerging technologies like nanotechnology has led to an increased demand for graduates who have sound knowledge of fundamental principles underlying the design and operation of complex systems. However, incorporation of such learning processes, including state-of-the-art technological tools and equipment, requires considerable time and financial resources. Keeping curricula and labs current with the rapid change of technology poses another challenge for academia. Academicians can address some of these challenges by using simulation and virtual experiments (Banky & Wong, 2007).

Advantages and Limitations of Simulation

In addition to cost savings, simulation offers a number of other advantages:

- Allowing the user to modify system parameters and observe the outcomes without any harmful side effects
- Eliminating component or equipment faults that affect outcomes
- Supporting users' progress at their own pace in discovery and understanding of concepts and issues
- Enhancing the presentation of "dry" concepts by integrating theory and practice
- Minimizing the risk of "What if something goes wrong?"

The use of simulation and virtual laboratories emulating nanotechnology can provide instructional opportunities in many modes: at campus or school (traditional on-site teaching), at home (warm-ups, post-class tasks, or self-learning), or through distance learning (online). Figures 15.1–15.3 illustrate examples of simulation. Realistic simulations (Cherner, Khan, Karim, & Mullett, 2011) enable learners to:

1. Observe the physical processes insightfully at different levels of detail (micro and macro)
2. Analyze the constraints between relevant parameters (relationships)
3. Push these parameters beyond normally allowed values to simulate infrequent operating conditions or casualty situations (cause and effect)
4. Run what-if scenarios (minimizing risk)
5. Acquire data from virtual experiments for detailed analysis and comparison to actual operating conditions in a theory-to-practice approach (This narrows the understanding gap between theory and practice.)

The screen shots in Figure 15.1 illustrate two simulations that model devices built using nanoscale technologies and materials. The sensor in the top measures a lead concentration in the air. Students are able to explore its microstructure and processes when lead atoms, gold nanoparticles, and DNAzymes interact with each other (middle right). The simulation of a device for diagnosis of diabetes mellitus by measuring the amount of acetone in a human breath is shown at the bottom.

However, simulation of "real processes" is not a panacea for the problems of keeping curricula current, as it has some limitations, too. For example, the use of software simulations of physical entities, such as electronic circuits,

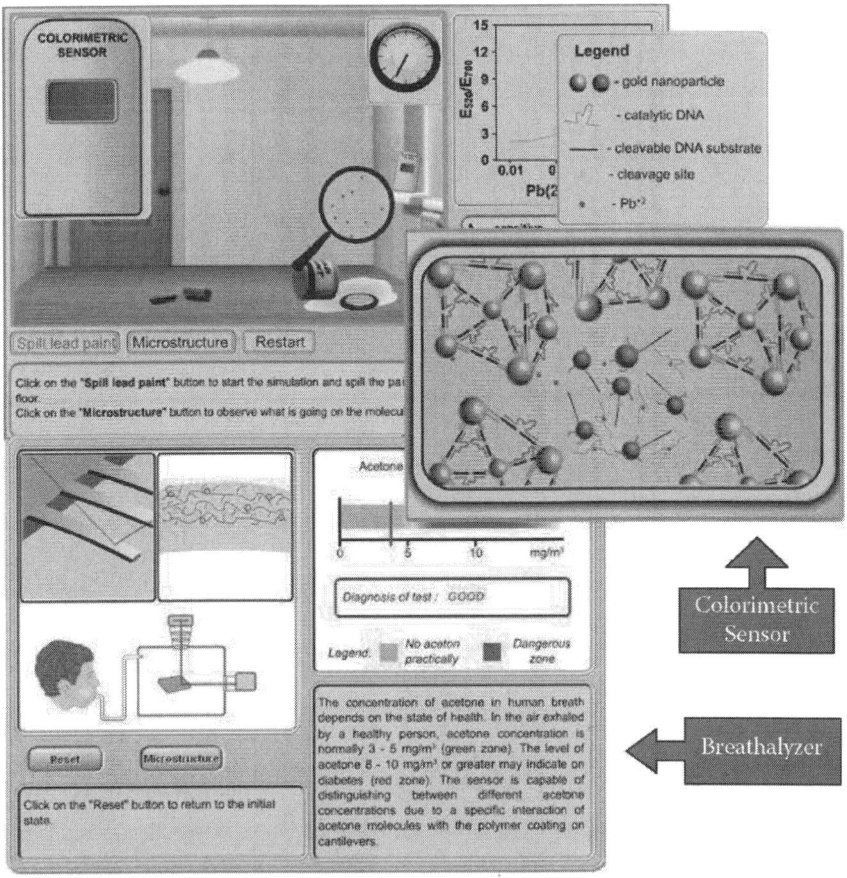

FIGURE 15.1 (See color insert)
Application of simulation to help students visualize the interaction between nanoparticles and DNAzymes. (Courtesy Atel LLC.)

denies the user a chance to physically handle the circuit components or construct the circuit (Bartelt, 2006). Given the limitations of software tools, it is also difficult to construct simulation models emulating complex systems and processes. On the other hand, simulation has another advantage in that a simulation is task- and learner-neutral; it models an object/system construction and operation or learning situation. Within simulation functionality, there are no restrictions on the student's actions. In the application of nanotechnology to food, health, and medicine, the use of simulation tools could provide insights into exploring risks and uncertainties, and thus help in developing risk management strategies.

The screen shots in Figure 15.2 show the quantum cascade laser (QCL) simulation that depicts the design of quantum-well heterostructures and the operating principles of QCL.

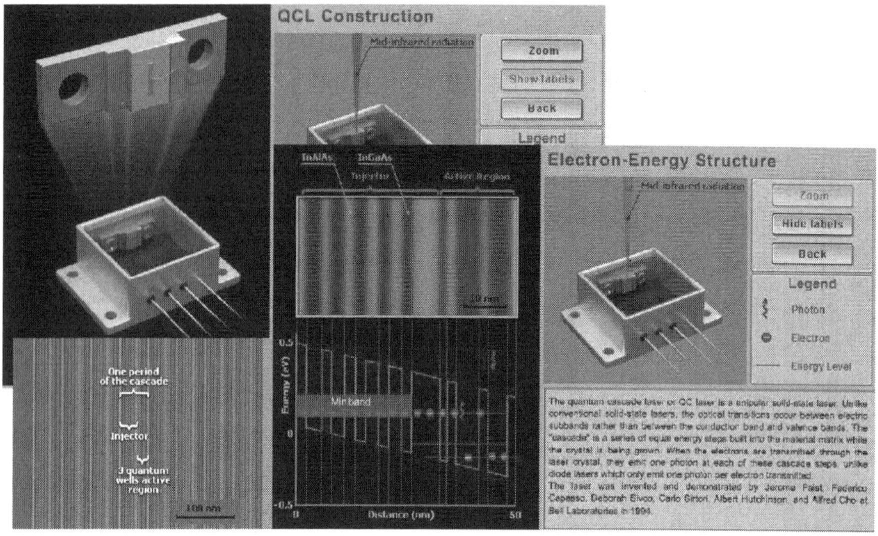

FIGURE 15.2 (See color insert)
Application of simulation to help students understand the principles of quantum-well devices. (Courtesy Atel LLC)

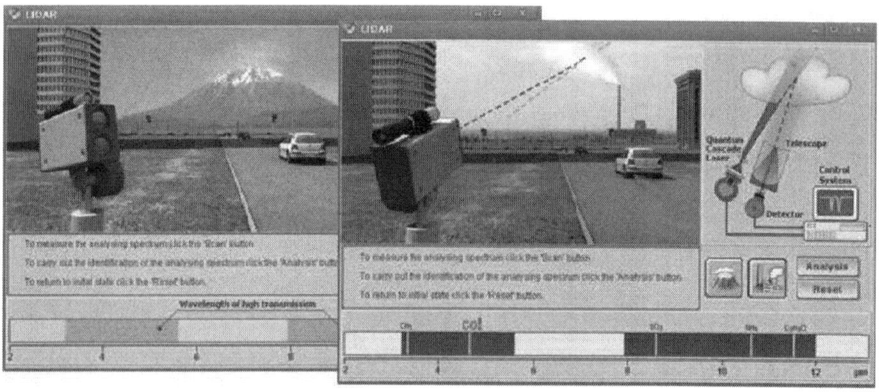

FIGURE 15.3 (See color insert)
Application of simulation to help students visualize and measure air pollution. (Courtesy Atel LLC.)

The screen shots in Figure 15.3 depict the light-detection and ranging (LIDAR) simulation operation of a device that measures the properties of scattered light to determine the composition of pollutants in the atmosphere near a natural (volcano, *left*) or industrial (*right*) source of air pollution. The diagram at the far right illustrates a LIDAR operational scheme. The LIDAR simulation enables students to first choose either a natural pollution source (volcano) or an industrial one (factory chimney), and then to use the LIDAR scan for determining the chemical composition of the air and clouds near the chosen object. Next the simulation can be triggered into the analysis mode to analyze the chemical spectrum. If any components exceed the allowed limits, the program marks them and generates a warning message.

Modes of Teaching

Recent studies suggest that students should first be exposed to theoretical concepts in the simulation environment and then required to perform a hands-on activity. This approach ensures progressive and sequenced learning in the form of scaffolding, an aspect of cognitive apprenticeship. Furthermore, to enhance student learning, the instructional design should consider three approaches. The first is to use simulation-based experiment in the first half of the course, followed by hands-on activity or experiments in the second phase (sequential design). The second approach is simultaneous use of simulation and hands-on activity or experiment (parallel design). The third approach is to use simulation and hands-on activity in an alternating mode (mixed design). All three approaches support a combinational approach or hybrid instructional delivery (Taher, 2009).

Questions

1. List advantages and disadvantages of the use of simulation in education.

Advantages (+)	Disadvantages (–)

2. Complete the following table by listing advantages and disadvantages of the use of simulation in various modes of teaching:

Use of simulation in teaching	Advantages (+)	Disadvantages (–)
Sequential design		
Parallel design		
Mixed design		

3. Discuss the advantages and disadvantages of using simulations to explore the risks and uncertainties posed by the application of nanotechnology in the domains of food, agriculture, and medicine. Can an effective risk management strategy be developed using simulation techniques?

4. Summarize your learning of new ideas and information presented in this chapter by completing the following flowchart:

Flow Chart

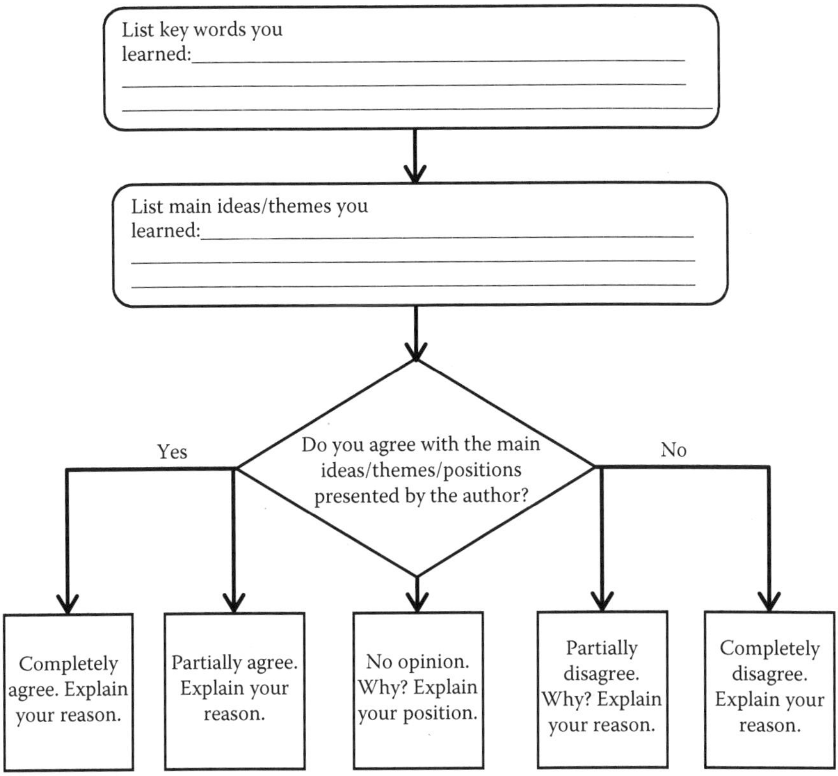
List key words you learned:
List main ideas/themes you learned:
Do you agree with the main ideas/themes/positions presented by the author?
Yes
No
Completely agree. Explain your reason.
Partially agree. Explain your reason.
No opinion. Why? Explain your position.
Partially disagree. Why? Explain your reason.
Completely disagree. Explain your reason.

References

Banky, G. P., & Wong, K. K. (2007). *Troubleshooting exercises using circuit simulator software: Support for deep learning in the study of electronic circuits.* Paper presented at ICEE International Conference on Engineering Education, Coimbra, Portugal.

Bartelt, T. (2006). *Instrumentation and process control.* Florence, KY: Delmar CENGAGE Learning.

Cherner, Y. E., Karim, A., Khan, A., Rubanchik, V., & Mullett, G. (2008). Using simulation-based hybrid and multilevel virtual labs for fiber optics, photonics, and telecom education. In *Proceedings of ASEE 115th Annual Conference.* Washington, DC: American Society for Engineering Education. Retrieved November 23, 2011, from http://soa.asee.org/paper/conference/paper-view.cfm?id=8907

Cherner, Y. E., Khan, A., Karim, A., & Mullett, G. (2011). Use of adaptable simulation-based virtual laboratories for teaching alternative energy and energy conservation in engineering and technology programs. In *Proceedings of ASEE 2011 Annual Conference.* Washington, DC: American Society for Engineering Education.

Taher, M. (2009). *Integration of computer simulation into traditional lecture-lab activities: Impact of using computer simulation on student learning in a technology-based course.* Unpublished doctoral dissertation, Northern Illinois University, DeKalb, IL.

16

Predicting the Future: Review of Public-Perception Studies of Nanotechnology

Michael Siegrist

CONTENTS

KEY WORDS: nanotechnology, risk perception, regulation, labeling

Introduction

Nanotechnology is assumed to be one of the key technologies of this century (Renn & Roco, 2006; Royal Society, 2004). The technology promises to deliver many useful products, but nanotechnology also poses risks (O'Brien and Cummins, 2008). Possible risks are unintended economic, societal, human health, and environmental impacts, which could lead to lack of public acceptance. For this reason, one of the first reports on the opportunities and risks of nanotechnology emphasized that public perception should be considered in risk assessments of nanotechnology (Royal Society, 2004). Studies examining public perception of nanotechnology, conducted in different countries and utilizing various methods, have been published in the meantime. Along with the narrative review presented in this chapter, implications for industry and science are outlined.

Why Is the Public Important?

Public perception may have a strong impact on the progress of a new technology (Köhler & Som, 2008). There may be both direct and indirect influences. Fueled by public fears, governmental agencies may be motivated to implement stricter regulations, which may result in higher production costs (Ayers, 1998). Consumers may also be unwilling to purchase products produced utilizing a certain technology (Siegrist, 2008). In the domain of gene technology, both effects have been observed. In some countries, retailers do not stock genetically modified (GM) foods because they expect the public to be opposed to such products. Elsewhere, there are regulations that do not allow farmers to use GM plants. Based on these past experiences with other technologies, industry and scientists are well advised to take public perception of nanotechnology into account at an early stage of technological development.

It is relatively easy to agree on the fact that public perception is important. More disagreement exists on the question of how public perception of a technology can be meaningfully measured before a public dialogue about it has started, and most people have no attitude toward it. Measuring public perception of a technology is most difficult when most needed, at an early stage in the life of that technology. It is most difficult to measure public perception at an early stage because most people have not yet developed strong attitudes, or may not even be familiar with the technology. Moreover, early surveys may not help predict people's acceptance of a technology at a later stage.

Surveys Measuring General Attitudes Toward Nanotechnology

Several surveys have examined public perception of nanotechnology in the United States and Europe. The results of these studies suggest that public awareness of nanotechnology is low and that knowledge about it is limited (Cobb & Macoubrie, 2004; Lee, Scheufele, & Lewenstein, 2005). At the same time, the public in the United States and Canada perceives nanotechnology in a rather positive light (Cobb & Macoubrie, 2004; Priest, 2006). Similar to the case of gene technology, Europeans seem to be less optimistic about nanotechnology than North Americans (Gaskell, Ten Eyck, Jackson, & Veltri, 2004, 2005). A study by Currall and colleagues addressed the question of how nanotechnology is perceived relative to other technologies (Currall, King, Lane, Madera, & Turner, 2006). The results of their Web survey suggested that nanotechnology is perceived as less risky and more beneficial than, for example, gene technology or pesticides. Nanotechnology, however, is perceived as more risky and less beneficial than vaccinations or hydroelectric power.

Providing information about general health benefits (positive framing) or health risks (negative framing) associated with nanotechnology influenced participants' perception of nanotechnology (Cobb, 2005). Positive framing resulted in higher benefit perception, and negative framing resulted in higher risk perception. If different types of benefits (e.g., economic or environmental) were described, they did not seem to influence risk perception (Schütz & Wiedemann, 2008).

Review of Public-Perception Studies of Nanotechnology

Surveys examining public perception of nanotechnology in the abstract provide information about how familiar people are with the term *nanotechnology* and which risk perceptions this term may evoke. It is an implicit assumption of this research that people have stable attitudes toward nanotechnology and that surveys measure these attitudes; however, there are good reasons for skepticism.

Simulating the Future: Providing Information

Most people do not have preferences about issues they are not very familiar with, but instead construct preferences when asked to respond to survey questions on unfamiliar topics (Slovic, 1995). It is highly likely that almost all those participating in surveys on nanotechnology construct their preferences. For this, they use whatever information they have available, even the questions in a questionnaire. Asked about possible risks of nanotechnology, people may infer that there must be some risks, because the question would not otherwise be asked. It is a challenge for social science research on nanotechnology to reduce the impact of irrelevant information on people's responses.

Providing specific information about nanotechnology applications is one approach to improving control over the information people use for assessing benefits and risks associated with nanotechnology. In such a research approach, participants are informed about the term *nanotechnology* and about possible risks and the benefits associated with various applications. Such information enables participants to form attitudes toward nanotechnology applications during the survey process (Siegrist, Cousin, Kastenholz, & Wiek, 2007a; Siegrist, Keller, Kastenholz, Frey, & Wiek, 2007b; Siegrist, Stampfli, Kastenholz, & Keller, 2008). Providing information implies that participants are primed in a certain way. It is important, therefore, to use realistic information that can be expected to be disseminated to the public in the near future. Such a process does not allow the expression of public risk perception in absolute terms, but it enables the identification of applications that could be more problematic and applications that most likely will not be in the focus of public attention in the future.

Results from this research suggest that nanotechnology applications in the food and health domains are perceived as more risky compared with other nanotechnology applications (Siegrist, Keller, et al., 2007b). Overall, this study demonstrated that risk and benefit perceptions are different for various applications. Therefore, it is important to measure public perception of concrete applications and not people's general attitude toward nanotechnology. Furthermore, nanotechnology food packaging seems to be more accepted by the public than nanotechnology foods (Siegrist et al., 2008). People's risk perception was influenced by the fact of whether nanoparticles are consumed or not. The fact that food packaging was relatively positively assessed suggests that participants have not taken into account the possible migration of nanoparticles from the food packaging to the food.

A further means of information dispersal would be public engagement workshops, which can be used to facilitate public discussion of emerging technologies. This technique involves several stages in which participants receive information about the technology and discuss various topics. Persons participating in a Swiss study showed a pragmatic attitude: They were neither scared nor did they show great enthusiasm (Burri & Bellucci, 2008). Results of U.K. and U.S. studies indicated that different applications are perceived differently (Pidgeon, Herr Harthorn, Bryant, & Rogers-Hayden, 2009). Energy applications were perceived more positively than health and human enhancement applications. Overall, even when presented with possible risks, benefits dominated participants' perception of nanotechnology. A similar finding was reported in a U.S. study, which found that respondents attending a program of citizen engagement about nanotechnology said positive things about nanotechnology to other people afterward (Besley, Kramer, Yao, & Toumey, 2008).

Factors Influencing Perception of Nanotechnology

Experts and lay people differ in their perception of nanotechnology: Experts perceive fewer risks associated with nanotechnology than do lay people (Siegrist, Keller, et al., 2007b). Experts and lay people also differ in respect to the kinds of risks they are most concerned about. The lay public seems to be more concerned about societal risks (e.g., loss of jobs), and experts more about health and environmental problems (Scheufele et al., 2007). One possible explanation for this finding is that participants in the study did not receive information about nanotechnology. As a result, they might not have had a clear idea of the domains and applications for which nanotechnology is used.

Most people lack an understanding of the fundamental concepts that are important for a full comprehension of nanotechnology (Batt, Waldron, & Broadwater, 2008; Castellini et al., 2007; Waldron, Spencer, & Batt 2006). The authors of these studies concluded, therefore, that scientists should inform the public about their research, and people should be informed about how nanotechnology works and what its benefits and risks are. A better-informed public, however, does not mean that there will be a higher acceptance of nanotechnology. Research in other fields suggests that knowledge is not a significant, but only a weak, predictor for the acceptance of a technology (Bredahl, 1999; Scholderer & Frewer, 2003). Further, the impact of knowledge may depend on people's values. It was shown, for example, that for highly religious respondents, knowledge had a weaker influence on support for federal funding of nanotechnology than it did for less religious respondents (Brossard, Scheufele, Kim, & Lewenstein, 2009). It should be emphasized, however, that the effect size of this interaction term was rather small.

Several research groups have suggested that values and trust influence acceptance of nanotechnology. People's cultural worldviews influence their interpretation of information about nanotechnology and, as a consequence, risk-benefit perception of the technology (Kahan, Braman, Slovic, Gastil, & Cohen, 2009). The same description of nanotechnology and its applications resulted in a more positive assessment in hierarchical and individualistic participants, and in a more negative assessment in participants with egalitarian and communitarian worldviews. A similar line of research found that membership in a political party influenced interpretation of information about nanotechnology (Smiley Smith, Hosgood, Michelson, & Stowe, 2008). Members of the Republican Party were more likely to have more positive attitudes, and members of the Democratic Party were more likely to have more negative attitudes toward nanotechnology after exposure to the same information. Religious beliefs have also been found to be negatively correlated with support for funding nanotechnology (Brossard et al., 2009).

Trust is another shortcut people can use to make a decision when they lack the knowledge for making an informed decision. People with trust in the industry using, and in the agencies regulating, nanotechnology assessed its applications more positively than did people who did not have trust (Siegrist, Cousin, et al., 2007a; Siegrist, Keller, et al. 2007b; Siegrist et al., 2008).

The Importance of Tangible Benefits

People interested in science seem to be very enthusiastic about nanotechnology, and they perceive many benefits associated with this technology (Bainbridge, 2002). Benefits are probably the most important factor influencing public acceptance of a technology. The results of several studies have demonstrated that people who perceive many benefits associated with nanotechnology applications perceived fewer risks and were more willing to accept nanotechnology compared with people who perceived few benefits associated with nanotechnology applications (Siegrist, Cousin, et al., 2007a, Siegrist, Keller, et al., 2007b; Siegrist et al., 2008). The future acceptance of nanotechnology may therefore depend on whether the public perceives tangible benefits associated with this technology. A lack of tangible benefits is probably one of the most important factors explaining why the European public is hesitant to accept this technology (Siegrist, 2008). Scientists should be cautious, however, to promise realistic benefits only. Promising more than a technology can deliver will result in a negative perception of that technology (Keller, 2007).

Few consumers have knowingly experienced tangible benefits of nanotechnology. As a consequence, the media are an important resource for informing people about benefits associated with nanotechnology. Since the media focus more on potential benefits and less on risks, users of science media possess more positive attitudes toward nanotechnology than do nonusers of science media (Scheufele & Lewenstein, 2005).

Analogies With Other Technologies

Biotechnology is an example of a technology for which public opposition, or at least lack of public acceptance, can be observed in many countries (Gaskell, Bauer, Durant, & Allum, 1999). Since both are enabling technologies and are used in very heterogeneous applications, it is not surprising that they are often compared (Einsiedel & Goldenberg, 2006; Mehta, 2006; Priest, 2008). We certainly agree that we should learn from the mistakes made in the case of biotechnology (Mehta 2006). There are, however, obvious differences in public perception between bio- and nanotechnology.

In the food domain, it has been shown that physical transformations are perceived as more natural than gene modifications (Rozin, 2005, 2006). The idea of tampering with nature (Sjöberg, 2000) seems to be relevant for the negative perception of biotechnology. This idea is less prominent in the case of nanotechnology because it involves physical processes on materials and not living organisms. It seems likely, therefore, that few people will have a moral impetus to oppose nanotechnology (Siegrist, 2008). Even though there has not been a serious incident or a clear indication of risks associated with biotechnology, the public in many countries is not willing to accept gene technology (Gaskell et al., 2000). For nanotechnology, such public opposition, in the absence of evidence for risks to consumers, seems very unlikely. A serious incident or evidence of health or environmental risks could change public perception of nanotechnology, of course. It should be in the interest of the industry, therefore, to make sure that such an incident does not occur. The industry has been called on to assess the risks of engineered nanomaterials before commercial products are well defined (Colvin, 2003). Unfortunately, according to the results of industry surveys, there are some doubts as to whether all companies address possible risks associated with nanotechnology as seriously as they should (Conti et al., 2008; Helland et al., 2008).

Risks for a Sustainable Development

Since the label *nanotechnology* embraces many different applications, one that turned out to be a risk for human health or the environment would result in a major challenge for the nanotechnology industry. The public would have to be educated to differentiate among types of nanotechnology application, because the impact depends on numerous factors such as surface characteristics (Stern & McNeil, 2008). A risky nanotechnology application could be a trigger for a social amplification process (Pidgeon, Kasperson, & Slovic, 2003), which could result in opposition to applications that are judged as low risks by experts.

Labeling is an issue that could become more important in the near future. The public is in favor of labeling nanotechnology products (Burri & Bellucci, 2008; Pidgeon & Rogers-Hayden, 2007). The question, however, is whether a nanotechnology label would help consumers to make informed decisions. Possible effects of nanoparticles depend not only on particle size, but also on other factors like surface properties (Brayner 2008; Maynard 2006). A label indicating only that the product contains nanoparticulate materials may therefore be of little use to consumers. One must not neglect, furthermore, the fact that new regulations, such as mandatory labeling, can have unwanted effects. The public could erroneously infer that a precautionary measure like labeling is a signal for risks associated with this technology (Wiedemann & Schütz, 2005). As a result, a nanotechnology label could increase public risk perception, instead of having a diminishing effect on risks perceived.

Research has shown that trust is an important factor influencing people's attitudes toward nanotechnology. One prominent trust model in the risk-perception literature suggests that trust is based on shared values (Siegrist, Earle, & Gutscher, 2003), that is, it is a judgment of similarity of intentions or values. Thus, an event with significant negative consequences must be avoided, because such an event could be interpreted as lack of concern for public welfare. Given the importance of trust for the acceptance of new technologies, studies showing that the public does not trust the government to manage nanotechnology risks should be viewed as a warning signal (Macoubrie, 2006). All stakeholders with an interest in the development of nanotechnology applications should be concerned about the level of trust people have. Scientists with different backgrounds and representing different values should reach out to the public in order to create trust and, as a result, foster public acceptance of nanotechnology.

Research Gaps

Public perception of nanotechnology is a very young research field. Nevertheless, some interesting and important knowledge has been accumulated, and, as observed by Currall (2009), public-perception research in the field of nanotechnology is becoming more elaborate and more theory-based. Future research should build on the findings reviewed here, and the following research questions might turn out to be fruitful pathways forward.

Labeling could become an important issue in the future. Therefore, expectations about and perception of nanotechnology labeling across different stakeholder groups should be examined. How labeling might influence public perception of nanotechnology products is a relevant question. Whether nanotechnology labeling is perceived in a similar way as the precautionary principle, leading to increased risk perception (Wiedemann & Schütz, 2005), should be examined. The impact of labeling on consumers' trust in industry and in governmental agencies is another question that requires examination. Nanotechnology is a technology most people have little knowledge about and have few stable preferences or attitudes toward. Research about public perception of nanotechnology could result in a better understanding of how people construct preferences when confronted with a new technology. Experimental designs in which participants receive different types of information about nanotechnology should be used more frequently. Such research could provide an answer as to whether variables associated with the technology (e.g., benefits, societal impact) or variables associated with persons (e.g., values) are more important for explaining people's perception of nanotechnology.

People may have misconceptions about nanotechnology, which may result in biased perception of the technology. Qualitative research might reveal associations and thinking patterns that are most likely to be evoked by the concept of nanotechnology. Such insights could be important for predicting which events and which future information might most likely result in increased risk perception.

Conclusion

Social science research on nanotechnology began about six years ago. Given this short span of time, an impressive body of research has accumulated. The risk-perception literature has addressed two research questions. First, factors have been identified that shape public perception of nanotechnology. The goal of this body of research is to better understand why the same technology is accepted by some people but not accepted by others. The second research question focuses on different perceptions of nanotechnology across applications. This body of research could be useful for identifying applications that may pose acceptance problems at an early stage of technological development.

In general, people seem not to have negative attitudes toward nanotechnology, although perceived benefits and level of acceptance vary across applications. Based on the reviewed literature, some risks for a sustainable development of nanotechnology can be identified. A major incident could mean a huge backlash for all domains in which nanotechnology is utilized. Regulations that minimize the probability of such a scenario should be regarded as in the interest of industry. Based on the findings of an industry survey suggesting that not all companies are similarly sensitive to risk assessments (Helland et al., 2008) or to issues of nanowaste disposal (Conti et al., 2008), the implementation of industry standards would appear to be indicated.

Biotechnology and nanotechnology are often compared. Based on the mistakes made in the field of biotechnology, there are certainly some lessons that can be learned for nanotechnology. One should not forget, however, that fundamental opposition toward nanotechnology is less likely than toward biotechnology. The latter is seen as tampering with nature, whereas the former does not evoke such associations.

Some possible avenues of future social science research have been outlined in the present chapter. Progress in such research related to nanotechnology is not only important for the social sciences, but also for the whole field of nanotechnology. More knowledge about lay people's perception of nanotechnology, their construction of preferences, and their attitudes toward nanotechnology is important for improving risk communication and for designing information better tailored to lay people.

Questions

1. What are some problems with measuring public perceptions?
2. How does positive and negative framing influence the results of a survey? Does framing induce a bias in inquiry?
3. How can public perceptions of technology be meaningfully measured?
4. Why do public perceptions differ for biotechnology and nanotechnology?
5. What factors influence public perception about the use of emerging technologies?
6. Develop a strategy for building and promoting trust between all stakeholders (public, researchers, industries, and regulatory bodies) in order to deal effectively with the risks and uncertainties of nanotechnology.
7. Using Internet resources, define the following terms:
   a. Alternate hypothesis
   b. Analysis of variance
   c. Association
   d. Confidence level

e. Construct validity
f. Content validity
g. Correlation
h. Criterion-related validity
i. Difference hypothesis
j. Effect size
k. Epistemology
l. Face validity
m. Hypothesis
n. Inferential statistics
o. Instrument validity
p. Mean
q. Measurement reliability
r. Median
s. Mode
t. Null hypothesis
u. Power of statistics
v. Reality
w. Research validity
x. Sampling
y. Theory
z. Validity

8. What are the limitations of using statistical tools to measure "'reality' vis-à-vis 'perceptions'"?
9. What is the effect of sample size on the results of a survey? What are the limitations of statistical tools in the research arena?
10. List examples of research where statistical methods were incorrectly used to arrive at results.
11. Summarize your learning of new ideas and information presented in this chapter by completing the following flowchart:

Flow Chart

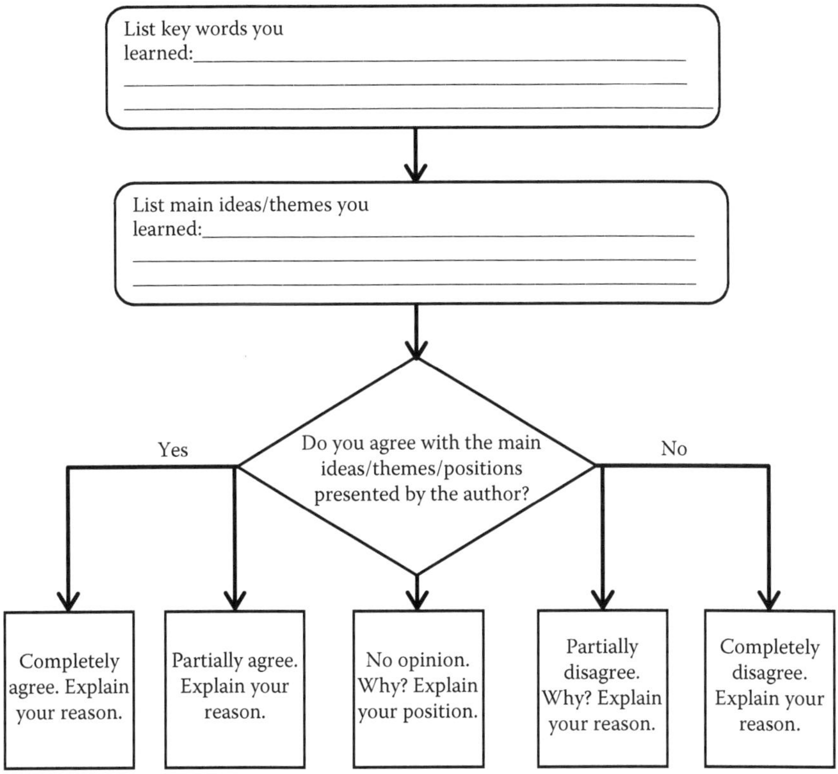

Acknowledgment

This chapter was originally published in *Human and Ecological Risk Assessment*, 16: 837–846, 2010. Reprinted by permission of Taylor & Francis, LLC.

References

Ayers, T. (1998). A tomato by any other name? U.S. and EC grapple with labeling. *Science, 281*, 714–715.

Bainbridge, W. S. (2002). Public attitudes toward nanotechnology. *J. Nanopart. Res., 4*, 561–570.

Batt, C. A., Waldron, A. M., & Broadwater, N. (2008). Numbers, scale and symbols: The public understanding of nanotechnology. *J. Nanopart. Res., 10*, 1141–1148.

Besley, J. C., Kramer, V. L., Yao, Q., & Toumey, C. (2008). Interpersonal discussion following citizen engagement about nanotechnology. *Sci. Commun., 30*, 209–235.

Brayner, R. (2008). The toxicological impact of nanoparticles. *Nano Today, 3*, 48–55.

Bredahl, L. (1999). Consumers' cognitions with regard to genetically modified foods: Results of a qualitative study in four countries. *Appetite, 33*, 343–360.

Brossard, D., Scheufele, D. A., Kim, E., & Lewenstein, B. V. (2009). Religiosity as a perceptual filter: Examining processes of opinion formation about nanotechnology. *Public Understanding Sci., 18*, 546–558.

Burri, R. V., & Bellucci, S. (2008). Public perception of nanotechnology. *J. Nanopart. Res., 10*, 387–391.

Castellini, O. M., Walejko, G. K., Holladay, C. E., Theim, T. J., Zenner, G. M., & Crone, W. C. (2007). Nanotechnology and the public: Effectively communicating nanoscale science and engineering concepts. *J. Nanopart. Res., 9*, 183–189.

Cobb, M. D. (2005). Framing effects on public opinion about nanotechnology. *Sci. Commun. 27*, 221–239.

Cobb, M. D., & Macoubrie, J. (2004). Public perceptions about nanotechnology: Risks, benefits and trust. *J. Nanopart. Res., 6*, 395–405.

Colvin, V. L. (2003). The potential environmental impact of engineered nanomaterials. *Nat. Biotechnol., 21*, 1166–1170.

Conti, J. A., Killpack, K., Gerritzen, G., Huang, L., Mircheva, M., Delmas, M., et al. (2008). Health and safety practices in the nanomaterials workplace: Results from an international survey. *Environ. Sci. Technol., 42*, 3155–3162.

Currall, S. C. (2009). New insights into public perceptions. *Nat. Nanotechnol., 4*, 79–80.

Currall, S. C., King, E. B., Lane, N., Madera, J., & Turner, S. (2006). What drives public acceptance of nanotechnology? *Nat. Nanotechnol., 1*, 153–155.

Einsiedel, E. F., & Goldenberg, L. (2006). Dwarfing the social? Nanotechnology lessons from the biotechnology front. In G. Hunt & M. D. Mehta (Eds.), *Nanotechnology: Risk, ethics and law* (pp. 213–221). London, UK: Earthscan.

Gaskell, G., Allum, N., Bauer, M., Durant, J., Allansdottir, A., Bonfadelli, H., et al. (2000). Biotechnology and the European public. *Nat. Biotechnol., 18*, 935–938.

Gaskell, G., Bauer, M. W., Durant, J., & Allum, N. C. (1999). Worlds apart? The reception of genetically modified foods in Europe and the U.S. *Science, 285*, 384–387.

Gaskell, G., Ten Eyck, T., Jackson, J., & Veltri, J. (2004). Public attitudes to nanotechnology in Europe and the United States. *Nat. Mater. 3*, 496.

Gaskell, G., Ten Eyck, T., Jackson, J., & Veltri, J. (2005). Imaging nanotechnology: Cultural support for technological innovation in Europe and the United States. *Public Understanding Sci., 14*, 81–90.

Helland, A., Scheringer, M., Siegrist, M., Kastenholz, H. G., Wiek, A., & Scholz, R. W. (2008). Risk assessment of engineered nanomaterials: A survey of industrial approaches. *Environ. Sci. Technol., 42*, 640–646.

Kahan, D. M., Braman, D., Slovic, P., Gastil, J., & Cohen, G. (2009). Cultural cognition of the risks and benefits of nanotechnology. *Nat. Nanotechnol., 4*, 87–90.
Keller, K. H. (2007). Nanotechnology and society. *J. Nanopart. Res., 9*, 5–10.
Köhler, A. R., & Som, C. (2008). Environmental and health implications of nanotechnology: Have innovators learned the lessons from past experiences? *Hum. Ecol. Risk Assess., 14*, 512–531.
Lee, C.-J., Scheufele, D. A., & Lewenstein, B. V. (2005). Public attitudes toward emerging technologies. *Sci. Commun., 27*, 240–267.
Macoubrie, J. (2006). Nanotechnology: Public concerns, reasoning and trust in government. *Public Understanding Sci., 15*, 221–241.
Maynard, A. D. (2006). Nanotechnology: Assessing the risks. *Nano Today, 1*, 22–33.
Mehta, M. D. (2006). From biotechnology to nanotechnology: What can we learn from earlier technologies? In G. Hunt & M. D. Mehta (Eds.), *Nanotechnology: Risk, ethics and law* (pp. 121–129). London, UK: Earthscan.
O'Brien, N., & Cummins, E. (2008). Recent developments in nanotechnology and risk assessment strategies for addressing public and environmental health concerns. *Hum. Ecol. Risk Assess., 14*, 568–592.
Pidgeon, N., Herr Harthorn, B., Bryant, K., & Rogers-Hayden, T. (2009). Deliberating the risks of nanotechnologies for energy and health applications in the United States and United Kingdom. *Nat. Nanotechnol., 4*, 95–98.
Pidgeon, N., Kasperson, R. E., & Slovic, P. (2003). *The social amplification of risk*. Cambridge, UK: Cambridge University Press.
Pidgeon, N., & Rogers-Hayden, T. (2007). Opening up nanotechnology dialogue with the publics: Risk communication or "upstream engagement"? *Health Risk Soc., 9*, 191–210.
Priest, S. (2006). The North American opinion climate for nanotechnology and its products: Opportunities and challenges. *J. Nanopart. Res., 8*, 563–568.
Priest, S. H. (2008). North American audiences for news of emerging technologies: Canadian and U.S. responses to bio- and nanotechnologies. *J. Risk Res., 11*, 877–889.
Renn, O., & Roco, M. C. (2006). Nanotechnology and the need for risk governance. *J. Nanopart. Res., 8*, 153–191.
Royal Society and Royal Academy of Engineering. (2004). *Nanoscience and nanotechnologies: Opportunities and uncertainties*. London, UK: Royal Society.
Rozin, P. (2005). The meaning of "natural." *Psychol. Sci., 16*, 652–658.
Rozin, P. (2006). Naturalness judgments by lay Americans: Process dominates content in judgments of food or water acceptability and naturalness. *Judgment and Decision Making, 1*, 91–97.
Scheufele, D. A., Corley, E. A., Dunwoody, S., Shih, T.-J., Hillback, E., & Guston, D. H. (2007). Scientists worry about some risks more than the public. *Nat. Nanotechnol., 2*, 732–734.
Scheufele, D. A., & Lewenstein, B. V. (2005). The public and nanotechnology: How citizens make sense of emerging technologies. *J. Nanopart. Res., 7*, 659–667.
Scholderer, J., & Frewer, L. (2003). The biotechnology communication paradox: Experimental evidence and the need for a new strategy. *J. Consumer Policy, 26*, 125–157.
Schütz, H., & Wiedemann, P. M. (2008). Framing effects on risk perception of nanotechnology. *Public Understanding Sci., 17*, 369–379.
Siegrist, M. (2008). Factors influencing public acceptance of innovative food technologies and products. *Trends Food Sci. Technol., 19*, 603–608.

Siegrist, M., Cousin, M.-E., Kastenholz, H., & Wiek, A. (2007a). Public acceptance of nanotechnology foods and food packaging: The influence of affect and trust. *Appetite, 49*, 459–466.

Siegrist, M., Earle, T. C., & Gutscher, H. (2003). Test of a trust and confidence model in the applied context of electromagnetic field (EMF) risks. *Risk Anal., 23*, 705–716.

Siegrist, M., Keller, C., Kastenholz, H., Frey, S., & Wiek, A. (2007b). Laypeople's and experts' perception of nanotechnology hazards. *Risk Anal., 27*, 59–69.

Siegrist, M., Stampfli, N., Kastenholz, H., & Keller, C. (2008). Perceived risks and perceived benefits of different nanotechnology foods and nanotechnology food packaging. *Appetite, 51*, 283–290.

Sjöberg, L. (2000). Perceived risk and tampering with nature. *J. Risk Res., 3*, 353–367.

Slovic, P. (1995). The construction of preference. *Am. Psychol., 50*, 364–371.

Smiley Smith, S. E., Hosgood, H. D., Michelson, E. S., & Stowe, M. H. (2008). Americans' nanotechnology risk perception: Assessing opinion change. *J. Ind. Ecol., 12*, 459–473.

Stern, S. T., & McNeil, S. E. (2008). Nanotechnology safety concerns revisited. *Toxicol. Sci., 101*, 4–21.

Waldron, A. M., Spencer, D., & Batt, C. A. (2006). The current state of public understanding of nanotechnology. *J. Nanopart. Res., 8*, 569–575.

Wiedemann, P. M., & Schütz, H. (2005). The precautionary principle and risk perception: Experimental studies in the EMF area. *Environ. Health Perspect., 113*, 402–405.

Epilogue
Nanotechnology: The Size of the Cosmos Versus the Size of Human Knowledge

Ahmed S. Khan

As the 21st century progresses, man has made great progress in the domains of science and technology. But when considered in context of the expanding cosmos, what is man's station, what is the size of man's knowledge? Let's ponder over the size of cosmos and size of human knowledge! We may gain a bit of wisdom!

Hubble Ultra Deep Field

A faint glimpse of the cosmos detected by the Hubble telescope. The image shows about 10,000 galaxies. The cosmos contains 12.7 million times more area than the Ultra Deep Field. (Hubble Ultra Deep Field Photo: NASA.)

Cosmos: 4.6% visible (known) and
95.4% dark matter and energy (unknown)

Size of Cosmos

		1	6	5	0	0	0	0	0	0	0	0	0	Light Years (LT)				Cosmic World
			1	0	0	0	0	0	0	0	0	0						1 LT = 9,500,000,000,000 km
				1	0	0	0	0	0	0	0							
					1	0	0	0	0	0								
						1	0	0	0									
							1	0										
								1										
								0	.1									
							0	0	.0	1								*Zeros have no significance without 1.* Rumi
						0	0	0	.0	0	1							
					0	0	0	0	.0	0	0	1						
				0	0	0	0	0	.0	0	0	0	1					
			0	0	0	0	0	0	.0	0	0	0	0	1				
		0	0	0	0	0	0	0	.0	0	0	0	0	0	1			
	0	0	0	0	0	0	0	0	.0	0	0	0	0	0	0	1		
0	0	0	0	0	0	0	0	0	.0	0	0	0	0	0	0	0	1	Nanoworld

Size of Nanoworld

Thirty-five Xenon atoms (IBM photo)

Size of Human Knowledge

The Library of Congress has 33 million books and printed items. Let's assume that the total accumulation of human knowledge over the past 5,000 years can be represented by 1,000 libraries of Congress; this would yield 33 billion books. Let's further assume that each book has 300 pages, with 450 words per page, and each word averages about 7 characters; if we digitize all the text using ASCII code (seven bits per character), this would yield 2.18295×10^{17} bits.

$$\text{Human knowledge in digitized form} = 2.18295 \times 10^{17}\ \text{bits}$$

And now let's transmit this total human knowledge in digital form using state-of-the-art telecommunications technology, fiber-optic communications with a transmission speed of 100 trillion bits per second.

The total human knowledge accumulated during the past 5,000 years can be transmitted to anywhere in the world in just 36.38 minutes!

To know what you know,
And know what you do not know—that is knowledge.

Confucius

It is a great advantage that man should know his station and not erroneously imagine that the whole universe exists only for him.

MAIMONIDES
DALALAT AL-HA'IRIN, PART III, CHAPTER XII, C. 1190

Index